AN INTRODUCTION
TO
ANALYTICAL GEOMETRY

AN INTRODUCTION TO ANALYTICAL GEOMETRY

by

A. ROBSON

Senior Mathematical Master at Marlborough College

VOLUME I

CAMBRIDGE

AT THE UNIVERSITY PRESS

1940

CAMBRIDGE UNIVERSITY PRESS
Cambridge, New York, Melbourne, Madrid, Cape Town, Singapore, São Paulo, Delhi

Cambridge University Press
The Edinburgh Building, Cambridge CB2 8RU, UK

Published in the United States of America by Cambridge University Press, New York

www.cambridge.org
Information on this title: www.cambridge.org/9780521116190

© Cambridge University Press 1940

First published 1940
This digitally printed version 2009

A catalogue record for this publication is available from the British Library

ISBN 978-0-521-06116-2 hardback
ISBN 978-0-521-11619-0 paperback

CONTENTS

PREFACE

This is an introduction to the use of coordinates and analytical methods in geometry. It is expected that the reader will usually have taken a previous course of elementary calculus and that during that course and from his study of graphs he will have gained some knowledge of rectangular cartesian coordinates.

The early chapters contain numerous exercises suitable for a beginner, so that a previous course of coordinate geometry is not necessary. The book is intended to be easy throughout. For the better students some harder questions are included in the illustrative examples and in the exercises.

The aim has been to introduce a large variety of methods and ideas. The use of parameters, envelope coordinates, and duality is emphasised, and vectors are used when this seems desirable.

The importance of the parabola, ellipse, and hyperbola must be recognised, although it has been exaggerated in the past. The analytical processes that are introduced can profitably be illustrated by applications to other curves as well as the conics, and many of the curves have an interest of their own. It has been found that the necessary knowledge of the conics is not easily acquired from an analytical course alone; therefore in this book some of the important properties are dealt with by pure geometry. Enough about the conics for the beginner is contained in Chapter 6, and it is recommended that chapters 13, 14, 15 should not be taken before any of the previous chapters; these three chapters contain sufficient detail for those who require a treatment of analytical conics on traditional lines.

An attempt is made in Chapter 8 to justify the use of complex coordinates and points at infinity. It is felt that they certainly ought not to be used without some justification,

and further that the important idea of an abstract geometry
is one which should be presented to ordinary students.

There is nowadays no definite line to be drawn between
pure and analytical geometry. Much of the bookwork of what
is usually called elementary projective geometry is included
in Volume I or Volume II. But the book is intended to be
mainly analytical and it will be necessary for the student to
supplement his reading by working sets of examples from some
text-books on pure geometry on such subjects as cross-ratio,
homography, involution, and inversion. Even then he should
always feel at liberty to apply analytical methods when they
are the most convenient. He needs to learn to choose for
himself the most suitable method for particular geometrical
problems.

The distinction between metrical and projective geometry
has been kept in mind in writing the book, with a view to
making it a good introduction to the work which will be done
later at the university. It is a distinction which the teacher
himself will do well to emphasise in the schools.

My thanks are due to Mr J. C. Manisty for the help he
has given me in the production of the book.

                                                A. R.

January 1940

CONVENTIONS AND ABBREVIATIONS

The following conventions or abbreviations are used in this book:

0·1. (a, b) means the point whose coordinates are a, b.

(a, b, c) means the point whose coordinates are a, b, c.

P_n means (x_n, y_n) or (x_n, y_n, z_n) according to the context.

0·2. "Line" means "straight line", and lines are supposed to be unlimited except when it is otherwise stated.

$[a, b]$ means the line whose coordinates are a, b.

$[a, b, c]$ means the line whose coordinates are a, b, c.

p_n means $[X_n, Y_n]$ or $[X_n, Y_n, Z_n]$ according to the context.

0·3. If the angle ω between the cartesian axes is relevant, it is assumed to be $\frac{1}{2}\pi$ unless the contrary is stated. When it is to be specially noted that the axes are oblique, the symbol $\{\omega\}$ or $\{\omega = \alpha\}$ is used.

0·41. The word "respectively" is omitted unless the omission is likely to lead to misunderstanding.

0·42. The words "whose equation is" are often omitted in such expressions as "the line whose equation is $ax + by + c = 0$".

0·43. The word "wo" means "with respect to".

0·5. If in an equation $x_1 : x_2 = a_1 : a_2$ it happens that $a_1 = 0$, the equation is taken to mean that $x_1 = 0$; if $a_2 = 0$, it is taken to mean that $x_2 = 0$; if $a_1 = a_2 = 0$, the equation has no meaning.

If $x_1 : x_2 : x_3 = a_1 : a_2 : a_3$ and, for example, $a_3 = 0$, the equation is taken to mean $x_1 : x_2 = a_1 : a_2$ and $x_3 = 0$;

if $a_2 = a_3 = 0$, it is taken to mean that $x_2 = x_3 = 0$;

if $a_1 = a_2 = a_3 = 0$, the equation has no meaning.

Similar conventions are made for equations of the same type with n variables.

0·6. A knowledge of the notation and elementary properties of determinants is assumed.

The determinant $\begin{vmatrix} a & h & g \\ h & b & f \\ g & f & c \end{vmatrix}$ which arises in connexion with the quadratic forms

$$ax^2 + 2hxy + by^2 + 2gx + 2fy + c, \quad ax^2 + 2hxy + by^2 + 2gxz + 2fyz + cz^2$$

is denoted by δ. Also it is assumed that if

$$A = bc - f^2, \quad B = ca - g^2, \quad C = ab - h^2,$$
$$F = gh - af, \quad G = hf - bg, \quad H = fg - ch,$$

then
$$\delta = Aa + Hh + Gg = Hh + Bb + Ff = Gg + Ff + Cc,$$
$$0 = Ah + Hb + Gf = Ha + Bh + Fg = Ga + Fh + Cg$$
$$0 = Ag + Hf + Gc = Hg + Bf + Fc = Gh + Fb + Cf,$$

and
$$\begin{vmatrix} A & H & G \\ H & B & F \\ G & F & C \end{vmatrix} = \delta^2 = \Delta.$$

Also
$$BC - F^2 = a\delta, \quad CA - G^2 = b\delta, \quad AB - H^2 = c\delta,$$
$$GH - AF = f\delta, \quad HF - BG = g\delta, \quad FG - CH = h\delta.$$

0·7. In mathematics the phrase "in general" is often used to qualify a statement. It means that the statement is true unless the "constants" involved satisfy some special condition.

For example:

(1) In general $ax + b = 0$ is true for just one value of x.

(2) In general two lines in a plane have just one common point.

In (1) for the special value $a = 0$, $ax + b = 0$ for no value of x unless also $b = 0$ when it is true for all values of x. In (2) if the lines are distinct and parallel, they have no common point; if they are coincident, they have an unlimited number of common points.

The "constants" may be explicit algebraic constants as in (1) or they may be implicit. In (2) the "constants" might be the coefficients in the equations of the lines or they might be implicit in geometrical conditions determining the lines.

0·8. The following abbreviations are used for references in the text:

P.M. *Pure Mathematics*, Hardy (Cambridge).

A.A. *Advanced Algebra*, 3 vols., Durell and Robson (Bell).

A.T. *Advanced Trigonometry*, Durell and Robson (Bell).

E.C. *Elementary Calculus*, 2 vols., Durell and Robson (Bell).

M.G. *Modern Geometry*, Durell (Macmillan).

P.G. *Projective Geometry*, Durell (Macmillan).

H.M. *A Short Account of the History of Mathematics*, W. W. R. Ball (Macmillan).

Chapter 1

COORDINATES

1·1. Geometry of One Dimension

1·11. If O is a fixed point on the line $X'OX$, the position of a point P on the line is determined by one coordinate x which takes positive and negative values and determines the displacement OP of P from O. P is called the point (x).

If the length of OP is l units, the coordinate of P is $+l$ or $-l$ according as P is on the same side of O as X or X'.

O is called the *origin*. It has the coordinate zero.

If A and B are the points (a) and (b), we write

$$AB = b - a, \tag{1}$$

and then, since the distance between the points is $|a-b|$ units, AB gives the distance or minus the distance according as $a < b$ or $a > b$.

Since P_n denotes the point (x_n), the formula (1) may be written

$$P_1 P_2 = x_2 - x_1.$$

1·12. EXAMPLE. Verify that

$$BC \cdot AD + CA \cdot BD + AB \cdot CD = 0.$$

Take D as origin and let the coordinates of A, B, C be a, b, c. Then

$$BC \cdot AD + CA \cdot BD + AB \cdot CD$$
$$= (c-b)(-a) + (a-c)(-b) + (b-a)(-c) = 0.$$

EXERCISE 1A

[In this exercise the points are in one line]

In Nos. 1–4 verify

1. $BC + CA + AB = 0.$
2. $AB + BC = AC.$
3. $P_0 P_1 + P_1 P_2 + P_2 P_3 + \ldots + P_{n-1} P_n = P_0 P_n.$
4. $AD^2 \cdot BC + BD^2 \cdot CA + CD^2 \cdot AB = CB \cdot AC \cdot BA.$

5. If $P_1P = \kappa PP_2$, find the coordinate (x) of P in terms of κ and the coordinates x_1, x_2 of P_1, P_2.

6. If A and B are fixed points and κ is a constant, show that a point P such that $PA^2 + PB^2 = \kappa AB^2$ has 2, 1, 0 possible positions according as $\kappa >, =, < \frac{1}{2}$.

In Nos. 7–9, the points P_1, P_2 vary so that their coordinates satisfy the given condition. Find the positions in which P_1 coincides with P_2.

7. $x_1x_2 + 4x_1 - 3x_2 - 12 = 0.$ **8.** $4x_1x_2 - 7x_1 - 5x_2 + 9 = 0.$

9. $x_1x_2 + 9x_1 - 8x_2 + 1 = 0.$

10. Verify that

$$EA^2 . BC . CD . DB + EB^2 . CA . AD . DC$$
$$+ EC^2 . AB . BD . DA + ED^2 . CB . BA . AC = 0.$$

1·13. The ratio $AP : PB$ is called the ratio in which P divides AB. It is positive if and only if P is between A and B.

For example, if $AB = 2BP$, then $AP : PB = 3 : -1$, and P divides AB in the ratio $3 : -1$. P may be said to divide AB externally in the ratio $3 : 1$.

If x is the coordinate of the point P which divides P_1P_2 in the ratio $\kappa_2 : \kappa_1$,

$$AP : PB = \kappa_2 : \kappa_1.$$

$$\therefore \ \kappa_1(x - x_1) = \kappa_2(x_2 - x),$$

$$\therefore \ x = \frac{\kappa_1 x_1 + \kappa_2 x_2}{\kappa_1 + \kappa_2}.$$

This important formula applies whether κ_2/κ_1 is positive or negative, but $\kappa_1 + \kappa_2$ must not be zero.

EXERCISE 1B

[The points are in one line and A, B are the points (a), (b)]

In Nos. 1–7, AB is divided at P in the given ratio and the coordinate of P is to be found.

1. $a = 3, b = 8; 3 : 2.$ **2.** $a = 3, b = -2; 2 : 3.$

3. $a = -5, b = 3; 3 : 1.$ **4.** $a = 5, b = -3; 1 : 7.$

5. $a = -7, b = 5; 9 : -5.$ **6.** $a = 3, b = -11; -2 : 9.$

7. $a = x_1, b = x_2; \kappa_2 : \kappa_1$ externally.

8. If $DA = AB = 2BC$, give the coordinates of C and D.

9. If $\frac{1}{2}PA = AB = \frac{1}{3}BQ$, give the coordinates of P and Q.

10. If $\lambda RA = \mu AB = \nu BS$, give the coordinates of R and S.

11. If AB is divided internally at P and externally at Q in the ratio $\lambda : 1$, find the length of PQ in terms of that of AB.

1·14. The position of a point P in a line may be determined wo two fixed points A and B instead of wo a single origin O.

The value of AP/PB, $\equiv \lambda$, is called the *ratio-coordinate* of P. The position of P is uniquely determined by λ.

Instead of using λ, it is possible to use two coordinates κ_1, κ_2 such that $AP : PB = \kappa_2 : \kappa_1$, calling P the point (κ_1, κ_2). These two coordinates are effectively equivalent to one. It is only their ratio that is relevant. The points $(3, 2)$, $(30, 20)$, $(3c, 2c)$ are all the same provided that c is not zero. The pair of coordinates $0, 0$ corresponds to no point.

The coordinates of A and B are 1, 0 and 0, 1, and the point (κ_1, κ_2) is the centre of mass of κ_1 at A and κ_2 at B. For this reason the coordinates are sometimes called *barycentric* coordinates.

EXERCISE 1 c

[The points are in one line and A, B are the fixed points of reference]

1. If $EA = 2AC = 2CB = BD$, what are the ratio-coordinates of C, D, E, A?

2. What point has no ratio-coordinate, and what number λ is the coordinate of no point?

3. Sketch a graph to show how λ varies when P describes the whole line.

4. In No. 1, give the barycentric coordinates of A, B, C, D, E.

5. Does every point of the line have barycentric coordinates?

6. What barycentric coordinates correspond to no point of the line?

7. A, B, C, D are points of a line having coordinates a, b, c, d measured from a point O in the line, and ratio-coordinates λ, μ, ν, ρ referred to any two fixed points in the line. Evaluate

$$(AB.CD)/(AD.CB)$$

in terms of a, b, c, d and in terms of λ, μ, ν, ρ.

1·2. Geometry of Two Dimensions

The position of a point P in a plane is determined with reference to two lines $X'OX$, $Y'OY$ by two coordinates $x \equiv ON$ and $y \equiv OM$ which are the one-dimensional co-ordinates (referred to O as origin) of points N, M on $X'OX$, $Y'OY$ such that $ONPM$ is a parallelogram.

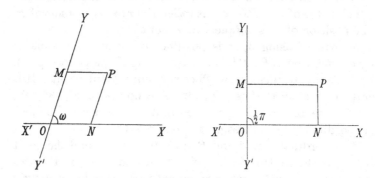

x and y are called cartesian coordinates. They were introduced by Descartes (1596–1650).

O is called the *origin*. The line $X'OX$ is called the *axis of x*, or, since every point on it has its y-coordinate zero, it may be called the *line y* = 0. Similarly $Y'OY$ is called the *axis of y* or the *line x* = 0.

It is assumed that the reader is familiar with elementary numerical applications of cartesian coordinates for axes $X'OX$, $Y'OY$ at right angles to one another. These co-ordinates are particularly convenient for the investigation of problems in metrical geometry, i.e. problems in which distances are involved.

There is an advantage for certain problems in using oblique axes ($\angle XOY = \omega \neq \frac{1}{2}\pi$) and many formulae are as easily obtained for oblique as for rectangular axes, but rectangular axes are often used in applications of coordinate geometry. See 0·3.

1·3. Displacements and Vectors

1·31. Instead of using two coordinates to determine the position of a point P we may determine it by the *displacement* **OP** from the origin to the point.

This displacement has a magnitude or amount which is the length of OP, a direction namely that of OP, and a sense (from O towards P). It is an example of a vector: a vector is a number associated with a direction.

If $ABCD$ is a parallelogram, the displacements AB, DC are equivalent. Vectors having this property are called *free* vectors; they are to be contrasted with *localised* vectors, such as forces, for which AB, DC are not equivalent.

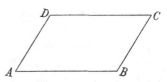

A displacement from A to B followed by a displacement from B to C is equivalent to a displacement from A to C. This is denoted by

$$\mathbf{AB} + \mathbf{BC} = \mathbf{AC}, \tag{1}$$

where + and = have new meanings.

We also write

$$\mathbf{P} + \mathbf{Q} = \mathbf{R}, \tag{2}$$

where **P**, **Q**, **R** are any displacements equivalent to **AB**, **BC**, **AC**.

(2) is, in effect, the definition of the sum of two free vectors. The equivalence of **AB** + **BC** and **AD** + **DC**,

$$\text{i.e. } \mathbf{P} + \mathbf{Q} = \mathbf{Q} + \mathbf{P}, \tag{3}$$

is the commutative law of addition for free vectors.

The identity $\mathbf{P} + (\mathbf{Q} + \mathbf{R}) = (\mathbf{P} + \mathbf{Q}) + \mathbf{R}$ can be illustrated geometrically. In virtue of the result, either expression may be denoted by $\mathbf{P} + \mathbf{Q} + \mathbf{R}$, and by (3) the order of the terms **P**, **Q**, **R** can be changed.

$\mathbf{P} - \mathbf{Q}$ denotes the vector **S** such that

$$\mathbf{P} = \mathbf{Q} + \mathbf{S}.$$

$-\mathbf{Q}$ denotes the vector **T** with the same amount as **Q** but the opposite direction.

Thus

$$\mathbf{P} + \mathbf{T} = \mathbf{P} - \mathbf{Q}.$$

1·32. The *magnitude* or *amount* of a vector **AB** is the length of AB. When a vector is denoted by a clarendon symbol **P**, its amount is denoted by the corresponding italic letter P.

A vector of magnitude 1 is called a *unit* vector.

If k is a positive number and **P** is a vector, k**P** denotes a vector of amount kP with the same direction as **P**. If k is negative, k**P** has amount $-kP$ and the direction opposite to **P**.

1·33. Unit vectors in the axes OX, OY are denoted by **i**, **j**. Hence if P is a point (x, y),

$$\mathbf{OP} = x\mathbf{i} + y\mathbf{j}.$$

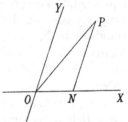

1·34. *Scalar Product.* The scalar product of two vectors **R**$_1$, **R**$_2$ is defined to be $R_1 R_2 \cos\theta$, where R_1, R_2 are the amounts of the vectors and θ is an angle from the direction of **R**$_1$ to that of **R**$_2$. It is denoted by **R**$_1$. **R**$_2$.

Another meaning with which we shall not be concerned in this book is given to **R**$_1$ × **R**$_2$.

Since $\cos(2n\pi + \theta) = \cos\theta$, it is immaterial in the above definition which angle from **R**$_1$ to **R**$_2$ is taken. And since $\cos(-\theta) = \cos\theta$

$$\mathbf{R}_2 . \mathbf{R}_1 = \mathbf{R}_1 . \mathbf{R}_2. \tag{1}$$

R^2 denotes **R** . **R** and

$$\mathbf{R}^2 = R^2. \tag{2}$$

In particular $\mathbf{i}^2 = \mathbf{j}^2 = 1$, $\mathbf{i} . \mathbf{j} = \mathbf{j} . \mathbf{i} = \cos\omega$.

If **R** = **R**$_1$ + **R**$_2$, the projection of **R** on any line is equal to the sum of the projections of **R**$_1$, **R**$_2$ on the same line. This is equivalent to Exercise 1A, No. 2. Hence, if **u** is a unit vector in the line,

$$\mathbf{R} . \mathbf{u} = \mathbf{R}_1 . \mathbf{u} + \mathbf{R}_2 . \mathbf{u}.$$

Multiplication by S gives

$$\mathbf{R} . \mathbf{S} = \mathbf{R}_1 . \mathbf{S} + \mathbf{R}_2 . \mathbf{S}. \tag{3}$$

By means of (1), (2), (3), relations between vectors analogous to ordinary algebraic formulae can be proved. For example:

$$(X+Y)^2 = (X+Y).(X+Y)$$
$$= X.(X+Y) + Y.(X+Y)$$
$$= (X+Y).X + (X+Y).Y$$
$$= X.X + Y.X + X.Y + Y.Y,$$
$$\therefore \quad (X+Y)^2 = X^2 + 2X.Y + Y^2. \tag{4}$$

EXERCISE 1 D

1. Simplify $2GX + GA$ and $VA + VB + VC$, where ABC is a triangle, G is its centroid, X is the mid-point of BC, and V is any point.

2. If $P + Q = R$, show that $kP + kQ = kR$, and interpret this when $k = \frac{1}{2}$.

3. Give the geometrical interpretation of 1·31(3).

4. Give the value of $OP_1.OP_2$ when $\angle P_1 OP_2 = \frac{1}{2}\pi$.

5. What conclusion can be drawn from $P.Q = 0$?

In Nos. 6–8 evaluate the product.

6. $i.(2i+3j)$.   **7.** $(i+j).(i-j)$.   **8.** $(3i-4j)^2$.

9. Verify that 1·34(4) is equivalent to the trigonometrical formula

$$a^2 = b^2 + c^2 - 2bc \cos A.$$

1·4. Distance between Two Points

Two points P_1 and P_2 may be represented by coordinates x_1, y_1 and x_2, y_2, or by vectors OP_1, OP_2, where

$$OP_1 = x_1 i + y_1 j, \qquad OP_2 = x_2 i + y_2 j.$$

Hence
$$P_1 P_2 \equiv OP_2 - OP_1$$
$$= (x_2 - x_1) i + (y_2 - y_1) j.$$
$$\therefore \quad P_1 P_2{}^2 = P_1 P_2{}^2 = \{(x_2 - x_1) i + (y_2 - y_1) j\}^2$$
$$= (x_1 - x_2)^2 + (y_1 - y_2)^2 + 2(x_1 - x_2)(y_1 - y_2) \cos \omega,$$

and for rectangular axes

$$P_1 P_2{}^2 = (x_1 - x_2)^2 + (y_1 - y_2)^2.$$

These results are equivalent to the theorem of Pythagoras and its extensions. The vector method is applicable also in geometry of three dimensions (1·9).

EXERCISE 1 E

In Nos. 1–4, state the distance OP in terms of the coordinates x, y of P for the given value of ω.

1. $\frac{1}{2}\pi$.   **2.** $\frac{1}{3}\pi$.   **3.** $\frac{2}{3}\pi$.   **4.** $-\frac{1}{4}\pi$.

In Nos. 5–12, find the distance between the pair of points.

5. $(2, 5)$, $(5, 9)$.   **6.** $(3, 5)$, $(2, 7)$.

7. (a, b), $(c, -d)$.   **8.** $(-4, 1)$, $(-5, 7)$.

9. $(4, -2)$, $(-3, -1)$.   **10.** $(a+c, b-d)$, $(0, 0)$.

11. (a, b), $(a+60, b-91)$.   **12.** $(p+12, q-1)$, $(p-12, q-8)$.

13. If the distance between $(k, 8)$ and $(-5, 3)$ is 13, find the value of k.

14. Prove that the distance between $(a \cos \alpha,\ a \sin \alpha)$ and $(a \cos \beta,\ a \sin \beta)$ is $\mid 2a \sin \frac{1}{2}(\alpha - \beta) \mid$, and verify the result geometrically.

15. Find the lengths of the sides of the triangle $(-1, 7)$ $(3, 10)$ $(13, 0)$.

16. Calculate the perimeter of the convex quadrilateral whose vertices are $(36, 50)$, $(98, 2)$, $(71, 38)$, $(-19, 2)$.

17. Prove that the points $(121, 0)$, $(-71, 56)$, $(39, -164)$, $(4, 81)$ lie on a circle centre $(4, -44)$.

18. Find the centre of the circle through $(0, 1)$, $(1, 2)$, $(2, 2\frac{1}{2})$.

In Nos. 19–21, find the distance between $(1, 2)$ and $(-3, 4)$ for the given value of ω.

19. $\frac{1}{3}\pi$.   **20.** $\frac{1}{4}\pi$.   **21.** $-\frac{2}{3}\pi$.

22. If the distance between $(4, 1)$ and $(7, -9)$ is $\sqrt{139}$, find ω.

1·5. Point dividing $P_1 P_2$ in a given Ratio

1·51. If the point $P(x, y)$ divides $P_1 P_2$ in the ratio $\kappa_2 : \kappa_1$, then, since in the figure

$$N_1 N : N N_2 = P_1 P : P P_2 = \kappa_2 : \kappa_1$$

and P and N have the same x-coordinate, therefore by 1·13

$$x = \frac{\kappa_1 x_1 + \kappa_2 x_2}{\kappa_1 + \kappa_2}. \tag{1}$$

Similarly $$y = \frac{\kappa_1 y_1 + \kappa_2 y_2}{\kappa_1 + \kappa_2}.$$

This proof may be expressed more concisely by means of

vectors. If $\mathbf{r}_1$, $\mathbf{r}_2$, $\mathbf{r}$ are the vectors $\mathbf{OP}_1$, $\mathbf{OP}_2$, $\mathbf{OP}$, then

$$\frac{\mathbf{r}-\mathbf{r}_1}{\kappa_2} = \frac{\mathbf{r}_2-\mathbf{r}}{\kappa_1}.$$

$$\therefore \; \mathbf{r} = \frac{\kappa_1\mathbf{r}_1+\kappa_2\mathbf{r}_2}{\kappa_1+\kappa_2}.$$

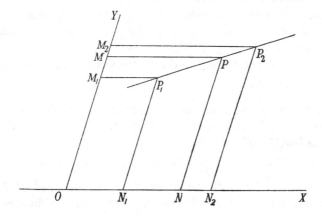

This includes the two results of (1). It may also be expressed in the form

$$\kappa_1\mathbf{OP}_1+\kappa_2\mathbf{OP}_2 = (\kappa_1+\kappa_2)\,\mathbf{OP}. \tag{2}$$

1·52. If the line joining the point P in 1·51 to P_3 is divided at Q in the ratio $\kappa_3 : \kappa_1+\kappa_2$, then by (2)

$$(\kappa_1+\kappa_2)\,\mathbf{OP}+\kappa_3\mathbf{OP}_3 = (\kappa_1+\kappa_2+\kappa_3)\,\mathbf{OQ},$$

and therefore by (2) again

$$\kappa_1\mathbf{OP}_1+\kappa_2\mathbf{OP}_2+\kappa_3\mathbf{OP}_3 = (\kappa_1+\kappa_2+\kappa_3)\,\mathbf{OQ}.$$

If QP_4 is divided at R in the ratio $\kappa_4 : \kappa_1+\kappa_2+\kappa_3$, it may be proved in the same way that

$$\kappa_1\mathbf{OP}_1+\kappa_2\mathbf{OP}_2+\kappa_3\mathbf{OP}_3+\kappa_4\mathbf{OP}_4 = (\kappa_1+\kappa_2+\kappa_3+\kappa_4)\,\mathbf{OR},$$

and so on. If there are n points P_1, P_2, ..., P_n and n numbers κ_1, κ_2, ..., κ_n associated with them, the final point $G(x, y)$ reached by the above process is called the *centroid* of κ_1 at P_1, κ_2 at P_2, ..., κ_n at P_n, and

$$\Sigma(\kappa\mathbf{OP}) = (\Sigma\kappa)\,\mathbf{OG}.$$

This is equivalent to

$$x = \frac{\Sigma \kappa x}{\Sigma \kappa}, \qquad y = \frac{\Sigma \kappa y}{\Sigma \kappa}.$$

The symmetry of the result shows that the points P_1, P_2, ..., P_n can be taken in any order provided that each P_r is associated with its assigned κ_r.

The values of the κ's need not be positive, but $\Sigma \kappa$ must not be zero. When $\kappa_1 = \kappa_2 = \ldots = \kappa_n$, the point G is called simply the *centroid* of P_1, P_2, ..., P_n.

If the κ's are positive, G is the centre of mass of κ_1 at P_1, κ_2 at P_2, ..., κ_n at P_n.

1·53. EXAMPLE. Find the incentre of the triangle $(-2, 47)$ $(-30, -49)$ $(70, 26)$, and the ecentre opposite to $(-2, 47)$.

The lengths of the sides are

$$\sqrt{(100^2 + 75^2)} = 125, \quad \sqrt{(72^2 + 21^2)} = 75, \quad \sqrt{(28^2 + 96^2)} = 100.$$

Hence $(A.T.$ p. 10) the incentre is the centroid of 5 at $(-2, 47)$, 3 at $(-30, -49)$, 4 at $(70, 26)$ and its coordinates are

$$x = \frac{5(-2) + 3(-30) + 4(70)}{5 + 3 + 4} = 15,$$

$$y = \frac{5(47) + 3(-49) + 4(26)}{5 + 3 + 4} = 16.$$

The ecentre is the centroid of -5 at $(-2, 47)$, 3 at $(-30, -49)$, 4 at $(70, 26)$ and its coordinates are

$$x = \frac{-5(-2) + 3(-30) + 4(70)}{-5 + 3 + 4} = 100,$$

$$y = \frac{-5(47) + 3(-49) + 4(26)}{-5 + 3 + 4} = -139.$$

EXERCISE 1F

In Nos. 1–4, state the coordinates of the mid-point of the line joining the given points.

1. $(2, 3)$, $(5, 9)$.   **2.** $(3, -7)$, $(-1, 5)$.

3. (a, b), $(-a, 2b)$.   **4.** $(a-c, b-d)$, $(a+c, b+d)$.

5. Prove that the line joining $(0, 0)$ to $(3, -4)$ is bisected by the line joining $(6, 1)$ to $(-3, -5)$.

In Nos. 6–11, find the point dividing the line joining the given points in the given ratio.

6. $(5, 2)$, $(8, 11)$; $2 : 1$.   **7.** $(-1, 2)$, $(5, 8)$; $2 : 3$.

8. $(-7, -4)$, $(2, -13)$; $5 : 4$.   **9.** $(9, 4)$, $(7, 2)$; $-7 : 5$.

10. $(6, 9)$, $(-4, 14)$; $6 : -11$.   **11.** (a, b), (b, a); $a+b : a-b$.

12. Find the points that divide the line joining $(3, -7)$ to $(-6, -3)$ externally in the ratios $2 : 1$ and $1 : 2$.

13. Find the points of trisection of the line joining $(2, 8)$ to $(-4, -1)$.

14. Find the points that divide the line joining $(2, 81)$ to $(-4, 11)$ into four equal parts.

15. The centre of a circle is $(3, 5)$ and one end of a diameter is $(2, -1)$. Find the other end.

16. A is $(3, 7)$, B is $(-1, 4)$, C is the point of trisection of AB nearer to A. D is the point in AB produced such that $5BD = AB$. Find the coordinates of C and D.

17. Find the ratio in which $(7, 37)$ divides the line joining $(35, 21)$ to $(-14, 49)$.

18. Find the ratio in which $(14, 7)$ divides the line joining $(-1, -8)$ to $(8, 1)$.

19. Find the ratio in which the line joining $(5, 11)$ to $(13, 2)$ is divided by OX.

20. Show that the line joining $(-1, 13)$ to $(9, -7)$ is divided at $(5, 1)$ and $(29, -47)$ internally and externally in the same ratio.

In Nos. 21–25, find the centroid.

21. $(15, -17)$, $(1, 6)$, $(-1, 11)$.

22. $(2, 5)$, $(-7, 3)$, $(4, -6)$, $(-1, -2)$, $(7, 15)$.

23. 3 at $(1, 2)$, 4 at $(5, 6)$, 7 at $(0, -8)$.

24. -2 at $(4, -6)$, 1 at $(-3, 5)$, -3 at $(7, 0)$

25. $t_2 t_3$ at (t_1^2, t_1), $t_3 t_1$ at (t_2^2, t_2), $t_1 t_2$ at (t_3^2, t_3).

26. Find the incentre of $(3, 1)$ $(-5, -7)$ $(2, -6)$.

27. Find the ecentres of $(8, 37)$ $(-20, -59)$ $(80, 16)$.

28. If AC is divided by BD at E in the ratio $2 : 5$, show that the centroids of 2 at A, 7 at B, 5 at C, 7 at D

and 1 at A, 1 at B, 1 at C, 1 at D, -1 at E

are the same.

29. P_1, P_2, P_3, P_4 are any four points and M_{rs} denotes the mid-point of $P_r P_s$. Prove that $M_{12} M_{34}$, $M_{13} M_{24}$, $M_{14} M_{23}$ have the same mid-point.

30. Prove that if the centroid of κ_1 at P_1, κ_2 at P_2, ..., κ_n at P_n is P, and that of λ_1 at Q_1, λ_2 at Q_2, ..., λ_m at Q_m is Q, then the centroid of $\Sigma\kappa$ at P and $\Sigma\lambda$ at Q is the same as the centroid of κ_1 at P_1, κ_2 at P_2, ..., κ_n at P_n, λ_1 at Q_1, λ_2 at Q_2, ..., λ_m at Q_m.

1·6. Polar Coordinates

1·61. We have seen that a point P in two dimensions is determined by the vector $\mathbf{OP}$ or by cartesian coordinates x, y. It is also uniquely determined by the two coordinates $r \equiv OP$ and $\theta \equiv \angle XOP$, which are called *polar coordinates*.

O is called the *pole* and OX the *initial line*.

θ is measured from OX to OP with the usual sign convention of trigonometry.

If r is restricted to be positive and θ is restricted to satisfy $-\pi < \theta \leqslant \pi$, the polar coordinates of a point P are unique. [They are the same as the modulus and principal value of the amplitude of the complex number represented by P in the Argand Diagram.]

In geometry it is usually convenient to leave r and θ unrestricted. The point $(+c, \alpha)$ is then the same as $(+c, \alpha + 2n\pi)$ and $(-c, \alpha + \pi + 2n\pi)$, where n is any integer or zero.

If x, y are the rectangular cartesian coordinates and r, θ are the polar coordinates of the same point P, the equations

$$\cos\theta : \sin\theta : 1 = x : y : r$$

give x, y uniquely in terms of r, θ in the form

$$x = r\cos\theta, \qquad y = r\sin\theta,$$

but they do not give r, θ uniquely in terms of x, y unless the restrictions are imposed.

1·62. The distance between the points $P_1(r_1, \theta_1)$ and $P_2(r_2, \theta_2)$ is given by

$$P_1P_2{}^2 = \mathbf{P_1P_2}^2 = (\mathbf{OP_2 - OP_1})^2$$
$$= OP_1{}^2 + OP_2{}^2 - 2OP_1OP_2 \cos(\theta_1 - \theta_2)$$
$$= r_1{}^2 + r_2{}^2 - 2r_1r_2 \cos(\theta_1 - \theta_2).$$

1·63. *Bipolar Coordinates.* Occasionally the position of a point is fixed by its distances r_1, r_2 from two fixed points A_1, A_2. These are called *bipolar* coordinates, and they are unique; but given coordinates correspond to two points P, images of one another in $A_1 A_2$. The values of r_1, r_2 must satisfy

$$r_1 \sim r_2 \leqslant A_1 A_2 \leqslant r_1 + r_2.$$

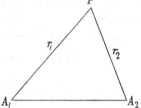

EXERCISE 1 G

1. Show in a figure the points whose polar coordinates are

$(1, 0)$, $(2, \frac{1}{2}\pi)$, $(3, \frac{5}{6}\pi)$, $(1, \pi)$, $(-1\cdot1, \pi)$, $(1\cdot1, -\pi)$, $(\frac{1}{2}, \frac{3}{2}\pi)$, $(1, -\frac{3}{2}\pi)$.

In Nos. 2–9, state the restricted polar coordinates of the point whose cartesian coordinates are:

2. $(1, \sqrt{3})$. **3.** $(1, -\sqrt{3})$. **4.** $(-1, -\sqrt{3})$. **5.** $(-1, \sqrt{3})$.

6. $(a^2, -b^2)$. **7.** $(-a^2, b^2)$. **8.** $(-a^2, -b^2)$. **9.** (x, y).

In Nos. 10–18, state the cartesian coordinates of the points whose polar coordinates are:

10. $(2\sqrt{2}, \frac{1}{12}\pi)$. **11.** $(2\sqrt{2}, \frac{11}{12}\pi)$.

12. $(2\sqrt{2}, -\frac{1}{12}\pi)$. **13.** $(5, \tan^{-1}\frac{3}{4})$.

14. $(5, \pi + \tan^{-1}\frac{3}{4})$. **15.** $(13, -\pi + \tan^{-1}\frac{5}{12})$.

16. $(101, \pi - \tan^{-1}\frac{20}{99})$. **17.** (ρ^2, ϕ).

18. $(-\rho^2, \phi)$.

In Nos. 19–22, find the distance between the points whose polar coordinates are:

19. $(2, \frac{3}{4}\pi)$, $(3, \frac{11}{12}\pi)$. **20.** $(-3, \frac{1}{4}\pi)$, $(5, \frac{7}{12}\pi)$.

21. $(-\rho, \alpha)$, $(2\rho, \alpha + \pi)$. **22.** $(5, 0)$, $(7, \sec^{-1} 7)$.

23. Show in a figure the points with the following bipolar coordinates, taking $A_1 A_2 = 10$:

$$(2, 12), \; (12, 2), \; (11, 11), \; (6, 8), \; (5, 5), \; (0, 10).$$

In Nos. 24 and 25, find the mid-point of the line joining the points whose polar coordinates are:

24. $(3, 70°), \; (3, 110°).$ **25.** $(r_1, \theta_1), \; (r_2, \theta_2).$

1·7. Area of Triangle $P_1 P_2 P_3$

1·71. If the polar coordinates of P_1, P_2 are (r_1, θ_1), (r_2, θ_2), the area of the triangle OP_1P_2 is $\frac{1}{2}r_1r_2 \sin(\theta_1 - \theta_2)$ or $\frac{1}{2}r_1r_2 \sin(\theta_2 - \theta_1)$, whichever is positive.

$$\pm \triangle OP_1 P_2 = \frac{1}{2}r_1 r_2 (\sin\theta_2 \cos\theta_1 - \sin\theta_1 \cos\theta_2)$$

$$= \frac{1}{2}(x_1 y_2 - x_2 y_1)$$

$$= \frac{1}{2} \begin{vmatrix} x_1 & y_1 \\ x_2 & y_2 \end{vmatrix}.$$

If new axes are chosen parallel to OX, OY and passing through P_3, the new coordinates of P_1, P_2 are $(x_1 - x_3, y_1 - y_3)$, $(x_2 - x_3, y_2 - y_3)$.

$$\therefore \; \triangle P_1 P_2 P_2 = \pm \frac{1}{2} \begin{vmatrix} x_1 - x_3 & y_1 - y_3 \\ x_2 - x_3 & y_2 - y_3 \end{vmatrix}$$

$$= \pm \frac{1}{2} \begin{vmatrix} x_1 & y_1 & 1 \\ x_2 & y_2 & 1 \\ x_3 & y_3 & 1 \end{vmatrix}.$$

The sign is chosen so that the formula gives a positive area. This assumes that no convention has been made to distinguish between positive and negative areas.

The formula can be applied to find the area of other rectilinear figures by dividing them into triangles. It can also be used as a test of the collinearity of three points.

1·72. *Collinearity.* Three points (x_1, y_1), (x_2, y_2), (x_3, y_3) are collinear if and only if the area of the triangle $P_1 P_2 P_3$ is zero. Hence by 1·71 the necessary and sufficient condition for the collinearity of the three points is

$$\begin{vmatrix} x_1 & y_1 & 1 \\ x_2 & y_2 & 1 \\ x_3 & y_3 & 1 \end{vmatrix} = 0.$$

1·73. EXAMPLE. Find the area of the convex quadrilateral whose four vertices are $(0, 3)$, $(6, 2)$, $(4, 6)$, $(5, 1)$.

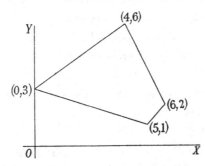

A rough figure shows that the quadrilateral is composed of the triangles $(0, 3)$ $(4, 6)$ $(6, 2)$ and $(0, 3)$ $(5, 1)$ $(6, 2)$. The areas of these triangles are the absolute values of

$$\tfrac{1}{2}\begin{vmatrix} 0 & 3 & 1 \\ 4 & 6 & 1 \\ 6 & 2 & 1 \end{vmatrix} = \begin{vmatrix} 0 & 0 & 1 \\ 2 & 3 & 1 \\ 3 & -1 & 1 \end{vmatrix} = -11$$

and

$$\tfrac{1}{2}\begin{vmatrix} 0 & 3 & 1 \\ 5 & 1 & 1 \\ 6 & 2 & 1 \end{vmatrix} = \tfrac{1}{2}\begin{vmatrix} 0 & 0 & 1 \\ 5 & -2 & 1 \\ 6 & -1 & 1 \end{vmatrix} = 3\tfrac{1}{2}.$$

Hence the area of the quadrilateral is $14\tfrac{1}{2}$.

EXERCISE 1H

In Nos. 1–10, find the area of the triangle.

1. $(0, 0)$ $(3, 0)$ $(0, -5)$.
2. $(5, 3)$ $(10, 8)$ $(12, 1)$.
3. $(8, 11)$ $(-6, -9)$ $(0, 0)$.
4. $(3, 5)$ $(7, -4)$ $(2, 1)$.
5. $(9, 8)$ $(-10, 3)$ $(-4, -1)$.
6. $(0, 1)$ $(\sqrt{3}, 0)$ $(\sqrt{3}, 2)$.
7. $(4, -7)$ $(18, 25)$ $(3, 11)$.
8. $(-a, b)$ $(a, -b)$ $(0, c)$.
9. $(bc, b+c)$ $(ca, c+a)$ $(ab, a+b)$.
10. (m, m^2) (n, n^2) (p, p^2).

In Nos. 11–14, find the area of the quadrilateral.

11. $(2, 1)$ $(7, 2)$ $(5, 3)$ $(6, -1)$.   **12.** $(0, 0)$ $(5, 0)$ $(0, 3)$ $(9, 7)$.

13. $(1, 0)$ $(2, 3)$ $(1\frac{1}{2}, 4)$ $(0, 2)$.

14. $(9, 7)$ $(31, 27)$ $(43, -13)$ $(-5, -21)$.

15. Find the area of the pentagon $(2, -1)$ $(9, 1)$ $(11, 7)$ $(2, 10)$ $(-3, 6)$.

In Nos. 16–21, verify the collinearity of

16. $(4, 7)$, $(3, 5)$, $(2, 3)$ **17.** $(1, 3)$, $(5, -1)$, $(-3, 7)$.

18. $(k + 2, k - 3)$, $(5, 7)$, $(2k - 1, 2k - 13)$.

19. $(-3, -4)$, $(-1, 2)$, $(7, 26)$, $(2, 11)$.

20. $(3p + 2, 1 - 2p)$, $(p + 3, 0)$, $(2p + 5, -2 - p)$.

21. $(a, b + c)$, $(b, c + a)$, $(c, a + b)$.

In Nos. 22, 23, find k, given that the points are collinear.

22. $(3k + 1, k)$, $(k - 2, k + 1)$, $(7k, k - 1)$.

23. $(k + 11, 6)$, $(2 - 3k, 5 - 2k)$, $(14, 13 - k)$.

24. Prove that $OP_1 OP_2 \cos P_1 OP_2 = x_1 x_2 + y_1 y_2$.

25. Prove that $OP_1 OP_2 \sin P_1 OP_2 = \pm (x_1 y_2 - x_2 y_1)$.

In Nos. 26 and 27, find the areas of the triangle

26. $\{\omega\}$. $(a, 0)$ $(0, b)$ $(0, 0)$.   **27.** $\{\omega\}$. (a, b) $(c, 0)$ $(0, 0)$.

28. $\{\omega\}$. Show that the area $OP_1 P_2$ is $\pm \frac{1}{2} r_1 r_2 \sin(\theta_1 - \theta_2)$, where

$$r_1 \cos \theta_1 = x_1 + y_1 \cos \omega, \quad r_1 \sin \theta_1 = y_1 \sin \omega, \quad \text{etc.},$$

and deduce that the area of $P_1 P_2 P_3$ is

$$\pm \tfrac{1}{2} \begin{vmatrix} x_1 & y_1 & 1 \\ x_2 & y_2 & 1 \\ x_3 & y_3 & 1 \end{vmatrix} \sin \omega.$$

29. Find the area of the triangle $(2, 60°)$ $(2, 30°)$ $(\cos 15°, 45°)$.

30. Find the area of the quadrilateral

$$(3, -20°) (8, 10°) (5, 40°) (2, 100°).$$

1·8. Homogeneous Coordinates

1·81. *Trilinear Coordinates.* If ABC is a fixed triangle, the position of a point P in its plane may be determined by its perpendicular distances α, β, γ from the sides. The convention is made that the first coordinate α is given a + or − sign according as the point is on the same side of BC as A or the opposite side. Similar conventions are made for β and γ.

ABC is called the *triangle of reference* and the coordinates are called *trilinear* coordinates.

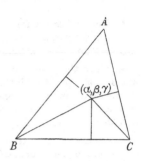

The three trilinear coordinates are effectively equivalent to only two. When two of them are known, the third can be found from the relation now to be proved.

Let P be a point (α, β, γ) inside the triangle of reference. Then

$$\triangle BPC = \tfrac{1}{2}\alpha BC = \tfrac{1}{2}a\alpha.$$

Similarly $\triangle CPA = \tfrac{1}{2}b\beta$ and $\triangle APB = \tfrac{1}{2}c\gamma.$

Hence $a\alpha + b\beta + c\gamma = 2\triangle ABC, = 2\triangle$, say.

See also Exercise 1ɪ, No. 2.

This relation shows that the coordinates α, β, γ cannot assume arbitrary values. The position of P is determined by two of the coordinates. In general it is most convenient to use the ratios $\alpha : \beta : \gamma$ instead of the actual values of two of α, β, γ. When the ratios are known, the actual values can be found from $a\alpha + b\beta + c\gamma = 2\triangle$.

1·82. Example. Prove that the bisectors of the angles of a triangle are concurrent.

Take the triangle as triangle of reference.

For any point (α, β, γ) on the bisector of B, $\alpha = \gamma$, and for any point on the bisector of C, $\alpha = \beta$. Hence for the point of intersection of these two bisectors, $\beta = \gamma$. This point therefore lies on the bisector of A.

This proof is equivalent to the usual proof by elementary geometry but it is expressed in a new language.

EXERCISE 1ɪ

1. Mark in a diagram the signs of α, β, γ in the seven regions of the plane.

2. Verify that the formula $a\alpha + b\beta + c\gamma = 2\triangle$ is true for all positions of (α, β, γ) by considering points in regions 2 and 3 of the figure.

3. Show that the trilinear coordinates of the point of intersection of the medians of the triangle of reference are in the ratios $bc : ca : ab$, and find their actual values.

4. Find the trilinear coordinates of the circumcentre of ABC.

5. Find the trilinear coordinates of the orthocentre of ABC.

6. For what points is $\alpha = 0$? What are the coordinates of A? What are the coordinates of the mid-point of BC?

7. If $a = 5$, $b = 4$, $c = 3$, find the actual trilinear coordinates of a point if they are proportional to 6, 7, 8.

8. If (α, β, γ) is on the circumcircle of the triangle of reference, prove that $a\beta\gamma + b\gamma\alpha + c\alpha\beta = 0$.

9. Find the coordinates of the points of intersection of the tangents at A, B, C to the circle ABC with BC, CA, AB.

10. What is the moment about (α, β, γ) of the resultant of forces P, Q, R along BC, CA, AB?

11. If the trilinear coordinates of a point on the line of action of the resultant of forces P, Q, R along BC, CA, AB are proportional to α, β, γ, prove that $P\alpha + Q\beta + R\gamma = 0$.

12. If P is the point (α, β, γ), find its cartesian coordinates x, y referred to oblique axes BC, BA. Verify that if α, β, γ satisfy a simple equation $l\alpha + m\beta + n\gamma = 0$, then x, y satisfy a simple equation

$$px + qy + r = 0.$$

1·83. *Areal Coordinates.* This is another system of coordinates for which a triangle of reference is used. The areal coordinates of a point P for a triangle of reference ABC are

$$x_1 \equiv \frac{\triangle BPC}{\triangle ABC}, \qquad x_2 \equiv \frac{\triangle CPA}{\triangle ABC}, \qquad x_3 \equiv \frac{\triangle APB}{\triangle ABC}$$

with the same sign conventions as for trilinear coordinates.

The areal coordinates satisfy

$$x_1 + x_2 + x_3 = 1.$$

This relation determines the three areal coordinates when their ratios $x_1 : x_2 : x_3$ are known. The ratios are generally used instead of the actual coordinates; thus the point $(2, 4, 1)$ means the point whose actual coordinates are $(\frac{2}{7}, \frac{4}{7}, \frac{1}{7})$.

The point $P(x_1, x_2, x_3)$ is the centre of mass of x_1 at A, x_2 at B, and x_3 at C. For, in the figure,

$$BX : XC = \triangle BAP : \triangle CAP = x_3 : x_2.$$

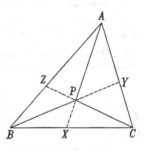

Therefore X is the centre of mass of x_2 at B and x_3 at C. Hence the centre of mass of x_1 at A, x_2 at B, and x_3 at C lies on AX. Similarly it lies on BY. Hence P is this centre of mass. Thus the areal coordinates are the two-dimensional barycentric coordinates. See 1·14.

1·84. *General Homogeneous Coordinates.* Trilinear or areal coordinates can occasionally be used to prove special properties of the triangle. Areal coordinates are to be preferred because of the simplicity of the relation $x_1 + x_2 + x_3 = 1$.

It is chiefly for general geometry that is not concerned with measurement that coordinates of this sort are used, and then it is immaterial which system is adopted. The actual trilinear coordinates of the incentre of the triangle of reference are (r, r, r) and the actual areal coordinate of the median point are $(\frac{1}{3}, \frac{1}{3}, \frac{1}{3})$. Thus, if only the ratios of the coordinates are considered, $(1, 1, 1)$ represents the incentre or the median point according as the coordinates are trilinear or areal. By using areal coordinates instead of trilinear the proof in 1·82 can be made to show that the medians are concurrent.

Other coordinates can be defined such that the orthocentre, circumcentre, or any other specified point has coordinates proportional to 1, 1, 1.

All these coordinates are special cases of general homogeneous coordinates which are introduced from another point of view in Volume II. They are mentioned in this chapter only as illustrations of different kinds of coordinates.

EXERCISE 1 J

1. Verify that the relation $x_1 + x_2 + x_3 = 1$ between areal coordinates holds for all positions of (x_1, x_2, x_3) in the plane.

In Nos. 2–6, find the ratios of the areal coordinates of the given points of the triangle of reference and deduce the actual coordinates.

2. Incentre. **3.** Ecentres. **4.** Circumcentre. **5.** Orthocentre.

6. For what points is $x_1 = 0$? What are the areal coordinates of A? What are the coordinates of the mid-point of BC?

7. If (α, β, γ) are the actual trilinear coordinates and (x_1, x_2, x_3) are the actual areal coordinates of the same point, prove that

$$x_1 = a\alpha/2\triangle, \quad x_2 = b\beta/2\triangle, \quad x_3 = c\gamma/2\triangle.$$

8. Give the algebraic sum of the moments of forces P, Q, R along BC, CA, AB about the point whose actual areal coordinates are (x_1, x_2, x_3).

9. If (x_1, x_2, x_3) lies on the line of action of the resultant of forces al_1, bl_2, cl_3 acting along BC, CA, AB, prove that $l_1x_1 + l_2x_2 + l_3x_3 = 0$.

1·9. Geometry of Three Dimensions

1·91. Three coordinates are required to determine the position of a point in space. These may be the cartesian coordinates referred to axes $X'OX$, $Y'OY$, $Z'OZ$. The coordinates of P are then the one-dimensional coordinates x, y, z of the points L, M, N where the axes meet the planes through P parallel to YOZ, ZOX, XOY. O and P are opposite corners of a parallelepiped having OL, OM, ON as edges. In general the angles YOZ $(=\lambda)$, ZOX $(=\mu)$, XOY $(=\nu)$ need not be right angles.

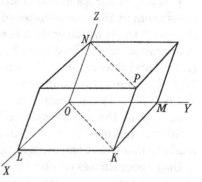

The position of P is also completely determined by a single vector, namely the displacement OP of P from the origin.

If unit vectors in OX, OY, OZ are denoted by $\mathbf{i}$, $\mathbf{j}$, $\mathbf{k}$, and P is the point (x, y, z), then

$$\mathbf{OP} = \mathbf{OM} + \mathbf{MK} + \mathbf{KP}$$

$$= x\mathbf{i} + y\mathbf{j} + z\mathbf{k}.$$

The results about vectors in a plane given in 1·3 are applicable to vectors in space.

1·92. Let $\lambda = \mu = \nu = \frac{1}{2}\pi$. The first two of the coordinates x, y, z are sometimes replaced by polar coordinates $\rho = OK$ and $\phi = XOK$ shown in the figure, so that P is then determined by ρ, ϕ, z. These are called *cylindrical* coordinates.

1·93. Again, in the plane $ZOKP$, ρ and z may be replaced by polar coordinates $r = OP$ and $\theta = ZOP$, so that P is determined by r, θ, ϕ. These are called *spherical polar* coordinates. If the sphere centre O and radius r is regarded as the Earth, and XOY as the plane of the equator, and XOZ as the plane of the meridian of Greenwich, then θ is the complement of the latitude and ϕ is the longitude of P.

1·94. A point in space may also be fixed wo a tetrahedron of reference by coordinates analogous to trilinear, areal, or general homogeneous coordinates in two dimensions. Four coordinates x_1, x_2, x_3, x_4 are then used and there is a relation between them, so that they are equivalent to three independent coordinates. A special system of coordinates can be chosen to make a specified point (not on a face of the tetrahedron of reference) have coordinates proportional to 1, 1, 1, 1. When metrical properties are not involved, the general coordinates are left arbitrary.

1·95. *Coordinates in general.* The idea of representation by means of coordinates is not restricted to points. For instance, in geometry of two dimensions a circle may be represented by three coordinates; these can be the radius of the circle and the cartesian coordinates of its centre. Again, in space a sphere can be represented by four coordinates.

It is important to find suitable coordinates to represent a line. In a plane, the number of coordinates needed to represent a line is two. These might be the intercepts OA, OB made by the line on the axes; but every line through the origin would then have coordinates 0, 0. Actually it is a modification of these intercepts that suggests the coordinates used in Chapter 4. Another possibility is to use the coordinates of any two points of the line to determine it; a line does not then have definite coordinates; its four coordinates are not independent.

EXERCISE 1κ

1. Use a figure of a parallelepiped to illustrate the equivalence of

$$P + (Q + R), \quad (P + Q) + R, \quad (R + P) + Q, \text{ etc.}$$

2. If $ABCD$ is a tetrahedron, and V is any point, interpret geometrically the equivalence of expressions such as

$$(VA + VB + VC) + VD \quad \text{and} \quad (VA + VB) + (VC + VD).$$

3. In the figure of 1·91 use the vector $OL + OM + ON$ to show that the plane LMN trisects OP.

4. Express OP^2 in terms of the cartesian coordinates (x, y, z) of P by squaring the vector OP. State the formula for $P_1 P_2^2$.

5. If $P_1 O P_2$ is a right angle, prove that $x_1 x_2 + y_1 y_2 + z_1 z_2 = 0$ by using $P_1 P_2^2 = OP_1^2 + OP_2^2$ and by using $OP_1 . OP_2 = 0$.

6. What points have their first cartesian coordinate (x) zero, and what points have $y = z = 0$?

In Nos. 7, 8, describe the figure formed by points whose rectangular cartesian coordinates are:

7. $(\pm 1, \pm 1, \pm 1)$.   **8.** $(\pm 1, 0, 0), (0, \pm 1, 0), (0, 0, \pm 1)$.

9. Express the cylindrical coordinates ρ, ϕ in terms of the cartesian coordinates x, y, z.

10. What points have (i) $\rho = k$, (ii) $\phi = \alpha$, (iii) $\rho = k$ and $\phi = \alpha$, where k, α are constants?

11. Express the spherical polar coordinates r, θ, ϕ in terms of the rectangular cartesian coordinates x, y, z.

12. What points have (i) $r = k$, (ii) $\theta = \alpha$, (iii) $\phi = \beta$, (iv) $\theta = \alpha$ and $\phi = \beta$, (v) $r = k$ and $\phi = \beta$, (vi) $r = k$ and $\theta = \alpha$?

13. If M_{rs} is the mid-point of the edge $P_r P_s$ of a tetrahedron $P_1 P_2 P_3 P_4$, prove that $M_{12} M_{34}$, $M_{13} M_{24}$ have the same mid-point $(\frac{1}{4}(x_1 + x_2 + x_3 + x_4), \frac{1}{4}(y_1 + y_2 + y_3 + y_4))$. Prove also that this point lies on the line joining P_1 to the centroid of P_2, P_3, P_4 and on the three similar lines.

14. Show that the four coordinates of a point in space which are its perpendicular distances from the faces of a tetrahedron of reference satisfy an equation $a_1 x_1 + a_2 x_2 + a_3 x_3 + a_4 x_4 = 3V$, where V is the volume of the tetrahedron, and interpret a_1, a_2, a_3, a_4.

15. Define the barycentric coordinates of a point in space.

Chapter 2

EQUATIONS AND LOCI

2·1. Graphs

2·11. *Cartesian Graphs.* Points whose coordinates satisfy an equation $y = f(x)$ or $g(x, y) = 0$ can be plotted by giving values to x and calculating the corresponding values of y. This method is particularly convenient when $f(x)$ is a one-valued function of x. For an equation $x = f(y)$ it is convenient to give values to y and calculate the values of x.

If the axes are rectangular, the graphs are usually drawn with the help of graph paper. Graph paper suitable for oblique axes can be obtained or can easily be made.

The equation of a graph may be given in another form, x and y being given as functions of a single variable t. For example, a graph can be drawn by giving values to t in the equations $x = t^2$, $y = t^3$. This graph is the same as that given by $y^2 = x^3$, but the use of a single variable will be seen to be important.

2·12. *Polar Graphs.* From an equation $r = f(\theta)$ a curve can be plotted by giving values to θ, just as a cartesian graph is plotted from $y = f(x)$.

Polar equations may also occur in the form $\theta = f(r)$ or $g(r, \theta) = 0$.

Sometimes it is convenient to use polar graph paper. This consists of concentric circles, on each of which r is constant, and their radii, on each of which θ is constant, just as ordinary graph paper consists of lines on which x is constant and lines on which y is constant.

Polar equations are rarely used in the general theory of algebraic curves, but an algebraic equation can always be expressed in polar form and this form is useful for particular problems.

The transformation from cartesian coordinates to polar coordinates or from polar coordinates to cartesian coordinates involves the use of the relations $x = r\cos\theta$, $y = r\sin\theta$.

An algebraic equation in one system will therefore not be transformed into an algebraic equation in the other system. For example, the polar equation $r = \theta$ is transformed into the cartesian equation $y = x \tan \sqrt{(x^2 + y^2)}$.

2·13. *Bipolar Graphs.* Points can be plotted for which the bipolar coordinates r_1, r_2 satisfy a given equation. For this purpose paper ruled with two sets of concentric circles centres A_1, A_2 would be useful.

EXERCISE 2A

In Nos. 1–6, sketch and compare the graphs of the equation for $\omega = \frac{1}{2}\pi$ and $\omega = \frac{1}{3}\pi$.

1. $y = x^2.$ **2.** $y^3 = x.$ **3.** $x^2 + y^2 = 1.$

4. $xy = 24.$ **5.** $x^2 - xy + y^2 = 1.$ **6.** $y = \sin x.$

In Nos. 7–12, by giving values to t, plot points (x, y) such that

7. $x = t,\ y = t^2.$ **8.** $x = t + 3,\ y = 2t + 5.$

9. $x = t^2,\ y = t^3.$ **10.** $x = \cos t,\ y = \sin t.$

11. $x = t^2,\ y = 1/t.$ **12.** $x = t^6,\ y = t^5.$

13. Sketch the graph of $r = \theta$ if (i) r, θ are both unrestricted; (ii) $r > 0$, θ unrestricted; (iii) $r > 0$, $-\pi < \theta \leqslant \pi$.

In Nos. 14–22, sketch the graph of the polar equation.

14. $r = \sin \theta.$ **15.** $r = \sec \theta.$ **16.** $r = \tan \theta.$

17. $r = 1.$ **18.** $\theta = 1.$ **19.** $r = \sin \theta + 1.$

20. $r = 4 - k \cos \theta$ for $k = 1, 2, 3, 4, 8.$

21. $1/r = 1 - k \cos \theta$ for $k = \frac{1}{2}, 1, 2.$ **22.** $r = 1 - 1/\theta.$

In Nos. 23–26, transform the cartesian equation into a polar equation.

23. $x^2 + y^2 = ax.$ **24.** $(x - a)(y - b)(y - cx) = 0.$

25. $(x^2 + y^2)^3 = c^2 xy(x^2 - y^2).$ **26.** $(x^2 + y^2)^2 - 2ax(x^2 + y^2) = a^2 y^2.$

In Nos. 27–34, transform the polar equation into a cartesian equation.

27. $r = a.$ **28.** $\theta = \alpha.$ **29.** $r \sin \theta = a.$

30. $r = a \cos \theta.$ **31.** $r^2 = a^2 \sin 2\theta.$ **32.** $r\theta = \pi.$

33. $1/r^2 = (\cos^2 \theta)/a^2 + (\sin^2 \theta)/b^2.$ **34.** $r = a \tan \theta + b \sec \theta.$

In Nos. 35–40, describe or sketch the graph given by the bipolar equation, taking $A_1A_2 = 10$.

35. $r_1 = r_2$. **36.** $r_1 + r_2 = 15$. **37.** $r_1 - r_2 = 3$.

38. $r_1^2 + r_2^2 = 68$. **39.** $r_1^2 - r_2^2 = 30$. **40.** $r_1 = 2r_2$.

2·2. Degree of Freedom

2·21. We have seen that in geometry of one dimension a point has one coordinate, that in two dimensions it has two, and that in three dimensions it has three. In fact, the number of dimensions is the same as the number of coordinates required to determine a point.

This is also expressed by saying that an arbitrary point in a line has one degree of freedom, a point in a plane has two degrees of freedom, and a point in space has three degrees of freedom.

For an arbitrary point (x, y) in two dimensions there is a double choice, the choice of x and the choice of y. For points in a plane which are not arbitrary but are restricted, say, to lie on the graph of $y = x^3$, there is no longer the double choice: when x has been chosen, y is determined by the equation. An arbitrary point of such a curve is therefore said to have only one degree of freedom: it needs only one coordinate to fix the position of a point on the curve.

Curves are not always given by equations of the form $y = f(x)$, $x = g(y)$, or $r = h(\theta)$. An important kind of equation is illustrated in 2·11. To determine a point of a curve given by equations like $x = t^2$, $y = t^3$, there is a single choice, the choice of t; a point of this curve has one degree of freedom.

The equation of a curve, whether of the form $y = f(x)$, $x = g(y)$, $r = h(\theta)$, $\theta = k(r)$, or $\{x = f(t), y = g(t)\}$, restricts the position of the point and reduces its freedom from two degrees to one.

A point on a curve $f(x, y) = 0$ or $f(r, \theta) = 0$ is also said to have one degree of freedom, although it is not true that, for a given value of x, the equation $f(x, y) = 0$ necessarily determines a unique value of y. There will usually be a finite number of values, and the effect of the equation is to reduce the freedom of (x, y) from two degrees to one.

2·22. Instead of saying that the geometry of the line is one-dimensional or that a point on a line has one coordinate or one degree of freedom, we shall sometimes say that there are ∞^1 points on a line. In the same sense there are ∞^1 points on a curve. And since a point of a plane has two coordinates or two degrees of freedom we say that there are ∞^2 points in a plane. Similarly there are ∞^3 points in space.

There is no such number as ∞, ∞^1, ∞^2, or ∞^3. The statement that there are ∞^2 points in a plane only implies that a point in a plane has two coordinates. Nevertheless this form of words is convenient.

2·23. When a curve is defined by equations

$$x = f(t), \quad y = g(t),$$

t is called a *parameter*, and the equations are called parametric equations of the curve.

In Exercise 2A, Nos. 7–12, it is easy to discover, by eliminating t, what are the corresponding cartesian equations. For example, in No. 7, $y = x^2$; in No. 8, $2x - y = 1$; and in No. 10, $x^2 + y^2 = 1$. The opposite process of discovering parametric equations is more important and more difficult. It is easy to guess parametric equations for $y^2 = x^3$, $xy^2 = 1$, $x^5 = y^6$; these are given in Nos. 9, 11, 12. But it is not always possible to find a suitable parametric representation. The curve $x^3 + y^3 = 1$ might be represented by

$$x = t, \quad y = \sqrt[3]{(1 - t^3)}$$

but cannot be represented by rational parametric equations.

The most useful parametric equations are of the form

$$x = \frac{f(t)}{g(t)}, \qquad y = \frac{h(t)}{k(t)},$$

where $f(t)$, $g(t)$, $h(t)$, $k(t)$ are polynomials in t. Also it is desirable that each point of the curve should determine just one value of the parameter.

For example, $x = t$, $y = t^2$ is a suitable representation for $y = x^2$; but $x = t^2$, $y = t^4$ is not suitable because $t = t_1$, $t = -t_1$ give the same point.

Sometimes parametric equations are at once evident: if the equation of the curve is $y = f(x)$, it may be represented by $\{x = t,\ y = f(t)\}$ in which x itself is, in effect, the parameter. Similarly, $x = g(y)$ may be represented by $\{x = g(t),\ y = t\}$.

A simple equation $ax + by = c$ can be replaced by parametric equations $x = bt,\ y = -at + c/b$.

EXERCISE 2B

1. State the number of degrees of freedom of a circle in a plane, a line in a plane, a sphere in space, a circle in space, a line in space.

2. Verify that $x = 3t - 4$, $y = 2t + 1$ are parametric equations of $2x - 3y + 11 = 0$.

3. Verify that $x = (1 - t^2)/(1 + t^2)$, $y = 2t/(1 + t^2)$ are parametric equations of $x^2 + y^2 = 1$.

In Nos. 4–13, find the x, y equation.

4. $x = t + 1,\ y = t - 1$.   **5.** $x = 3t - 5,\ y = 4 - 2t$.

6. $x = (4t - 3)/(t + 1),\ y = (t + 2)/(t + 1)$.   **7.** $x = 2t^2,\ y = 1/t$.

8. $x = 2at^3,\ y = 3at^4$.   **9.** $x = \sec t,\ y = \tan t$.

10. $x = 4 \sin t - 3,\ y = 3 \cos t + 4$.   **11.** $x = at^2 + bt,\ y = ct$.

12. $x = a \cos^3 t,\ y = a \sin^3 t$.   **13.** $x : y : 1 = 1 + t^2 : 2t : 1 - t^2$.

In Nos. 14–22, give parametric equations.

14. $y = x^3$.   **15.** $y^3 = ax^2$.   **16.** $x^2 + 4y^2 = 9$.

17. $2x + 3y = 4$.   **18.** $5x - 6y = 7$.   **19.** $ax + by + c = 0$.

20. $xy = 1$.   **21.** $x^2 y = a^3$.   **22.** $y = x(x - 1)$.

2·3. Cartesian Equations and Loci in Two Dimensions

2·31. The study of the curves that can be represented by given cartesian equations is an important part of analytical geometry. It begins with the theorem that any simple equation represents a line, and continues with the study of second degree equations. We shall not in this book attempt the general discussion of curves whose equations are of the third degree or of higher degree. Newton's classification of cubic curves, published in 1704, gave more than seventy varieties of these curves, many of which are illustrated in Miscellaneous Exercise

C. See *H.M.* 3rd edition, p. 350. There are more than 200 varieties of quartic curve. The precise numbers of cubics and quartics depend upon the principle of classification that is adopted.

The solution of a locus problem by cartesian methods consists essentially of three parts: the defining property of the locus is expressed algebraically; the algebraic relations are reduced to a single equation; finally the equation is interpreted. It is for the final interpretation that we need to know what is represented by a given equation. Unless we have made some progress in that branch of the subject, our solutions are necessarily incomplete. Nevertheless a locus problem can be regarded as partially solved when the cartesian equation of the locus has been found. Certain conclusions can be drawn about the curve from the degree and form of the equation even if the complete interpretation is not known. For example, properties of symmetry can often be recognised (2·32). When the graph of one function is known, the graphs of related functions can be deduced (2·33).

Another important part of analytical geometry is the discussion of the properties of special curves. For this purpose, axes are chosen so as to simplify as much as possible the equation of the curve under discussion. The properties of three important curves are studied in this way in Chapters 13, 14, 15. Other curves are illustrated in Miscellaneous Exercise D and elsewhere.

2·32. EXAMPLE. Discuss the symmetry of the curves

$$\text{(i) } x^2 - y^2 = x^4 y, \quad \text{(ii) } x^3 - y^3 = x^2 y^2.$$

(i) Since only even powers of x occur in the equation, it follows that if (a, b) lies on the curve, so also does $(-a, b)$; the condition being $a^2 - b^2 = a^4 b$ in each case. Thus the curve is symmetrical about OY. But a different condition is required for $(a, -b)$ to lie on the curve: the curve is not symmetrical about OX. Nor is it symmetrical through the origin, since $a^2 - b^2 = a^4 b$ does not secure that $(-a, -b)$ lies on the curve.

(ii) This curve is not symmetrical about OX, OY, or through O. But if (a, b) lies on the curve, so also does $(-b, -a)$. Hence there is symmetry about the bisector of $\angle XOY'$.

2·33. EXAMPLE. Sketch the curve $y^2 = 1 - \dfrac{x^2}{4} + \dfrac{1}{x-1}$.

The curves $y = 1$, $y = -\tfrac{1}{4}x^2$, $y = 1/(x-1)$ are drawn in figures (a), (b), (c). By adding the ordinates of these three curves the curve $y = 1 - \tfrac{1}{4}x^2 + 1/(x-1)$ is found. See figure (d). But the ordinate of $y^2 = 1 - \tfrac{1}{4}x^2 + 1/(x-1)$ is plus or minus the square root of the ordinate of the curve in figure (d) and hence figure (e) can be sketched.

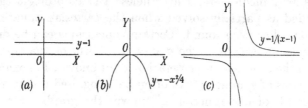

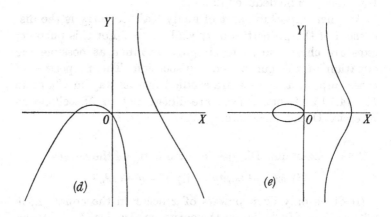

EXERCISE 2c

1. When the graph of $y = f(x)$ is given, how can the graphs of $y = f(x) + c$, $y = f(x + c)$, $y = cf(x)$ be deduced?

2. Sketch in the same figure the graphs of x^2, $x^2 + 2$, $(x + 2)^2$, $2x^2$.

3. Sketch in the same figure the graphs of $\sin x$, $\sin x + 1$, $\sin(x + 1)$, $2 \sin x$, $x \sin x$.

4. When the polar graph of $r = f(\theta)$ is given, how can the graphs of $r = f(\theta) + c$, $r = f(\theta + \gamma)$, $r = cf(\theta)$ be deduced?

5. What is the relation of the graph of $r = -f(\theta)$ to that of $r = f(\theta)$? Answer the same question for $r = f(-\theta)$ and $r = f(\theta)$.

6. Sketch in the same figure the polar graphs of
$$r = \sec\theta, \quad r = \sec\theta + 1, \quad r = \sec(\theta + 1).$$

7. Sketch in the same figure the polar graphs of $r = \theta$, $r = \theta + 2$, $r = 2\theta$ for values of θ from 0 to 2π.

In Nos. 8–16, state any obvious symmetry of the curve.

8. $y^3 = x^5$.   **9.** $y^2 = x^7$.   **10.** $x^5 - xy + y^5 = 0$.

11. $(x^2 + y^2)^2 = x^2 - y^2$.   **12.** $x^5 - y^5 = 5xy$.

13. $x^6 - y^6 = 6x^3y^3$.   **14.** $x = t$, $y = t^5$.

15. $x = t^2$, $y = t^3 + t$.   **16.** $y = \cos x$.

17. State how to tell from an equation $f(x, y) = 0$ whether there is symmetry about the bisector of $\angle XOY$.

18. If a curve has symmetry about both axes or about both bisectors of the angles between the axes, show that it has symmetry about the origin.

19. Sketch the graphs of $x + 1/x$, $x - x^3$, $x + \log x$, $x \log x$.

20. Sketch the graphs of $y = x + 1/x^2$, $y^2 = x + 1/x^2$.

In Nos. 21–26, state what is represented by the equation.

21. $x - 3 = 0$.   **22.** $x^2 - x = 6$.   **23.** $f(x) = 0$.

24. $f(y) = 0$.   **25.** $f(x/y) = 0$.   **26.** $xy = 0$.

27. Sketch the graph of $y = 1 - \frac{1}{4}x^2 + \epsilon/(x - 1)$, where ϵ is a small positive constant.

[Further examples of graphs are given in Miscellaneous Exercise C.]

2·4. Analytical Solution of Locus Problems

2·41. EXAMPLE. Find the locus of a point P whose distance from the origin is twice its distance from A (1, 0).

Let (x, y) be an arbitrary point on the locus.

Then $$OP^2 = 4OA^2.$$
$$\therefore \quad x^2 + y^2 = 4\{(x - 1)^2 + y^2\},$$
$$\therefore \quad x^2 + y^2 - \tfrac{8}{3}x + \tfrac{4}{3} = 0.$$

This is the equation of the locus. It may be written
$$(x-\tfrac{4}{3})^2+y^2 = \tfrac{16}{9}-\tfrac{4}{3} = (\tfrac{2}{3})^2,$$
and this expresses that the distance from the point (x,y) of the locus to the fixed point $(\tfrac{4}{3},0)$ is $\tfrac{2}{3}$. Hence the locus is the circle centre $(\tfrac{4}{3},0)$ and radius $\tfrac{2}{3}$.

2·42. EXAMPLE. Find the locus of a point which is equidistant from the point $(-5,4)$ and the line $y=2$.

Let (x,y) be an arbitrary point on the locus.

Then $\qquad \sqrt{\{(x+5)^2+(y-4)^2\}} = |\,y-2\,|.$

$\qquad\therefore\quad (x+5)^2+(y-4)^2 = (y-2)^2,$

$\qquad\qquad\therefore\quad (x+5)^2 = 4y-12.$

This is the equation of the locus. It is considered more fully in 2·51.

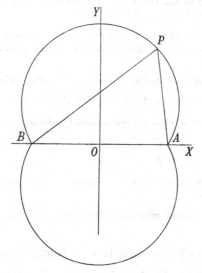

2·43. EXAMPLE. A and B are the points $(a,0)$ and $(-a,0)$. Find the locus of the point P such that $\angle APB$ is $\tfrac{1}{3}\pi$.

Take an arbitrary point $P(x,y)$ on the locus.

If $y>0$,
$$\tan XAP = y/(x-a), \quad \tan XBP = y/(x+a),$$
$$\tfrac{1}{3}\pi = \tan^{-1}\{y/(x-a)\}-\tan^{-1}\{y/(x+a)\}.$$

If $y < 0$, similarly

$$\tfrac{1}{3}\pi = \tan^{-1}\{-y/(x-a)\} - \tan^{-1}\{-y/(x+a)\}.$$

From these equations may be deduced

$$\tan\frac{\pi}{3} = \frac{\dfrac{y}{x-a} - \dfrac{y}{x+a}}{1 + \dfrac{y^2}{x^2-a^2}} = \frac{2ya}{x^2+y^2-a^2},$$

i.e. $\qquad\qquad x^2+y^2-a^2 = 2ay/\surd 3$: a circle;

and $\qquad\qquad \tan\dfrac{\pi}{3} = \dfrac{-2ya}{x^2+y^2-a^2},$

i.e. $\qquad\qquad x^2+y^2-a^2 = -2ay/\surd 3$: another circle.

The steps of the solution in 2·43 are not reversible, and so the solution does not prove that every point on the circles belongs to the locus. Geometrical considerations show that the locus consists of the major arcs of the circles shown in the figure.

In 2·41, 2·42, 2·44, 2·45 the steps are reversible.

2·44. Example. A rod AB of length 5 inches moves with A and B on fixed perpendicular lines. Find the locus of the point P in AB 2 inches from B.

Take the lines as axes of coordinates and let an arbitrary position of P be (x, y).

1st method. In the figure $OA = \tfrac{5}{2}x$ and $OB = \tfrac{5}{3}y$.

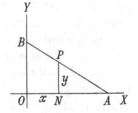

Also $\quad OA^2 + OB^2 = AB^2.$

$$\therefore \quad (\tfrac{5}{2}x)^2 + (\tfrac{5}{3}y)^2 = 5^2,$$

$$\therefore \quad 9x^2 + 4y^2 = 36.$$

This is the equation of the locus. It is a curve symmetrical about both axes meeting them at $(\pm 2, 0)$, $(0, \pm 3)$; these correspond to the positions of the rod with the ends at the origin.

2nd method. In the arbitrary position of the rod, let $\angle OAB = t$. The corresponding position of P is $(2\cos t, 3\sin t)$.

$$\therefore \quad x = 2\cos t, \quad y = 3\sin t.$$

These are parametric equations of the locus. Elimination of t gives $\frac{1}{4}x^2 + \frac{1}{9}y^2 = 1$, which is the same as the result in the first solution. To sketch the curve, it is simplest to give values to t in the parametric equations. This curve is called an ellipse.

2·45. EXAMPLE. Q is any point on the curve $y = x^3 + 1$, and P is the point of trisection of OQ nearer to O. Find the equation of the locus of P.

Let an arbitrary position of P be (x_1, y_1) and let the corresponding position of Q be (x_2, y_2).

Then $\qquad y_2 = x_2{}^3 + 1, \quad x_1 = \frac{1}{3}x_2, \quad y_1 = \frac{1}{3}y_2.$

$$\therefore \quad x_2 = 3x_1, \quad y_2 = 3y_1,$$

$$\therefore \quad 3y_1 = (3x_1)^3 + 1 = 27x_1{}^3 + 1.$$

$$\therefore \quad P \text{ lies on the curve } 3y = 27x^3 + 1.$$

2·46. In each of the locus problems that has been solved, the first step has been to take an arbitrary position of the point whose locus was required and to denote it by (x, y). This is a necessity: for the required equation is the relation which holds between the coordinates x, y of an arbitrary point on the locus.

If the arbitrary point is represented otherwise, it must also be represented by (x, y). In 2·44, 2nd method, the arbitrary point is $(2\cos t, 3\sin t)$ but it is also denoted by (x, y). Sometimes it is necessary to use (x_1, y_1) instead of (x, y) as a temporary measure to avoid confusion with the x and y coordinates of other points involved in the solution. This happens in 2·45. Even then x and y have eventually to be substituted for x_1 and y_1.

The relation which holds between the coordinates x, y of the arbitrary point P on the locus must hold no matter which point of the locus is chosen as P. This is what is meant by the use of the word 'arbitrary'. Sometimes the word 'any' is used instead of 'arbitrary'. It must be interpreted in the same way.

EXERCISE 2D

In Nos. 1–27, find the locus of P.

1. P is equidistant from $(0, 7)$ and $(-4, 0)$.

2. P is equidistant from $(3, 1)$ and $(-5, 0)$.

3. The distance of P from $(3, 0)$ is twice its distance from $(-3, 0)$.

4. The distance of P from $(3, 5)$ is twice its distance from $(-2, 6)$.

5. The distances of P from $(2, 0)$ and $(-3, 0)$ are in the ratio $3 : 2$.

6. The distances of P from $(5, 7)$ and $(-3, -1)$ are in the ratio $5 : 3$.

7. P is equidistant from $(5, 0)$ and OY.

8. P is equidistant from $(-2, 5)$ and OX.

9. The distance of P from OY is three times its distance from $(1, 1)$.

10. P is equidistant from $(3, 1)$ and $y + 5 = 0$.

11. P is equidistant from $(a, 0)$ and $x + a = 0$.

12. The sum of the squares of the distances of P from $(1, 1)$ and $(-2, 0)$ is 16.

13. The square of the distance of P from $(2, -7)$ exceeds the square of its distance from $(0, 11)$ by 10.

14. The difference between the squares of the distances of P from $(3, 4)$ and $(11, -2)$ is 7.

In Nos. 15–20, A, B are the fixed points $(a, 0)$, $(-a, 0)$ and c is constant.

15. $PA^2 + PB^2 = 2c^2$. **16.** $PA^2 - PB^2 = 2c^2$.

17. $PA = cPB$. **18.** $\angle PAB = 2 \angle PBA$.

19. $\tan PAB = 2 \tan PBA$. **20.** AB subtends $45°$ at P.

In Nos. 21–27, L, M are variable points on OX, OY and c is constant.

21. $LM = c$. P is the mid-point of LM.

22. $OL + OM = c$. P is the mid-point of LM.

23. $LM = c$. P divides LM in the given ratio $l : m$.

24. $OL . OM = c^2$. P is the mid-point of LM.

25. $\{\omega\}$. $LM = 5c$. P divides LM in the ratio $3 : 2$.

26. $OL + OM = c$. P is the foot of the perpendicular from O to LM.

27. $\{\omega\}$. $OL + OM = c$. PL, PM are perpendicular to OX, OY.

28. An equilateral triangle of given side has two vertices on fixed perpendicular lines. Find the locus of the third vertex.

29. Find the locus of a point P such that $PR^2 - PS^2 = 4c^2$, where R, S are the fixed points (c, c), $(-c, -c)$.

30. Q is any point on $2x + 3y = 4$ and OQ is produced to P so that $QP = OQ$. Find the locus of P.

31. Find the locus of P if OP is produced to Q so that $PQ = OP$ and Q lies on $x^2 + y^2 = 2x$.

32. A is $(1, 1)$, AB cuts OX at B, AC is perpendicular to AB and cuts OY at C. Find the locus of the mid-point of BC.

2·5. Change of Axes

2·51. *Change of Origin.* If P is a point whose coordinates referred to axes OX, OY are x, y and if new axes are taken parallel to OX, OY through the point O' (f, g), then the coordinates of P referred to these new axes are x', y' such that

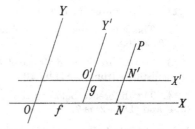

$$x' = x - f, \quad \text{or} \quad x = x' + f,$$
$$y' = y - g, \quad\quad\quad y = y' + g.$$

If the coordinates of a fixed point referred to axes OX, OY are p, q, then its coordinates referred to $O'X'$, $O'Y'$ are $p - f$, $q - g$. The figure is drawn for positive values of f and g. The reader should see that the result is true for all cases.

If the coordinates x, y of a variable point P referred to axes OX, OY satisfy an equation $f(x, y) = 0$, it follows that $f(x' + f, y' + g) = 0$ is the equation satisfied by the coordinates of P referred to axes $O'X'$, $O'Y'$.

The way in which a discussion can be simplified by a change of origin is illustrated by 2·41 and 2·42.

In 2·41, after reaching the equation $(x - \frac{4}{3})^2 + y^2 = (\frac{2}{3})^2$ the new equation $x'^2 + y'^2 = (\frac{2}{3})^2$ is obtained by change of origin to $(\frac{4}{3}, 0)$. This equation represents a circle centre the new origin and radius $\frac{2}{3}$.

In 2·42, the equation of the locus is $(x + 5)^2 = 4(y - 3)$. By change of origin to $(-5, 3)$ the new equation $x'^2 = 4y'$, or

$y' = \frac{1}{4}x'^2$ is obtained. The curve may be plotted by giving values to x'.

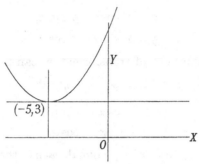

2·52. EXAMPLE. Show that there is no essential difference between the curves $x^2 - 2y^2 = 1$ and $x^2 - 2y^2 - 4x - 12y = 15$.

The second equation can be written

$$x^2 - 4x + 4 - 2(y^2 + 6y + 9) = 15 + 4 - 18,$$

i.e. $$(x-2)^2 - 2(y+3)^2 = 1,$$

and if the origin is moved to the point $(2, -3)$ the new equation of this curve is $x^2 - 2y^2 = 1$.

Hence the second equation gives the same curve relative to axes $O'X'$, $O'Y'$ as the first gives relative to OX, OY.

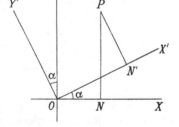

2·53. *Rotation of Rectangular Axes.* If new axes OX', OY' are taken, without change of origin, and if $\angle XOX' = \alpha$, $\angle XOY' = \alpha + \frac{1}{2}\pi$, then the new coordinates x', y' of a point $P(x, y)$ are given by

$x' = $ the projection of OP on OX'

$\quad =$ the sum of the projections of ON, NP on OX'

$\quad = x\cos\alpha + y\sin\alpha,$

$y' = $ the projection of OP on OY'

$\quad =$ the sum of the projections of ON, NP on OY'

$\quad = -x\sin\alpha + y\cos\alpha.$

By solving these equations for x, y in terms of x', y', or by using $-\alpha$ instead of α:

$$x = x' \cos\alpha - y' \sin\alpha,$$

$$y = x' \sin\alpha + y' \cos\alpha.$$

The four equations of transformation can be read off from the scheme

	x'	y'
x	$\cos\alpha$	$-\sin\alpha$
y	$\sin\alpha$	$\cos\alpha$

They make $x^2 + y^2$ and $x'^2 + y'^2$ equal, each of them being equal to OP^2.

2·54. Change of Oblique Axes. For a change from axes OX, OY inclined at an angle ω to new axes OX', OY' such that $\angle XOX' = \alpha$, $\angle YOY' = \beta$, the relations between the

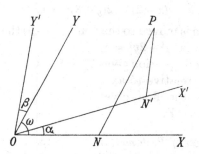

coordinates x, y and x', y' may be found by equating the expressions for the projection of OP in two directions. For example, taking directions perpendicular to OX, OY,

$$y \sin\omega = x' \sin\alpha + y' \sin(\omega + \beta),$$

$$x \sin\omega = x' \sin(\omega - \alpha) - y' \sin\beta.$$

These are of the form

$$x = ax' + by',$$

$$y = cx' + dy',$$

where $$ad - bc = \sin(\omega - \alpha + \beta) \operatorname{cosec}\omega.$$

By combining the results of 2·51 and 2·54 it is proved that

for any change of axes, the equations connecting the co-ordinates of a point relative to the two sets of axes are

$$x = ax' + by' + e,$$
$$y = cx' + dy' + f.$$

Since $ad \neq bc$ the equations can be solved for x', y' and thus can be expressed in the form

$$x' = Ax + By + E,$$
$$y' = Cx + Dy + F.$$

The equations in 2·51, 2·53, 2·54 can also be used in another way to define the position of a new point P' whose coordinates x', y' are expressed in terms of the known coordinates x, y of a given point P, the same axes being used for both. See Exercise 2E, Nos. 18–29.

EXERCISE 2E

1. If the origin is moved to $(4, 5)$, state the new coordinates of $(3, 6)$, $(11, -2)$, (a, b), and the old origin.

2. If the origin is moved to $(3, -3)$, state the new coordinates of $(8, 4)$, $(3, 3)$, $(-4, -4)$, (a, b).

3. If the origin is moved to $(1, 1)$, state the new equations of the line $2x + 3y = 1$ and the curve $y = x^2$.

4. If the origin is moved to $(-2, 3)$, state the new equations of the line $x + y = 1$ and the curve $(x + 2)^2 + (y + 3)^2 = 1$.

5. To what point must the origin be moved so that
   (i) $(3, 7)$ becomes $(9, 6)$?   (ii) $(-3, 4)$ becomes $(3, -4)$?
   (iii) (a, b) becomes (c, d)?

6. To what point must the origin be moved so that
   (i) $(y - 2)^2 = 4(x + 3)$ becomes $y'^2 = 4x'$?
   (ii) $x^2 + y^2 = 1$ becomes $(x' + 1)^2 + (y' - 1)^2 = 1$?
   (iii) $x^2 + 2y^2 + 3x - 4y + 7 = 0$ becomes of the form $x^2 + 2y^2 + c = 0$?

7. Find the new coordinates of $(1, \sqrt{3})$, $(1, -1)$, (a, b) when the axes are turned through $\frac{1}{6}\pi$.

8. Find the new coordinates of $(\sqrt{3}, 1)$, $(1, -1)$, (c, d) when the axes are turned through $-\frac{1}{6}\pi$.

9. If the axes are turned through $\frac{1}{2}\pi$, find new equations of the line $x = y$ and the curve $x^2 + 2xy + 2y^2 = 1$.

10. If the axes are turned through $\frac{2}{3}\pi$, find the new equations of the line $2x + 2\sqrt{3}y + 3 = 0$ and the curve $x^2 = 4y$.

11. If the axes are turned through $\frac{1}{4}\pi$, find the new equations of $y - 3x = 1$, $x^2 - y^2 = a^2$, $xy = c^2$, $7x^2 + 2xy + 7y^2 = 24$.

12. Find the angle through which the axes must be turned so that $11x^2 + xy - 9y^2 = 0$ may assume the form $ax'^2 + by'^2 = 0$.

13. $\{\omega = \frac{1}{4}\pi.\}$ If the axes are turned through $\frac{1}{4}\pi$, what are the new coordinates of (a, b) and the new equation of $y = x^2$?

14. If the origin is moved to $(1, -3)$ and the axes are then rotated through $\tan^{-1}\frac{3}{4}$, find the new coordinates of the old origin and the new equation of $36x^2 + 24xy + 29y^2 + 150y + 45 = 0$.

15. By moving the origin and then rotating the axes, reduce $x^2 + xy + y^2 - 2x - y - 4 = 0$ to the form $ax^2 + by^2 + c = 0$.

16. The axes are changed twice by the substitutions

	x'	y'		x''	y''
x	$\cos\alpha$	$-\sin\alpha$	x'	$\cos\beta$	$-\sin\beta$
y	$\sin\alpha$	$\cos\alpha$	y'	$\sin\beta$	$\cos\beta$

What single substitution would produce the same effect?

17. If $ax^2 + 2hxy + by^2$ becomes $a'x'^2 + 2h'x'y' + b'y'^2$ by rotating the axes through an angle α, prove that $a' + b' = a + b$ and $a'b' - h'^2 = ab - h^2$.

In Nos. 18–29, the coordinates of P and P' referred to the same rectangular axes are x, y and x', y'. In Nos. 18–21, P lies on a network of lines parallel to the axes at integral distances from them. Show in a diagram on what lines P' lies.

18. $x' = 2x$, $y' = y$.   **19.** $x = x'$, $y = 3y'$.

20. $x' = x + 2y$, $y' = y$.   **21.** $5x' = 3x + 4y$, $5y' = 4x - 3y$.

In Nos. 22–29, P lies on a given curve. State how to derive from this curve the curve on which P' lies.

22. $x' = y$, $y' = -x$.   **23.** $x' = y$, $y' = x$.

24. $x' = x$, $y' = -y$.   **25.** $x' = -y$, $y' = x$.

26. $x' = 3x$, $y' = y$.   **27.** $x' = x$, $y' = -2y$.

28. $x' = cx$, $y' = cy$.   **29.** $x' = ax + b$, $y' = ay + c$.

2·6. Equations and Loci in Space

2·61. When the cartesian coordinates x, y, z of a point in space are connected by a single relation $f(x, y, z) = 0$, the point is restricted to lie on a locus called a *surface*, and $f(x, y, z) = 0$

is called the equation of the surface. A surface contains ∞^2 points. The simplest example of a surface is a plane.

The equations of the planes YOZ, ZOX, XOY are $x = 0$, $y = 0$, $z = 0$.

2·62. Another kind of locus in space consists of ∞^1 points and is called a *curve*. The simplest example is the line. The coordinates of a point in three dimensions must be subject to a double restriction to confine the point to a curve.

The axis OX can be represented by $y = z = 0$.

2·63. A line or curve in space may also be represented by parametric equations

$$x = f(t), \quad y = g(t), \quad z = h(t),$$

which leave the point only one degree of freedom.

Consider, for example, the points given by

$$x = \sin t, \quad y = \cos t, \quad z = \tan t.$$

These points are all such that $x^2 + y^2 = 1$ and this expresses that they are at unit distance from OZ. They lie on a cylinder with OZ for axis; $x^2 + y^2 = 1$ is the equation of this cylinder. The points also lie on another surface whose equation is $x = yz$. They do not lie entirely in any one plane.

In general the parameter t can be eliminated from

$$x = f(t), \quad y = g(t), \quad z = h(t)$$

in two independent ways, giving results

$$\phi(x, y, z) = 0, \quad \psi(x, y, z) = 0,$$

which are the equations of two surfaces passing through the curve. These equations may be called the equations of the curve; but it must not be assumed that every curve consists of the common points of two surfaces found in this way. See Exercise 2F, No. 20.

2·64. From equations $x = f(u, v)$, $y = g(u, v)$, $z = h(u, v)$, which express the coordinates of a point in terms of two parameters u and v, there can be obtained ∞^2 points by giving values to the parameters. These points constitute a surface. The equation of the surface is found in the form $\phi(x, y, z) = 0$ by elimination of u and v from the three parametric equations.

2·65. Thus in space a locus may be a curve-locus consisting of ∞^1 points or a surface-locus consisting of ∞^2 points. Both a curve-locus and a surface-locus can sometimes be represented parametrically.

EXERCISE 2F

In Nos. 1–8, state what surface is represented by the equation.

1. $y = x$. **2.** $xz = 0$. **3.** $xyz = 0$.

4. $x = 2$. **5.** $2z + 3 = 0$. **6.** $y^2 = 4$.

7. $x^2 + y^2 + z^2 = 1$. **8.** $x^2 + y^2 = a^2$.

In Nos. 9–15, state what curve is represented by the equations.

9. $x = 0, y = 0$. **10.** $x = y, z = 0$. **11.** $x = y, z = 1$.

12. $x = y = z$. **13.** $y^2 + z^2 = 1, x = 0$.

14. $x^2 + y^2 + z^2 = 1, x = y$. **15.** $xz = 0, yz = 0$.

In Nos. 16–18, state what you can about the form of the surface represented by the equation.

16. $f(y, z) = 0$. **17.** $f(x) = 0$. **18.** $f(x/z, y/z) = 0$.

19. Describe the curve given by $x = a \cos t, y = a \sin t, z = bt$.

20. Verify that the curve given by $x = t^3, y = t^2, z = t$ lies on the surfaces $y^2 = zx, yz = x$. Show also that the point $(0, 0, k)$ lies on these surfaces for all values of k, and only lies on the curve for $k = 0$.

In Nos. 21–23, find the x, y, z equation of the surface given by the parametric equations.

21. $x = s^2, y = t^2, z = st$. **22.** $x = s + t, y = st, z = s^2 + t^2$.

23. $x = a \cos s \cos t, y = b \sin s \cos t, z = c \sin t$.

In Nos. 24–27, state the x, y, z equations of two surfaces on which the curves with the given parametric equations lie.

24. $x = t^2, y = t^3, z = t^4$. **25.** $x = 2 + 3t, y = 3 - 4t, z = t + 1$.

26. $x = \sec t, y = \tan t, z = \sin t + \cos t$.

27. $x = t^2 + 3t, y = t + 2, z = 5t^2 + 4$.

28. What is represented by the equations $x = t + 1, y = t - 1$ when t is (i) a constant, (ii) a parameter?

Chapter 3

THE POINT AND LINE

3·1. Equation of Line

3·11. The graph of $ax+b$, where a and b are constants, passes through the point $B(0,b)$. Any other point P on the graph has coordinates $k, ak+b$, and therefore BP makes with OX an angle $\tan^{-1}a$. Hence the graph is a line with gradient a. $ax+b$ is called a linear function.

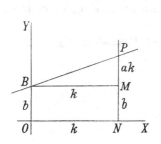 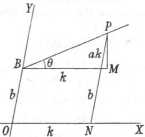

When a line is said to make an angle θ with OX, this is to be taken to mean that an angle from OX to the line, measured in the positive direction of rotation, is θ.

When the axes are oblique, the graph is still a line. This line makes with OX an angle θ given by

$$\tan \theta = \frac{MP}{BM} = \frac{ak \sin \omega}{k+ak \cos \omega} = \frac{a \sin \omega}{1+a \cos \omega}.$$

Any linear equation $ux+vy+w = 0$ in which u,v are not both zero is of the form $y = ax+b$ or else $x = c$. It therefore represents a line.

3·12. EXAMPLE. Find the equation of the line through $(-1, 2)$ perpendicular to the line joining $(2, 3)$ to $(9, 8)$.

The gradient of the joining line is $\frac{5}{7}$. Hence that of the required line is $-\frac{7}{5}$. Therefore the equation is of the form

$$5x + 7y = \dots.$$

The right side is a constant, and its value is found from the fact that the equation must be satisfied by $x = -1$, $y = 2$. Hence the constant is $5(-1) + 7.2$. In practice the result is written down as follows:

$$5x + 7y = 5(-1) + 7.2 = 9.$$

EXERCISE 3A

In Nos. 1–6, state the gradient of the line.

1. $2x - y = 7$.   2. $4x - 5y = 6$.   3. $7x + 8y = 9$.

4. $y = 3$.   5. $x = 2$.   6. $ax + by + c = 0$.

In Nos. 7–22, find the equation of the line through the given point satisfying the given condition, and state whether the result holds when $\omega \neq \frac{1}{2}\pi$.

7. $(1, 3)$, gradient 2.   8. $(-2, 1)$, gradient $-\frac{2}{3}$.

9. $(4, -2)$, parallel to $3x + 5y = 7$.

10. $(3, 1)$, parallel to $2x - y + 5 = 0$.

11. $(5, -7)$, parallel to $x = 4$.

12. $(5, -1)$, perpendicular to $2x - y = 3$.

13. $(2, 3)$, perpendicular to $4x - 7y + 8 = 0$.

14. O, parallel to $3x - 7y = 2$.

15. O, perpendicular to $ax + by + c = 0$.

16. (h, k), perpendicular to $y = tx$.

17. (p, q), parallel to $ax + by + c = 0$.

18. $(4, 7)$, parallel to the y-axis.

19. O, making an angle of $60°$ with the x-axis.

20. $(0, -1)$, making an angle of $135°$ with the x-axis.

21. $(2, 3)$, making an angle of $30°$ with OY.

22. $(3, -4)$, making an angle of $\tan^{-1} \frac{2}{5}$ with OX.

In Nos. 23–31, find the equation of the line satisfying the given conditions and state whether the result holds when $\omega \neq \frac{1}{2}\pi$.

23. Through $(0, 0)$ and (p, q).   24. Through $(3, 5)$ and $(-2, 7)$.

25. Through $(5, 7)$ and $(5, -3)$.

26. Perpendicular to $5x + 6y = 1$ at the point where it meets OX.

27. Parallel to $3x + 4y + 1 = 0$ and $3x + 4y - 6 = 0$ and midway between them.

28. Through $(2, 0)$ perpendicular to the line joining $(4, -8)$ and $(0, 3)$.

29. Through (a, b) parallel to the line joining (c, d) to the origin.

30. Find the perpendicular bisector of the line joining $(0, 0)$ and $(3, 7)$.

31. Find the perpendicular bisector of the line joining $(4, 1)$ and $(-5, 8)$.

32. Prove that the lines joining $(-4, 1)$ to $(2, 4)$ and $(-4, 3)$ to $(-3, 1)$ are at right angles.

33. Find the equations of the lines through $(1, 1)$ making $45°$ with $2x + y = 7$.

3·2. Point of Intersection of Two Lines

If two lines have a common point, the coordinates of this point satisfy the equations of the lines and are found by solving these simultaneous equations.

(i) If the equations are $x + 2y = 7$, $3x - 4y = 1$, the solution is $x = 3$, $y = 2$. The point of intersection is $(3, 2)$.

(ii) If the equations are $3x + 2y = 5$, $6x + 4y = 7$, there is no solution. The equations are inconsistent. The lines have no common point: they are parallel.

(iii) If the equations are $3x + 2y = 5$, $6x + 4y = 10$, the equations are identical and are satisfied by an unlimited number of values of x, y. The lines are coincident.

In the general case of $a_1 x + b_1 y + c_1 = 0$ and $a_2 x + b_2 y + c_2 = 0$ the coordinates of any common point satisfy

$$b_2(a_1 x + b_1 y + c_1) = b_1(a_2 x + b_2 y + c_2)$$

and $\qquad a_2(a_1 x + b_1 y + c_1) = a_1(a_2 x + b_2 y + c_2),$

and therefore

$$\begin{vmatrix} a_1 & b_1 \\ a_2 & b_2 \end{vmatrix} x = \begin{vmatrix} b_1 & c_1 \\ b_2 & c_2 \end{vmatrix}, \quad \begin{vmatrix} a_1 & b_1 \\ a_2 & b_2 \end{vmatrix} y = - \begin{vmatrix} a_1 & c_1 \\ a_2 & c_2 \end{vmatrix}. \quad (1)$$

Put $\qquad \begin{vmatrix} b_1 & c_1 \\ b_2 & c_2 \end{vmatrix} = A, \quad \begin{vmatrix} a_1 & c_1 \\ a_2 & c_2 \end{vmatrix} = -B, \quad \begin{vmatrix} a_1 & b_1 \\ a_2 & b_2 \end{vmatrix} = C.$

Then, unless $C = 0$,

$$x = A/C, \quad y = B/C.$$

It is easy to verify that these values actually satisfy the two equations. They are the coordinates of the point of intersection of the lines, and this exists when C is not zero.

No exception arises from $A = 0$ or $B = 0$, but if the solution is written

$$x:y:1 = A:B:C, \tag{2}$$

this must be interpreted by the convention of 0·5.

Thus if $A = 0$, (2) means $x = 0$ and $y = B/C$,

and   if $B = 0$, it means $y = 0$ and $x = A/C$,

and   if $A = B = 0$, it means $x = y = 0$.

If $C = 0$ and if the equations have a solution, it follows from (1) that $A = B = 0$. Hence there is no point common to the lines when $C = 0$ unless also $A = B = 0$.

The lines are parallel when $C = 0$ and coincident when $A = B = C = 0$.

3·25. Concurrence

Suppose that the lines

$$a_1 x + b_1 y + c_1 = 0 \quad \text{and} \quad a_2 x + b_2 y + c_2 = 0$$

have a single common point, so that $C \neq 0$.

Then by 3·2 (1), the third line $a_3 x + b_3 y + c_3 = 0$ will also pass through this point if

$$a_3 \begin{vmatrix} b_1 & c_1 \\ b_2 & c_2 \end{vmatrix} - b_3 \begin{vmatrix} a_1 & c_1 \\ a_2 & c_2 \end{vmatrix} + c_3 \begin{vmatrix} a_1 & b_1 \\ a_2 & b_2 \end{vmatrix} = 0,$$

i.e.
$$\Delta \equiv \begin{vmatrix} a_1 & b_1 & c_1 \\ a_2 & b_2 & c_2 \\ a_3 & b_3 & c_3 \end{vmatrix} = 0.$$

If $\Delta = 0$, the point $x = A/C$, $y = B/C$ lies on

$$a_3 x + b_3 y + c_3 = 0$$

as well as on the other lines because

$$a_3 A/C + b_3 B/C + c_3 = \Delta/C = 0.$$

Although this argument breaks down when $C = 0$, a similar

argument would then apply, with the lines in a different order, unless also

$$\begin{vmatrix} a_2 & b_2 \\ a_3 & b_3 \end{vmatrix} = 0 \quad \text{and} \quad \begin{vmatrix} a_3 & b_3 \\ a_1 & b_1 \end{vmatrix} = 0.$$

When $\begin{vmatrix} a_1 & b_1 \\ a_2 & b_2 \end{vmatrix}$, $\begin{vmatrix} a_2 & b_2 \\ a_3 & b_3 \end{vmatrix}$, $\begin{vmatrix} a_3 & b_3 \\ a_1 & b_1 \end{vmatrix}$ are all zero, the three lines are parallel.

Therefore when $\Delta = 0$ the lines are either concurrent or else parallel.

3·3. Angle of Intersection of Two Lines. $\{\omega\}$

3·31. Unit vectors in the axes being denoted by $\mathbf{i}$, $\mathbf{j}$, the vector $p\mathbf{i} + q\mathbf{j}$ has the direction of the line $x/p = y/q$. Therefore a vector which has the direction of the line $ax + by + c = 0$ is $b\mathbf{i} - a\mathbf{j}$. Hence the angle between two lines

$$a_1 x + b_1 y + c_1 = 0,$$

$$a_2 x + b_2 y + c_2 = 0$$

is the same as the angle between the two vectors

$$\cdot \quad \mathbf{R}_1 \equiv b_1 \mathbf{i} - a_1 \mathbf{j}, \quad \mathbf{R}_2 \equiv b_2 \mathbf{i} - a_2 \mathbf{j}.$$

$$\therefore \quad R_1 R_2 \cos\theta = (b_1 \mathbf{i} - a_1 \mathbf{j}).(b_2 \mathbf{i} - a_2 \mathbf{j})$$

$$= b_1 b_2 \mathbf{i}^2 + a_1 a_2 \mathbf{j}^2 - b_1 a_2 \mathbf{i}.\mathbf{j} - b_2 a_1 \mathbf{j}.\mathbf{i}$$

$$= a_1 a_2 + b_1 b_2 - (a_1 b_2 + a_2 b_1) \cos\omega. \tag{1,}$$

Also $R_1^2 = \mathbf{R}_1^2 = (b_1 \mathbf{i} - a_1 \mathbf{j})^2 = a_1^2 + b_1^2 - 2a_1 b_1 \cos\omega,$

$$R_2^2 = \mathbf{R}_2^2 = (b_2 \mathbf{i} - a_2 \mathbf{j})^2 = a_2^2 + b_2^2 - 2a_2 b_2 \cos\omega.$$

From the last three results, the value of $\cos\theta$ can be found. Putting $\cos\theta = 0$ in (1), the condition of perpendicularity is found to be

$$a_1 a_2 + b_1 b_2 = (a_1 b_2 + a_2 b_1) \cos\omega,$$

or, if $\omega = \tfrac{1}{2}\pi,$ $\qquad a_1 a_2 + b_1 b_2 = 0.$

In practice it is usually more convenient to use this result than to express that the product of the gradients is -1.

3·32. EXAMPLE. Find the angle between the lines

$$3x + 2y = 1, \quad 5x - y = 6,$$

when $\omega = \frac{1}{3}\pi$.

1st method. By 3·11, for the first line

$$a = -\tfrac{3}{2}, \quad \tan\theta_1 = \frac{-3\sin\omega}{2 - 3\cos\omega} = -3\sqrt{3},$$

and for the second line

$$a = +5, \quad \tan\theta_2 = \frac{5\sin\omega}{1 + 5\cos\omega} = \frac{5\sqrt{3}}{7}.$$

The angle between the lines

$$= \theta_1 - \theta_2 = \tan^{-1}\frac{-3\sqrt{3} - \tfrac{5}{7}\sqrt{3}}{1 - 3\sqrt{3}\tfrac{5}{7}\sqrt{3}} = \tan^{-1}\frac{13\sqrt{3}}{19}.$$

2nd method. By 3·31 (1),

$$R_1 = \sqrt{(3^2 + 2^2 - 2.3.2\cos\omega)} = \sqrt{7},$$

$$R_2 = \sqrt{(1^2 + 5^2 + 2.5\cos\omega)} \quad = \sqrt{31},$$

$$\sqrt{7}\sqrt{31}\cos\theta = 15 - 2 - 7\cos\omega = \tfrac{19}{2},$$

$$\therefore \quad \theta = \cos^{-1}\frac{19}{2\sqrt{217}}.$$

EXERCISE 3B

In Nos. 1–8, find the point of intersection of the lines.

1. $14x - 11y + 13 = 0, \ y = 5.$ 2. $x + 7y = 20, \ 4x + 5y = 11.$

3. $x + 2y = 0, \ 7x - 3y = 5.$ 4. $2x + 7y + 4 = 0, \ 5x + 9y + 3 = 0.$

5. $(a - 5)x + (a - 3)y = 2, \ 3x + (a - 1)y = 2a - 6.$

6. $m_1 y = x + am_1, \ m_2 y = x + am_2.$

7. $y = m_1 x + \dfrac{a}{m_1}, \ y = m_2 x + \dfrac{a}{m_2}.$

8. $x + yt_1^2 = 2at_1, \ x + yt_2^2 = 2at_2.$

9. Prove that $x + 2y = 8, \ 3x - 4y + 6 = 0, \ 5x - 7y + 11 = 0$ are concurrent.

10. Prove that $x + 4y + 7 = 0, \ 3x - 2y + 8 = 0, \ 11x + 2y + 38 = 0$ are concurrent.

11. If $ax + 3y + 4 = 0, \ x - 2y + 3 = 0, \ x = y$ are concurrent, find the value of a.

12. If $x+ky = 3$, $7x+5y = 4$, $5x+y+k = 0$ are concurrent, find the value of k.

13. Verify that the perpendiculars from the vertices to the opposite sides of the triangle $(4, 11)$ $(5, -6)$ $(2, -3)$ are concurrent.

14. Find the condition for the concurrence of the three distinct lines $3ax-2y = ka^3$, $3bx-2y = kb^3$, $3cx-2y = kc^3$.

In Nos. 15–18, find the area of the figure bounded by the lines.

15. $x+y = 7$, $11x-5 = 13y$, x-axis.

16. $2x-3y+11 = 0$, $13x-y = 58$, $9x+5y = 6$.

17. $4x-3y-13 = 0$, $4x-3y+37 = 0$, $3x+4y+9 = 0$,
$$3x+4y-16 = 0.$$

18. $x-7y+10 = 0$, $4x+3y-22 = 0$, $x-3y+17 = 0$, $4x-y+13 = 0$.

In Nos. 19–24, find the angle between the lines.

19. $x-3y+5 = 0$, $x+2y+7 = 0$. **20.** $x\sqrt{3}-y = 2$, $x+y\sqrt{3} = 0$.

21. $12x+20y = 15$, $x = 0$. **22.** $17x+11y = 53$, $4x+2y = 3$.

23. $(a+b)x+(a-b)y = c$, $(a-b)x+(a+b)y = d$.

24. $my/n = (m+n)x/(m-n)$, $nx/m = (m+n)y/(m-n)$.

25. Find the angles of the triangle formed by
$$2x+3y = 4, \quad 2x-3y = 2, \quad y+1 = 0.$$

In Nos. 26–32, find the angle between the lines.

26. $\{\omega = \tan^{-1}\frac{3}{4}\}$. $x = 2y$, OX.

27. $\{\omega = \tan^{-1}\frac{4}{3}\}$. $y = 2x$, OX.

28. $\{\omega = \frac{1}{4}\pi\}$. $x+2y = 3$, $2x-y = 1$.

29. $\{\omega\}$. $x-y = 1$, OX.

30. $\{\omega\}$. $3x-y+3 = 0$, $x+3y = 0$.

31. $\{\omega\}$. $3x+4y = 5$, $7x-6y+1 = 0$.

32. $\{\omega\}$. $x-a = y(\sin\omega - \cos\omega)$, $b-y = x(\sin\omega + \cos\omega)$.

33. $\{\omega\}$. State the condition of perpendicularity for
$$t_1 y = x+at_1^2, \quad t_2 y = x+at_2^2.$$

34. $\{\omega\}$. Write down the equations of the perpendiculars to the axes at the origin.

35. $\{\omega\}$. Find the equation of the perpendicular at $(1, 2)$ to $3x+y = 0$.

36. $\{\omega\}$. Find the equation of the perpendicular from (x_1, y_1) to $ax+by+c = 0$.

In Nos. 37–39, find the orthocentre of the triangle.

37. $(5, -7)$ $(8, 3)$ $(0, 0)$.

38. Sides $y = 2x - 3$, $2y = x + 2$, $2x + 3y = 24$.

39. Sides $t_1 y = x + at_1{}^2$, $t_2 y = x + at_2{}^2$, $t_3 y = x + at_3{}^2$.

40. Deduce the value of $\tan \theta$ from the value of $\cos \theta$ given in 3·31.

3·4. Length of Perpendicular

Let p denote the length, essentially positive, of the perpendicular from (x_1, y_1) to

$$ax + by + c = 0.$$

If ϕ is the angle between OX and the perpendicular, then the foot of the perpendicular is

$$(x_1 - p \cos \phi, \; y_1 - p \sin \phi)$$

or $(x_1 + p \cos \phi, \; y_1 + p \sin \phi)$.

Since this point lies on

$$ax + by + c = 0,$$

$$a(x_1 \mp p \cos \phi) + b(y_1 \mp p \sin \phi) + c = 0.$$

$$\therefore \quad \pm p = (ax_1 + by_1 + c)/(a \cos \phi + b \sin \phi).$$

But $\tan \phi = b/a$,

$$\therefore \quad p = \pm \frac{ax_1 + by_1 + c}{\sqrt{(a^2 + b^2)}}.$$

The sign is $+$ if $ax_1 + by_1 + c$ is positive and $-$ if $ax_1 + by_1 + c$ is negative.

When $\omega \neq \tfrac{1}{2}\pi$, the formula for the length of the perpendicular is

$$p = \pm \frac{(ax_1 + by_1 + c) \sin \omega}{\sqrt{(a^2 + b^2 - 2ab \cos \omega)}}.$$

A proof of this is indicated in Exercise 3c, No. 29.

3·45. Sign of $ax_1 + by_1 + c$

The expression $ax + by + c$, which is zero when (x, y) lies on the line $ax + by + c = 0$, is positive when (x, y) lies on one side of the line and negative when it lies on the other side.

For let P_1, P_2 be any two points, and let the point of intersection P of $P_1 P_2$ and $ax + by + c = 0$ be

$$\left(\frac{\kappa_1 x_1 + \kappa_2 x_2}{\kappa_1 + \kappa_2}, \frac{\kappa_1 y_1 + \kappa_2 y_2}{\kappa_1 + \kappa_2} \right).$$

See 1·51.

Then $\qquad\qquad P_1 P : PP_2 = \kappa_2 : \kappa_1.$

Since P lies on the line $ax + by + c = 0$,

$$a \frac{\kappa_1 x_1 + \kappa_2 x_2}{\kappa_1 + \kappa_2} + b \frac{\kappa_1 y_1 + \kappa_2 y_2}{\kappa_1 + \kappa_2} + c = 0.$$

$$\therefore \quad a(\kappa_1 x_1 + \kappa_2 x_2) + b(\kappa_1 y_1 + \kappa_2 y_2) + c(\kappa_1 + \kappa_2) = 0,$$

$$\therefore \quad \kappa_1(ax_1 + by_1 + c) = -\kappa_2(ax_2 + by_2 + c). \tag{1}$$

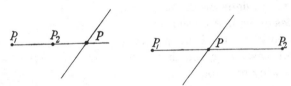

But κ_2/κ_1 is positive when P is between P_1 and P_2 and then P_1, P_2 are on opposite sides of $ax + by + c = 0$. (1) proves that $ax_1 + by_1 + c$ and $ax_2 + by_2 + c$ then have opposite signs.

Similarly, κ_2/κ_1 is negative when P_1, P_2 are on the same side of the line and $ax_1 + by_1 + c$, $ax_2 + by_2 + c$ then have the same sign.

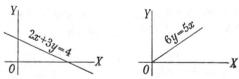

Consider for example the lines $2x + 3y = 4$ and $6y = 5x$.

When $x = y = 0$, $2x + 3y - 4$ is negative. Hence it is negative when x, y are the coordinates of a point below the line $2x + 3y = 4$, and positive whenever x, y are the coordinates of a point above the line.

When $x = 0$ and $y = 1$, $6y - 5x$ is positive. Hence it is positive when x, y are the coordinates of a point above the line $6y = 5x$ and negative when they are the coordinates of a point below it.

A line does not possess one side which is necessarily its positive side: the equation of the line $2x + 3y - 4 = 0$ can also be written $-2x - 3y + 4 = 0$.

3·5. Angle-bisectors

3·51. The equation of a bisector of an angle between two lines can be written down by equating the lengths of the perpendiculars to the lines from an arbitrary point (x, y). To distinguish between the two bisectors the signs of $ax + by + c$ must be considered. These signs can usually be determined from a rough figure as in the examples of 3·45. A method which may be used in a doubtful case is given in the following example.

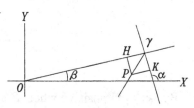

3·52. EXAMPLE. Find the equation of the bisector of the obtuse angle between the lines $11x = 60y$ and $99x + 20y = 1100$.

From the figure

$$\tan \gamma = \tan (\alpha - \beta) = \frac{\tan \alpha - \tan \beta}{1 + \tan \alpha \tan \beta} = \frac{-\frac{99}{20} - \frac{11}{60}}{1 - \frac{99}{20}\frac{11}{60}}.$$

And, since $20.60 > 99.11$, $\tan \gamma < 0$, thus γ is obtuse.

Now if P is (x, y) and PH, PK are the perpendiculars,

$$PH = \pm \frac{11x - 60y}{\sqrt{(11^2 + 60^2)}}, \qquad PK = \pm \frac{99x + 20y - 1100}{\sqrt{(99^2 + 20^2)}}.$$

The point $(1, 0)$ and the point P are on the same side of $11x = 60y$, and $11.1 - 60.0$ is positive. Therefore $11x - 60y$ is positive when x, y are the coordinates of P. Thus

$$PH = + (11x - 60y)/61.$$

Also the origin and P are on the same side of

$$99x + 20y = 1100.$$

Therefore $99x + 20y - 1100$ is negative when x, y are the co-ordinates of P. Thus

$$PK = -(99x + 20y - 1100)/101.$$

Hence the required equation is

$$101(11x - 60y) = -61(99x + 20y - 1100),$$

i.e. $$65x - 44y = 610.$$

In an example of this kind the figure should be used to check that the correct bisector has been found.

EXERCISE 3 c

In Nos. 1–8, find the length of the perpendicular from the point to the line.

1. $(0, 0)$, $3x + 4y = 2$.   **2.** $(5, 4)$, $4x + 3y = 11$.

3. $(1, -3)$, $y = 3x + 2$.   **4.** $(-6, 0)$, $\frac{1}{3}x - \frac{1}{2}y = 1$.

5. $(3, 7)$, $5(x - 4) = 12(y - 3)$.   **6.** $(a, 0)$, $y = 3x + \frac{1}{3}a$.

7. $(a, 0)$, $ty = x + at^2$.

8. (x_1, y_1), $x \cos \alpha + y \sin \alpha = p$. Also show how the sign is determined.

9. Find the distance between $3x + 4y + 7 = 0$ and $3x + 4y - 10 = 0$.

10. Prove that the lines $x = 65$, $63x + 16y = 4225$, $5x + 12y = 845$, $3x - 4y = 325$ touch a circle centre the origin.

11. Prove that any point on $9y = 7x$ is equidistant from $3x + 4y = 5$ and $12x + 5y = 13$.

12. Are $(1, 2)$ and $(3, -4)$ on the same or opposite sides of

$$60x + 20y = 99?$$

13. Which of the points $(3, -9)$, $(5, -11)$, $(4, 7)$, $(-3, 1)$, $(0, -1)$ are on the same side of $5x + 2y = 1$ as the origin?

14. For what values of c are $(8, -3)$ and $(3, c)$ on opposite sides of $12x + 5y = 70$?

15. Find the conditions for (f, g) to be inside the triangle formed by $3x - 2y = 6$, $x + 2y + 4 = 0$, $x + y = 5$.

In Nos. 16–19, find the bisectors of the angles between the lines.

16. $3x + 4y = 5$, $12x - 5y = 41$.   **17.** $4x + 3y = 0$, $y = 0$.

18. $x + y = 30$, $x - y = 4$.

19. $77x + 36y = 10$, $33x - 56y = 8$.

20. Find the bisector of the acute angle between $8x+15y = 7$ and $3x-4y = 13$.

21. Find the bisector of the obtuse angle between $12x-5y = 4$ and $3x+4y = 16$.

22. Find the bisectors of the angles between

$$x+y = 6 \quad \text{and} \quad y-x\sqrt{2}+1 = 0$$

and distinguish between them.

23. Find the bisectors of the angles between $60x = 11y$ and $7x+24y = 3$ and distinguish between them.

24. Find the bisector of that angle between $4x+3y = 12$ and $3x-4y = 3$ in which $(\tfrac{7}{2}, 2)$ lies.

25. Find the bisector of that angle between $8x-15y = 12$ and $5x-12y = -6$ in which the origin lies.

26. Find the internal bisectors of the angles of the triangle formed by $y = 7$, $12x-5y = 37$, $12x-9y = 57$. Verify that they are concurrent and find the incentre.

27. Find the incentre of the triangle formed by $x = y$, $7x-y = 2$, $x-7y = 14$.

28. Find the ecentres of the triangle formed by $4x-3y = 1$, $80x-39y = 41$, $40x-9y = 871$.

29. A and B are the points in which the line $ax+by+c = 0$ meets the axes. p is the length of the perpendicular from (x_1, y_1) to the line. By equating $\tfrac{1}{2}p \cdot AB$ to the area of the triangle P_1AB given by Exercise 1 H, No. 28, prove the formula stated in 3·4 for oblique axes.

30. $\{\omega\}$. Find the length of the perpendicular from $(5, -7)$ to $4x+13y+1 = 0$.

31. $\{\omega = \tfrac{2}{3}\pi\}$. Find the internal bisectors of the angles of the triangle formed by $3x+5y = 8$, $x = 4$, $7x+8y+28 = 0$.

$$\textbf{3·6.} \quad \frac{\mathbf{x}}{\mathbf{a}}+\frac{\mathbf{y}}{\mathbf{b}} = 1$$

3·61. By 3·11 the equation $x/a+y/b=1$ represents a line because it is of the first degree. It is satisfied by $x = a$, $y = 0$ and by $x = 0$, $y = b$. Hence the line passes through $A(a, 0)$ and $B(0, b)$. It cuts

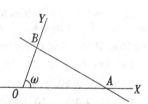

off intercepts a, b from the axes, due regard being paid to a sign convention.

3·62. EXAMPLE. ABC is a triangle with sides CB, CA along given lines and $1/CB + 1/CA$ is constant. Prove that AB passes through a fixed point.

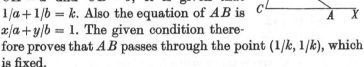

Take the given lines as axes. Then if $CA = a$ and $CB = b$, it is given that $1/a + 1/b = k$. Also the equation of AB is $x/a + y/b = 1$. The given condition therefore proves that AB passes through the point $(1/k, 1/k)$, which is fixed.

3·65. Polar Equation and x cos α + y sin α = p

Let the polar coordinates of an arbitrary point P on a given line be (r, θ) and let the polar coordinates of the foot of the perpendicular from the origin to the line be (p, α). Then in the figure

$$ON = OP \cos NOP,$$

i.e.

$$p = r \cos (\alpha - \theta),$$

which is therefore the polar equation of the line.

Since

$$r \cos (\theta - \alpha) = r \cos \theta \cos \alpha + r \sin \alpha \sin \theta$$

and the cartesian coordinates of P are $r \cos \theta \ (= x)$ and $r \sin \theta \ (= y)$,

$$p = x \cos \alpha + y \sin \alpha.$$

This is the cartesian equation of the line. Negative values of p can be avoided by the change of α into $\alpha \pm \pi$ when necessary. For example, $x + y\sqrt{3} + 4 = 0$ can be written

$$x \cos \tfrac{1}{3}\pi + y \sin \tfrac{1}{3}\pi = -2,$$

but it can also be written

$$x \cos\left(-\tfrac{2}{3}\pi\right) + y \sin\left(-\tfrac{2}{3}\pi\right) = +2.$$

Any equation $ax + by + c = 0$ can be reduced to the form $x \cos\alpha + y \sin\alpha = p$ by division of both sides by $\pm\sqrt{(a^2+b^2)}$.

EXERCISE 3D

In Nos. 1–4, express the equation in the form $x/a + y/b = 1$ and sketch the graph.

1. $3x - 4y = 12.$ **2.** $4x + 3y + 12 = 0.$

3. $2x + 3y = -6.$ **4.** $y = 3x - 7.$

In Nos. 5–15, express the equation in the form $x \cos\alpha + y \sin\alpha = 0$ $(p > 0)$.

5. $x + y\sqrt{3} = 4.$ **6.** $x - y = 2.$ **7.** $x + y + 1 = 0.$

8. $x = 3.$ **9.** $7x + 24y = 12.$ **10.** $y + 5 = 0.$

11. $2y - 3x = 4.$ **12.** $20x - 99y = 303.$

13. $x/a + y/b = 1,\ (a > 0, b > 0).$ **14.** $y = ax + b,\ (a > 0, b > 0).$

15. $ax + by + c = 0,\ (a > 0, b > 0, c > 0).$

In Nos. 16–24, write down the equation of the line.

16. Cutting off intercepts 5, 3 from the axes of x, y.

17. Cutting off intercepts $-2, \tfrac{3}{2}$ from the axes of x, y.

18. Joining $(8, 0)$, $(0, -4)$.

19. Through $(7, 11)$ making equal positive intercepts on the axes.

20. Through $(2, -3)$ having the part cut off between the axes bisected at that point.

21. Through (x_1, y_1) having the part cut off between the axes bisected at that point.

22. Through $(2, 5)$ having the part cut off between the axes divided at that point in the ratio $4 : 3$.

23. At a distance 3 from the origin such that the perpendicular from the origin makes 30° with OX.

24. At unit distance from the origin and making $\tfrac{1}{2}\pi + \tan^{-1}\tfrac{5}{12}$ with OX.

25. Write down the equation of the line joining $(p \cos t_1,\ p \sin t_1)$ to $(p \cos t_2,\ p \sin t_2)$ and the coordinates of the point of intersection of $x \cos t_1 + y \sin t_1 = p$ and $x \cos t_2 + y \sin t_2 = p$.

In Nos. 26–28, give the polar equation of the line.

   26. Through $(c, 0)$ perpendicular to the initial line.

   27. Through $(k, \tfrac{1}{2}\pi)$ parallel to the initial line.

   28. Through the pole and (r_1, θ_1).

In Nos. 29–34, show the position of the line in a diagram.

   29. $2 = r\cos(\theta - \tfrac{1}{3}\pi)$.   **30.** $-3 = r\cos(\theta + \tfrac{1}{4}\pi)$.

   31. $3 = 2r\sin\theta$.   **32.** $r = \sec(\theta + \tfrac{2}{3}\pi)$.

   33. $r = \operatorname{cosec}(\theta + \tfrac{7}{8}\pi)$.   **34.** $r\cos(\theta + \alpha) = p$.

In Nos. 35–37, give the polar equation of the line.

   35. Parallel to $\theta = -\tfrac{1}{4}\pi$ at unit distance from it.

   36. Through $(k, -\tfrac{1}{2}\pi)$ making $\tan^{-1}m$ with the initial line.

   37. Joining $(a, 0)$, $(b, \tfrac{1}{2}\pi)$.

   38. Find the point of intersection of the lines $r\cos(\theta - \alpha) = p$ and $r\cos(\theta - \beta) = q$.

   39. Show that $l/r + \cos\theta = \cos(\theta - \alpha)$ represents a line and find its distance from the pole.

   40. Find the point of intersection of the lines
$$l/r + \cos\theta = \cos(\theta - \alpha) \quad \text{and} \quad l/r + \cos\theta = \cos(\theta - \beta).$$

3·7. Line Joining Two Points

   3·71. Let P_1, P_2 be the points and let $P(x, y)$ be an arbitrary point on the line joining them. Then since

$$N_1 N : NN_2 = P_1 P : PP_2 = M_1 M : MM_2,$$

$$\kappa_1(x - x_1) = \kappa_2(x_2 - x),$$

$$\kappa_1(y - y_1) = \kappa_2(y_2 - y).$$

$$\therefore \quad \begin{vmatrix} x - x_1 & x_2 - x \\ y - y_1 & y_2 - y \end{vmatrix} = 0,$$

$$\therefore \quad \begin{vmatrix} x & y & 1 \\ x_1 & y_1 & 1 \\ x_2 & y_2 & 1 \end{vmatrix} = 0. \quad (1)$$

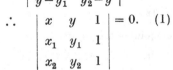

Alternatively this result follows from 1·71 by expressing that the area $P_1 P P_2$ is zero. (1) shows that the equation of a line is of the first degree.

In numerical examples the equation of the line joining two points can often be written down at sight. In particular this can always be done if one of the points is the origin. The line from $(0, 0)$ to $(3, 4)$ is $4x = 3y$ and the line from $(0, 0)$ to $(-2, 5)$ is $5x + 2y = 0$, because these equations are of the first degree and are satisfied by the given coordinates. Similarly the line joining $(1, 4)$ to $(3, 2)$ is $x + y = 5$.

The equation of the line joining two points whose polar coordinates are given is found as in the following example.

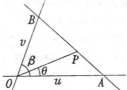

3·72. EXAMPLE. Find the polar equation of the line joining $(u, 0)$ to (v, β).

Let A, B be the given points and let P be an arbitrary point (r, θ) on the line joining them. Then in the figure

$$\triangle AOB = \triangle AOP + \triangle POB.$$

$$\therefore \quad \tfrac{1}{2}uv \sin \beta = \tfrac{1}{2}ur \sin \theta + \tfrac{1}{2}vr \sin (\beta - \theta),$$

i.e. $$(uv/r) \sin \beta = u \sin \theta + v \sin (\beta - \theta).$$

If P is in AB produced, then, in the above proof, $\triangle POB$ is replaced by $-\triangle BOP$ and this is $-\tfrac{1}{2}vr \sin (\theta - \beta)$. Hence the result is unaffected. It also remains the same when P is in BA produced.

By adopting a sign convention to distinguish between $\triangle BOP$ and $\triangle POB$, the same proof can be made to apply to all positions of P on the line.

EXERCISE 3E

In Nos. 1–8, find the equation of the line joining the points.

1. $(0, 0)$, $(5, -7)$.   **2.** $(5, 6)$, $(11, 12)$.   **3.** $(8, 12)$, $(-3, 7)$.

4. $(0, 0)$, (p, q).   **5.** $(0, p)$, $(-q, 0)$.   **6.** $(3, 11)$, $(-7, 20)$.

7. $(\tfrac{2}{3}, \tfrac{3}{4})$, $(-\tfrac{4}{5}, \tfrac{1}{2})$.   **8.** (am^2, am), (an^2, an).

9. A is $(0, 3)$, B is $(-2, 0)$, C is $(11, 5)$, D is $(4, 7)$. Find the equation of the line joining the mid-points of AB and CD.

10. Find the equations of the medians of the triangle $(4, -1)$ $(6, 7)$ $(11, 2)$ and verify that they are concurrent.

In Nos. 11–15, find the polar equation of the line joining the points.

11. (a, α), $(a, \pi - \alpha)$.   **12.** (b, β), $(b, 2\pi - \beta)$.

13. (c, γ), $(d, \gamma + \tfrac{1}{2}\pi)$.   **14.** $(2, 10°)$, $(3, 70°)$.

15. $(3, \tan^{-1}\tfrac{4}{3})$, $(-4, \tan^{-1}\tfrac{7}{24})$.

16. Prove that the line joining (r_1, θ_1) to (r_2, θ_2) is

$$(r_1 r_2/r) \sin(\theta_1 - \theta_2) = r_2 \sin(\theta - \theta_2) - r_1 \sin(\theta - \theta_1).$$

3·8. Coordinates of a Line

The various forms of the equation of a line

$$x/a + y/b = 1, \quad x \cos\alpha + y \sin\alpha = p,$$
$$r \cos(\theta - \alpha) = p, \quad y = ax + b$$

all contain two arbitrary constants.

In $ax + by + c = 0$ there are three constants, but they are not independent because

$$ax + by + c = 0 \quad \text{and} \quad \kappa ax + \kappa by + \kappa c = 0 \quad (\kappa \neq 0)$$

represent the same line. Two of the ratios $a:b:c$ are independent, and thus a, b, c are equivalent to two independent constants.

We therefore say that there are ∞^2 lines in a plane. A line in a plane can be represented either by one of the above equations or by two coordinates. The coordinates might be p, α or they might be the a, b of the intercept equation, but it is best to emphasise the duality that exists between the geometry of the point and of the line in two dimensions by defining coordinates X, Y of a line as follows:

When the equation of a line is written in the form

$$Xx + Yy + 1 = 0,$$

with the constant term equal to $+1$, the coefficients X, Y of x, y are called the coordinates of the line.

These coordinates are such that the intercepts made on the axes by the line are $-1/X$ and $-1/Y$. Thus the dimensions of X and Y are -1 in length.

We shall write line coordinates in square brackets. Thus $[X, Y]$ means the line whose coordinates are X, Y and whose equation is therefore $Xx + Yy + 1 = 0$. See 0·2.

EXERCISE 3F

In Nos. 1–10, state the coordinates of the line.

1. $4x - 5y + 2 = 0$.   **2.** $3x + 2y - 1 = 0$.   **3.** $2y - 7x + 3 = 0$.

4. $x = 5$.   **5.** $2y + 3 = 0$.   **6.** $ax + by + c = 0$.

7. $x/a + y/b = 1$.   **8.** $y = mx + c$.   **9.** $x \cos \alpha + y \sin \alpha = p$.

10. $x - y = \frac{1}{100}$.

In Nos. 11–16, state the equation of the line.

11. $[3, 4]$.   **12.** $[-1, 5]$.   **13.** $[1, -1]$.

14. $[4, 0]$.   **15.** $[\frac{1}{2}, \frac{2}{3}]$.   **16.** $[0, -\frac{5}{4}]$.

17. What lines have no coordinates?

18. What coordinates give no line?

19. If $[X, Y]$ passes through $(4, 5)$, what equation is satisfied by X, Y?

20. If $[X, Y]$ passes through $(-\frac{1}{2}, \frac{3}{4})$, what equation is satisfied by X, Y?

21. If $3X - 7Y + 1 = 0$, prove that $[X, Y]$ passes through a certain fixed point and state its coordinates.

22. If $4X + 5Y - 6 = 0$, prove that $[X, Y]$ passes through a certain fixed point and state its coordinates.

23. Prove that $[X, Y]$ and $[kX, kY]$ are parallel for all values of k except zero.

24. Prove that all lines $[X, Y]$ such that $2X + 3Y = 0$ are parallel.

25. Find the coordinates of the line through $(2, 3)$ parallel to $[4, 5]$.

26. State the condition for $[X_1, Y_1]$ and $[X_2, Y_2]$ to be parallel.

27. State the condition for $[X_1, Y_1]$ and $[X_2, Y_2]$ to be perpendicular.

28. What is known about $[X, Y]$ if $X^2 + Y^2 = a^{-2}$?

29. Find the tangent of the angle between $[X_1, Y_1]$ and $[X_2, Y_2]$.

30. Find the coordinates of the bisectors of the angles between $[3, 4]$ and $[5, -12]$.

31. Prove that $[X_1, Y_1], [X_2, Y_2]$ and

$$[(\kappa_1 X_1 + \kappa_2 X_2)/(\kappa_1 + \kappa_2),\ (\kappa_1 Y_1 + \kappa_2 Y_2)/(\kappa_1 + \kappa_2)]$$

are concurrent.

32. Find the coordinates of the line joining (x_1, y_1) to (x_2, y_2).

33. Find the coordinates of the point of intersection of $[X_1, Y_1]$ and $[X_2, Y_2]$.

34. Find the area of the triangle bounded by $[X_1, Y_1]$, $[X_2, Y_2]$, $[X_3, Y_3]$.

3·9. Duality in Two Dimensions

3·91. The idea of dual statements in two-dimensional geometry may be illustrated by:

There is one line	There is one point
and only one	and only one
which passes through	which lies on
two given points	two given lines
and it is called	and it is called
the *join*	the *meet*
of the two points.	of the two lines.

It is not asserted here that the dual of a true statement is necessarily true and there is in fact an exception to the statement on the right: the two given lines may be parallel.

A geometrical statement of a certain kind can be translated into another (dual) statement by using

point	line
lies on	passes through
join	meet
etc.	etc.

as a dictionary. Two-dimensional geometry can be built up in two different ways, with the point as the fundamental element as in the left column, or with the line as the fundamental element as in the right column.

3·92. The idea of duality also appears in analytical geometry:

When the coordinates x, y	When the coordinates X, Y
of a variable point P	of a variable line p
obey a linear relation	obey a linear relation
$lx + my + n = 0 \ (n \neq 0),$	$LX + MY + N = 0 \ (N \neq 0),$
the point (x, y) lies on	the line $[X, Y]$ passes through
a certain line, namely	a certain point, namely
$X = l/n, \quad Y = m/n.$	$x = L/N, \quad y = M/N.$

We call $lx + my + n = 0$ the equation of the line which may be thought of as generated by the variable point.	We call $LX + MY + N = 0$ the equation of the point which may be thought of as generated by the variable line.

When there is any danger of confusion, x, y are called the *locus* or *point coordinates* and X, Y are called the *envelope* or *line coordinates*. Also $lx + my + n = 0$ is called the *locus* or *point equation* of the line, and $LX + MY + N = 0$ is called the *envelope* or *line equation* of the point.

3·93. When $n = 0$, points such that $lx + my + n = 0$ still lie on a certain line, but this line passes through the origin and has no envelope coordinates.

When $N = 0$, lines such that $LX + MY + N = 0$, instead of passing through a fixed point, are parallel.

For if $[X_1, Y_1]$, $[X_2, Y_2]$ are two such lines,

$$LX_1 + MY_1 = 0 \quad \text{and} \quad LX_2 + MY_2 = 0,$$

and so

$$\begin{vmatrix} X_1 & Y_1 \\ X_2 & Y_2 \end{vmatrix} = 0,$$

which is the condition of parallelism by 3·2.

Thus lines $[X, Y]$ such that $LX + MY = 0$, where L, M are constants, form a set of parallel lines.

3·94. EXAMPLE. What is represented by $X = at + b$, $Y = ct + d$, where t is a parameter?

For any particular value of t, the equations give the coordinates of a line. When t varies, the line varies in such a way that its coordinates satisfy

$$\begin{vmatrix} X - b & a \\ Y - d & c \end{vmatrix} = 0,$$

i.e. $$cX - aY = cb - ad.$$

This is of the form $LX + MY + N = 0$ and is the envelope equation of a point unless $cb = ad$. The given equations are parametric equations of a point.

When $cb = ad$, the equations represent a set of parallel lines.

EXERCISE 3G

In Nos. 1–4, state the coordinates of the point.

1. $4X - 7Y + 1 = 0.$ **2.** $2X + 3Y = 5.$

3. $X + 7 = 0.$ **4.** $2Y = 3.$

In Nos. 5–9, state the equation of the point.

5. $(3, 4).$ **6.** $(-7, -6).$ **7.** $(2, 0).$ **8.** $(0, -\frac{1}{2}).$ **9.** $(-\frac{2}{3}, \frac{3}{2}).$

10. What point has no equation?

In Nos. 11–13, indicate the position of the point in a diagram.

11. $X + Y = 1.$ **12.** $3X - 4Y + 5 = 0.$ **13.** $2X = 7.$

14. Write down the distance between the points $a_1 X + b_1 Y + c_1 = 0$ and $a_2 X + b_2 Y + c_2 = 0.$

15. Write down the equation of the mid-point of the line joining $a_1 X + b_1 Y + c_1 = 0$ to $a_2 X + b_2 Y + c_2 = 0.$

In Nos. 16–19, state the coordinates of the line.

16. Through $2X + 3Y = 1$ parallel to $[4, -1].$

17. Through $aX + bY + c = 0$ parallel to $[X_1, Y_1].$

18. Through $3X - Y + 2 = 0$ perpendicular to $[1, 2].$

19. Through $aX + bY + c = 0$ perpendicular to $px + qy + r = 0.$

20. Find the condition for the lines joining O to $a_1 X + b_1 Y + c_1 = 0$ and $a_2 X + b_2 Y + c_2 = 0$ to be at right angles.

21. Prove that lines $[X, Y]$ such that $PX + QY = 0$ are parallel to the line joining the origin to $(P, Q).$

22. Find the relation satisfied by the coordinates of lines parallel to $px + qy = 0.$

23. Find the locus of the points given by $aX + bY = \lambda$ when λ varies.

24. Find the equation of the locus of the points given by

$$Y = \lambda^2(\lambda X + 1)$$

when λ varies.

Chapter 4

DUALITY AND DEGENERACY

4·1. Duality

4·11. When the coordinates x, y of a variable point P satisfy an equation

$$f(x, y) = 0,$$

the point generates a certain curve (locus) and $f(x, y) = 0$ is called the (locus) equation of the curve.

When the coordinates X, Y of a variable line p satisfy an equation

$$F(X, Y) = 0,$$

the line generates a certain curve (envelope) and $F(X, Y) = 0$ is called the (envelope) equation of the curve.

The words in brackets are omitted when there is no doubt about the meaning.

The simplest cases, when f and F are functions of the first degree in x, y and in X, Y, are given in 3·92. The line is the simplest form of a curve locus generated by a variable point. The point, when it is not taken as the fundamental element, occurs as the simplest form of curve envelope generated by a variable line: a point that occurs in this way may be called an envelope point.

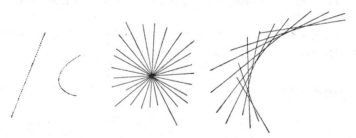

The locus equation of a curve is the relation which is satisfied by the coordinates of an arbitrary point of the

The envelope equation of a curve is the relation which is satisfied by the coordinates of an arbitrary line of the locus. Hence the solution of a locus problem usually begins with the words "Take an arbitrary point (x, y) of the locus...."

envelope. Hence the solution of an envelope problem usually begins with the words "Take an arbitrary line $[X, Y]$ of the envelope...."

4·12. *Tangents.* If A is a fixed point of a curve locus, and P is a variable point of it, the limit when P tends to A of the line through A and P (denoted by AP) is called the *tangent* at A to the curve locus.

AB is the join of the points A, B. A curve locus of a point Q has tangents, and these lines generate a curve envelope.

Contacts. If a is a fixed line of a curve envelope, and p is a variable line of it, the limit when p tends to a of the point on a and p (denoted by ap) is called the *contact* of a with the curve envelope.

ab is the meet of the lines a, b. A curve envelope of a line q has contacts, and these points generate a curve locus.

Thus there are two kinds of curve: a curve locus which is a collection of points, and a curve envelope which is a collection of lines. But a curve locus has tangents and these lines constitute a curve envelope, and a curve envelope has contacts and these points constitute a curve locus.

In Chapter 12 we start with a curve locus defined by an equation $f(x, y) = 0$ of the second degree. We prove that its tangents constitute a curve envelope whose equation $F(X, Y) = 0$ is also of the second degree, and that the contacts of this curve envelope are the points of the original curve locus. The name *conic* is given to this curve whether it is regarded as a locus or as an envelope. Other curves can be generated either as loci or as envelopes, but the degrees of their locus and envelope equations are not usually the same.

EXERCISE 4A

In Nos. 1–8, by drawing the lines $[X, Y]$ in various positions, show the curve envelope given by the equation.

1. $X + Y = 5$.   **2.** $XY = 24$.   **3.** $X = 0$.   **4.** $Y = X^2$.

5. $Y = 1$.   **6.** $5X = 7Y$.   **7.** $1/X + 1/Y = 1$.   **8.** $Y^2 = X^3$.

9. Prove that all lines such that $X^2 + Y^2 = \frac{1}{4}$ are at the same distance from the origin. Hence obtain the curve whose envelope equation is $X^2 + Y^2 = \frac{1}{4}$.

10. A is a fixed point and a is a fixed line not through A. Y is a variable point on a and p is the line through Y perpendicular to AY. By taking various positions of Y, show in a figure the envelope of p.

11. A piece of paper is in the form of a circle and A is a point inside the circle. Show in a figure the envelope of the crease formed by folding the paper so that the circumference passes through A.

In Nos. 12–15, state the duals.

12. Four points can be joined in pairs by six lines. These meet in three other points.

13. Five lines meet in pairs at ten points. These can be joined by fifteen other lines.

14. Three points A, B, C lie on a line p, and three points A', B', C' lie on a line q. The meets of BC' and CB', CA' and AC', AB' and BA' are collinear.

15. Two triangles which have their corresponding vertices joined by concurrent lines also have their corresponding sides intersecting in collinear points.

4·2. Degenerate Loci and Envelopes

4·21. The equations $x = 0$ and $y = 0$ represent the axes OY and OX respectively. The equation $xy = 0$ represents the pair of lines OX, OY; for it is satisfied by the coordinates of any point on either line.

More generally, $(ax + by + c)(lx + my + n) = 0$ represents the *line-pair* composed of the two lines $ax + by + c = 0$ and $lx + my + n = 0$; for it is satisfied by the coordinates of any point on either of these two lines.

Again, $(x^2 + y^2 - 1)(y - x) = 0$ represents a circle and line. And generally if $f(x, y) = 0$ and $g(x, y) = 0$ are any two loci, the equation
$$f(x, y) \cdot g(x, y) = 0$$
represents the locus composed of the separate curves $f(x, y) = 0$ and $g(x, y) = 0$. Such a composite locus is called a *degenerate locus.*

Dually an envelope equation
$$F(X, Y) \cdot G(X, Y) = 0$$

represents the envelope composed of the separate envelopes $F(X, Y) = 0$ and $G(X, Y) = 0$, and this is called a *degenerate envelope*. For example, the equation $(X+1)(X-1) = 0$ is satisfied by the coordinates of any line which passes through $(1, 0)$ or through $(-1, 0)$ and is the envelope equation of this *point-pair*.

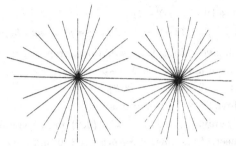

4·22. EXAMPLE. What is represented by the equation

$$(x^2 - 1)(x^2 + 2y^2) = 0?$$

This equation may be written $(x+1)(x-1)(x^2+2y^2) = 0$ and therefore represents the locus composed of the three loci

$$x = -1, \quad x = +1, \quad x^2 + 2y^2 = 0.$$

$x^2 + 2y^2 = 0$ is satisfied by $x = y = 0$ only. The equation therefore represents two lines $x = \pm 1$ together with the origin.

4·23. EXAMPLE. What is represented by the equation

$$(X-4)^2 + (Y^2-1)^2 = 0?$$

This equation is true if and only if $X = 4$ and $Y = \pm 1$. Therefore it represents the two lines $[4, 1]$ and $[4, -1]$.

EXERCISE 4B

In Nos. 1–20, state what is represented by the equation.

1. $x^2y = 0$.   2. $3x^2 + 2y^2 = 0$.
3. $(x+1)(y-2) = 0$.   4. $(X+1)(Y-2) = 0$.
5. $(X-A)^2 + (Y-B)^2 = 0$.   6. $X^2Y = Y$.
7. $(x^2-1)^2 + (y^2-4)^2 = 0$.   8. $xy + 3x - 2y - 6 = 0$.

9. $8x^2 + 30xy - 27y^2 = 0.$ **10.** $X^2 - XY + Y^2 = 0.$

11. $x^4 = y^4.$ **12.** $(X+Y)^2 = 4.$

13. $X^2 + Y^2 + 1 = 0.$ **14.** $x^4 - 5x^2y^2 + 4y^4 = 0.$

15. $X^3 + Y^3 = 0.$

16. $(x+1)^2 + (y-2)^2 = (x-3)^2 + (y+4)^2.$

17. $(x+y+a)^2 + (x-y-a)^2 = 0.$

18. $(X+Y+1)^2 + (X-Y-1)^2 = 0.$

19. $x^2 + 2xy + y^2 = 3x + 3y.$ **20.** $x^2 + y^2 = (ax+by)^2.$

In Nos. 21–26, give a single locus equation to represent:

21. The pair of lines $x + 2 = 0,\ y - 3 = 0.$

22. The pair of lines $[3, 1],\ [-2, 8].$

23. The three lines $x = 0,\ y = 0,\ x = y.$

24. $(3, -5)$ only.   **25.** $(0, 0)$ and $(1, 1)$ only.

26. The meet of the lines $a_1x + b_1y + c_1 = 0$ and $a_2x + b_2y + c_2 = 0$ only.

In Nos. 27–30, give a single envelope equation to represent:

27. The pair of points $X = 4,\ Y = -3.$

28. The pair of points $(4, 7),\ (-3, -2).$

29. $x + y = 1$ only.   **30.** $[1, 0]$ and $[0, 1]$ only.

4·3. The Locus $ax^2 + 2hxy + by^2 = 0$ and the Envelope $AX^2 + 2HXY + BY^2 = 0$

4·31. The locus equation may be written

$$(ax + hy)^2 - (h^2 - ab)y^2 = 0, \quad (a \neq 0).$$

If $h^2 > ab$, $h^2 - ab = k^2$, this becomes

$$(ax + hy + ky)(ax + hy - ky) = 0$$

and represents a pair of lines through the origin.

If $h^2 = ab$, it represents one line $ax + hy = 0$ and conventionally is said to represent two coincident lines.

If $h^2 < ab$, $h^2 - ab = -k^2$, the equation is

$$(ax + hy)^2 + k^2y^2 = 0.$$

This is only true if $ax + hy = 0$ and $y = 0$, i.e. if $x = y = 0$. Hence it represents the point $(0, 0)$ only.

The first two results hold even if a or b is zero.

4·32. *Orthogonal and Oblique Line-pairs.*

If $\qquad ax^2 + 2hxy + by^2 \equiv (l_1 x + m_1 y)(l_2 x + m_2 y),$

the condition for the lines represented by $ax^2 + 2hxy + by^2 = 0$ to be at right angles is

$$l_1 l_2 + m_1 m_2 = 0.$$

But $l_1 l_2 = a$ and $m_1 m_2 = b$.

$$\therefore \quad a + b = 0.$$

Thus the equation of a pair of perpendicular lines through the origin is

$$x^2 + 2cxy - y^2 = 0.$$

When the lines are not at right angles an angle between them is given by

$$\tan \theta = \frac{l_1 m_2 - l_2 m_1}{l_1 l_2 + m_1 m_2} = \frac{2\sqrt{(h^2 - ab)}}{a + b},$$

because

$$(l_1 m_2 - l_2 m_1)^2 = (l_1 m_2 + l_2 m_1)^2 - 4 l_1 l_2 m_1 m_2 = (2h)^2 - 4ab.$$

4·33.

For the envelope equation $AX^2 + 2HXY + BY^2 = 0$, as in 4·31, if $H^2 > AB$, the equation is of the form

$$(AX + HY + KY)(AX + HY - KY) = 0.$$

Hence $X:Y$ has one of two fixed values, and a line whose coordinates satisfy the equation belongs to one of two sets of parallel lines.

If $H^2 = AB$, the equation is satisfied by the coordinates of a line belonging to a set of parallel lines.

If $H^2 < AB$, the equation is satisfied by $X = Y = 0$ only, and there is no line which has coordinates $0, 0$.

When $H^2 > AB$, $H^2 - AB = K^2$, the two sets of parallel lines make angles $\tan^{-1}\{(H + K)/A\}$, $\tan^{-1}\{(H - K)/A\}$ with OX. Hence the angle between them is given by

$$\tan \theta = \frac{(H + K)/A - (H - K)/A}{1 + (H^2 - K^2)/A^2} = \frac{2AK}{A^2 + H^2 - K^2}$$

$$= \frac{2K}{A + B} = \frac{2\sqrt{(H^2 - AB)}}{A + B}.$$

The two sets of lines are at right angles if $A + B = 0$.

4·34. *Angle-bisectors of* $ax^2 + 2hxy + by^2 = 0$. It is assumed that $h^2 > ab$. Suppose as before that

$$ax^2 + 2hxy + by^2 \equiv (l_1 x + m_1 y)(l_2 x + m_2 y).$$

Let (x, y) be any point on either bisector. Then the sum of the angles made by the given lines with OX is $2 \tan^{-1}(y/x)$ or differs from this by a multiple of π. Hence

$$2 \tan^{-1} \frac{y}{x} = \tan^{-1} \frac{-l_1}{m_1} + \tan^{-1} \frac{-l_2}{m_2} + n\pi.$$

$$\therefore \quad \frac{2y/x}{1 - y^2/x^2} = \frac{-l_1/m_1 - l_2/m_2}{1 - l_1 l_2/m_1 m_2} = \frac{l_1 m_2 + l_2 m_1}{l_1 l_2 - m_1 m_2} = \frac{2h}{a - b},$$

i.e.
$$\frac{x^2 - y^2}{a - b} = \frac{xy}{h}.$$

This is an equation of the form

$$x^2 + 2cxy - y^2 = 0.$$

4·35. EXAMPLE. Find the angle between the lines

$$x^2 - 4xy - 3y^2 = 0$$

and find the equation of the angle-bisectors.

Let $\qquad x^2 - 4xy - 3y^2 = -3(y - px)(y - qx).$

Then $\qquad p + q = -\frac{4}{3}$ and $pq = -\frac{1}{3}.$

$$\therefore \quad (p - q)^2 = \tfrac{16}{9} + \tfrac{4}{3} = \tfrac{28}{9}.$$

The angle between the lines

$$= \tan^{-1} p - \tan^{-1} q$$

$$= \tan^{-1} \frac{p - q}{1 + pq} = \tan^{-1} \tfrac{1}{3}\sqrt{7}.$$

Also, if (x, y) is a point on either bisector,

$$\frac{2y/x}{1 - y^2/x^2} = \tan(\tan^{-1} p + \tan^{-1} q) = \frac{p + q}{1 - pq} = -1.$$

$$\therefore \quad x^2 + 2xy - y^2 = 0.$$

4·36. EXAMPLE. Find the equation of the lines through the origin perpendicular to the lines $ax^2 + 2hxy + by^2 = 0$.

If (x, y) lies on either of the required lines, then $(-y, x)$ lies on one of the given lines $ax^2 + 2hxy + by^2 = 0$.

$$\therefore \quad ay^2 - 2hxy + bx^2 = 0.$$

This is the required equation.

4·37. EXAMPLE. Find the orthocentre of the triangle formed by

$$x^2 + 3xy - 2y^2 = 0 \quad \text{and} \quad 4x - y + 2 = 0.$$

Let

$$x^2 + 3xy - 2y^2 \equiv (x + t_1 y)(x + t_2 y).$$

Then

$$t_1 + t_2 = 3, \quad t_1 t_2 = -2.$$

$x + t_1 y = 0$ meets $4x - y + 2 = 0$ at $\left(\dfrac{-2t_1}{1 + 4t_1}, \dfrac{2}{1 + 4t_1} \right)$ and the perpendicular from this point to $x + t_2 y = 0$ is

$$t_2 x - y = (-2t_1 t_2 - 2)/(1 + 4t_1).$$

The orthocentre is the point of intersection of this line and

$$t_1 x - y = (-2t_1 t_2 - 2)/(1 + 4t_2),$$

namely

$$x = \frac{2(t_1 t_2 + 1)}{t_2 - t_1} \left(\frac{1}{1 + 4t_2} - \frac{1}{1 + 4t_1} \right) = \frac{-8(t_1 t_2 + 1)}{1 + 4(t_1 + t_2) + 16 t_1 t_2} = \frac{-8}{19},$$

$$y = -\tfrac{1}{4}x = \tfrac{2}{19}.$$

Thus the orthocentre is $(-\tfrac{8}{19}, \tfrac{2}{19})$.

4·38. EXAMPLE. Find the equation of the lines through the origin whose perpendicular distance from (a, b) is c.

Let $px + qy = 0$ be either line.

Then

$$c = \pm \frac{pa + qb}{\sqrt{(p^2 + q^2)}},$$

$$c^2(p^2 + q^2) = (pa + qb)^2.$$

Eliminating $p : q$,

$$c^2(x^2 + y^2) = (ay - bx)^2.$$

EXERCISE 4c

In Nos. 1, 2, state for what values of k the locus is a line-pair and for what values of k the line-pair is orthogonal.

1. $kx^2 + 2xy + k^2y^2 = 0.$ **2.** $4x^2 - 2(k+3)xy - ky^2 = 0.$

In Nos. 3–8, find the angle between the lines.

3. $x^2 - 7xy + 10y^2 = 0.$ **4.** $2x^2 - 5xy - 12y^2 = 0.$

5. $7x^2 + 123xy - 7y^2 = 0.$ **6.** $3x^2 - 4xy - 10y^2 = 0.$

7. $x^2 + 2xy \sec\alpha + y^2 = 0.$ **8.** $(x^2 + y^2)\sin\alpha = 2xy.$

9. For what values of k does $kX^2 + XY - k^3Y^2 = 0$ represent two perpendicular sets of lines?

In Nos. 10–12, find the angle between the sets of lines.

10. $X^2 = 3Y^2.$ **11.** $7X^2 + 2XY - 3Y^2 = 0.$

12. $2X^2 + 3XY - 2Y^2 = 0.$

In Nos. 13–19, find the angle-bisectors of the line-pairs.

13. $y^2 - 4xy + 3x^2 = 0.$ **14.** $5x^2 - 11xy - 13y^2 = 0.$

15. $3x^2 = 5y^2.$ **16.** $x^2 + kxy - y^2 = 0.$

17. $x^2 + 2xy \operatorname{cosec}\alpha + y^2 = 0.$ **18.** $(x^2 + y^2)\cos\alpha = 2xy.$

19. $cx^2 + 2dxy + (c-d)y^2 = 0.$

In Nos. 20, 21, find the equation of the sets of parallel lines bisecting the angles between the given sets.

20. $X^2 - 5Y^2 = 0.$ **21.** $X^2 + 7XY + 12Y^2 = 0.$

In Nos. 22, 23, find the equation of the pair of lines through the origin perpendicular to the lines.

22. $x^2 - 3y^2 = 0.$ **23.** $7x^2 + 4xy - 5y^2 = 0.$

24. Find the orthocentre of the triangle formed by $x^2 - y^2 = 0$ and $2x + y = 2.$

25. Find the orthocentre of the triangle formed by
$$ax^2 + 2hxy + by^2 = 0 \quad \text{and} \quad lx + my + n = 0.$$

26. Prove that the centroid of the triangle formed by
$$ax^2 + 2hxy + by^2 = 0 \quad \text{and} \quad lx + my + n = 0$$
is given by
$$x : y : \tfrac{2}{3}n = hm - bl : hl - am : am^2 - 2hlm + bl^2.$$

4·4. Line-Pairs and Point-Pairs in general

4·41. When $ax^2 + 2hxy + by^2 + 2gx + 2fy + c$

has factors $l_1x + m_1y + n_1$, $l_2x + m_2y + n_2$,

the equation $ax^2 + 2hxy + by^2 + 2gx + 2fy + c = 0$

represents the pair of lines

$$l_1x + m_1y + n_1 = 0, \quad l_2x + m_2y + n_2 = 0.$$

At the same time $l_1x + m_1y$, $l_2x + m_2y$ are the factors of $ax^2 + 2hxy + by^2$. Hence $ax^2 + 2hxy + by^2 = 0$ represents the pair of lines through the origin parallel to the lines

$$ax^2 + 2hxy + by^2 + 2gx + 2fy + c = 0.$$

When $AX^2 + 2HXY + BY^2 + 2GX + 2FY + C$

has factors $L_1X + M_1Y + N_1$, $L_2X + M_2Y + N_2$,

the equation $AX^2 + 2HXY + BY^2 + 2GX + 2FY + C = 0$

represents the pair of points

$$L_1X + M_1Y + N_1 = 0, \quad L_2X + M_2Y + N_2 = 0.$$

At the same time $L_1X + M_1Y$, $L_2X + M_2Y$ are the factors of $AX^2 + 2HXY + BY^2$. Hence $AX^2 + 2HXY + BY^2 = 0$ represents the same as the equations

$$L_1X + M_1Y = 0, \quad L_2X + M_2Y = 0,$$

i.e. it represents two sets of parallel lines (3·93).

4·42. EXAMPLE. Find the equation of the lines through the origin parallel to the lines whose coordinates satisfy

$$AX^2 + 2HXY + BY^2 = 0.$$

Let $[X, Y]$ be one of the lines. The line through the origin parallel to it is $Xx + Yy = 0$. The coordinates of any point on this line satisfy $x:y = Y:-X$. But they also satisfy

$$AX^2 + 2HXY + BY^2 = 0. \quad \therefore \quad Ay^2 - 2Hxy + Bx^2 = 0.$$

This is the required equation because it is satisfied by the coordinates x, y of any point on either of the lines.

4·43. Example. Find λ such that

$$2x^2 + xy - y^2 + 5x + 2y + \lambda = 0$$

represents a line-pair.

Since $\qquad 2x^2 + xy - y^2 \equiv (x + y)(2x - y),$

put $\quad 2x^2 + xy - y^2 + 5x + 2y + \lambda \equiv (x + y + a)(2x - y + b).$

Then $\qquad 5 = 2a + b, \quad 2 = -a + b, \quad \lambda = ab.$

$$\therefore \quad a = 1, \quad b = 3, \quad \lambda = 3.$$

4·44. Example. Find μ such that

$$2X^2 + \mu XY - Y^2 + 5X + 2Y + 3 = 0$$

represents a point-pair.

Since $\qquad 2X^2 + 5X + 3 \equiv (X + 1)(2X + 3),$

put

$$2X^2 + \mu XY - Y^2 + 5X + 2Y + 3 \equiv (X + 1 + AY)(2X + 3 + BY).$$

Then $\quad \mu = 2A + B, \quad -1 = AB, \quad 2 = 3A + B.$

$$\therefore \quad A = 1, B = -1 \quad \text{or} \quad A = -\tfrac{1}{3}, B = 3,$$

$$\therefore \quad \mu = 1 \quad \text{or} \quad \mu = \tfrac{7}{3}.$$

4·45. When $l_1 x + m_1 y + n_1$, $l_2 x + m_2 y + n_2$ are the factors of

$$ax^2 + 2hxy + by^2 + 2gx + 2fy + c,$$

$$a = l_1 l_2 \qquad\qquad b = m_1 m_2 \qquad\qquad c = n_1 n_2$$

$$2f = m_1 n_2 + m_2 n_1 \quad 2g = n_1 l_2 + n_2 l_1 \quad 2h = l_1 m_2 + l_2 m_1.$$

But

$$\begin{vmatrix} l_1 & l_2 & 0 \\ m_1 & m_2 & 0 \\ n_1 & n_2 & 0 \end{vmatrix} \begin{vmatrix} l_2 & m_2 & n_2 \\ l_1 & m_1 & n_1 \\ 0 & 0 & 0 \end{vmatrix} = 0.$$

$$\therefore \quad \begin{vmatrix} 2l_1 l_2 & l_1 m_2 + l_2 m_1 & n_1 l_2 + n_2 l_2 \\ l_1 m_2 + l_2 m_1 & 2m_1 m_2 & m_1 n_2 + m_2 n_1 \\ n_1 l_2 + n_2 l_1 & m_1 n_2 + m_2 n_1 & 2n_1 n_2 \end{vmatrix} = 0,$$

$$\therefore \quad \begin{vmatrix} a & h & g \\ h & b & f \\ g & f & c \end{vmatrix} = 0.$$

Hence this is a necessary condition for

$$ax^2 + 2hxy + by^2 + 2gx + 2fy + c$$

to factorise. It is not a sufficient condition in real algebra: for it holds when $a = b = 1$, $c = f = g = h = 0$, whereas $x^2 + y^2$ does not factorise. When the factors exist, the equation

$$ax^2 + 2hxy + by^2 + 2gx + 2fy + c = 0$$

represents the pair of lines

$$l_1 x + m_1 y + n_1 = 0, \quad l_2 x + m_2 y + n_2 = 0.$$

4·46. As in 4·45, it can be proved that a necessary condition for $AX^2 + 2HXY + BY^2 + 2GX + 2FY + C$ to factorise is

$$\begin{vmatrix} A & H & G \\ H & B & F \\ G & F & C \end{vmatrix} = 0.$$

When the factors exist, the equation

$$AX^2 + 2HXY + BY^2 + 2GX + 2FY + C = 0$$

represents a pair of points. But either of these points may be replaced by a set of parallel lines as explained in 3·93.

4·47. EXAMPLE. If $x^2 + 2xy - 5y^2 + 7x + 8y + \lambda = 0$ represents a line-pair, find the value of λ and the angle between the lines.

From

$$\begin{vmatrix} 1 & 1 & \frac{7}{2} \\ 1 & -5 & 4 \\ \frac{7}{2} & 4 & \lambda \end{vmatrix} = 0, \quad \lambda = \tfrac{293}{24}.$$

If

$$x^2 + 2xy - 5y^2 \equiv (l_1 x + m_1 y)(l_2 x + m_2 y),$$

$$l_1 l_2 = 1, \quad l_1 m_2 + l_2 m_1 = 2, \quad m_1 m_2 = -5,$$

$$\therefore \quad (l_1 m_2 - l_2 m_1)^2 = 2^2 - 4 \cdot 1 \cdot (-5) = 24.$$

The angle is given by

$$\tan \theta = \frac{l_1 m_2 - l_2 m_1}{l_1 l_2 + m_1 m_2} = \tfrac{1}{2}\sqrt{6}.$$

4·48. EXAMPLE. Find what is represented by

$$X^2 + \mu Y^2 + X - Y = 0$$

when μ has the value for which the left side factorises.

From

$$\begin{vmatrix} 1 & 0 & \tfrac{1}{2} \\ 0 & \mu & -\tfrac{1}{2} \\ \tfrac{1}{2} & -\tfrac{1}{2} & 0 \end{vmatrix} = 0, \quad \mu = -1.$$

The equation is then

$$(X - Y)(X + Y + 1) = 0.$$

It represents the point $(1, 1)$ and the set of parallel lines making equal intercepts on the axes.

4·49. EXAMPLE. Write down the equation of the pair of lines through $(-2, 3)$ parallel to the lines

$$s \equiv 4x^2 - xy - y^2 = 0.$$

The equation is

$$s' \equiv 4(x+2)^2 - (x+2)(y-3) - (y-3)^2 = 0.$$

s' is found from s by changing x into $x+2$ and y into $y-3$. Hence if the factors of s are $l_1 x + m_1 y$, $l_2 x + m_2 y$, those of s' are $l_1(x+2) + m_1(y-3)$, $l_2(x+2) + m_2(y-3)$. Thus the lines $s' = 0$ are parallel to the lines $s = 0$. Also they both pass through $(-2, 3)$.

The equation is the same as that of the given lines referred to axes through $(2, -3)$ parallel to the original axes.

4·5. *The Locus* $a_0 x^n + a_1 x^{n-1} y + a_2 x^{n-2} y^2 + \ldots + a_n y^n = 0$

The function $f(x, y) \equiv a_0 x^n + a_1 x^{n-1} + \ldots + a_n y^n$ can be expressed as the product of r ($\leqslant n$) linear factors and $\tfrac{1}{2}(n-r)$ quadratic factors which are not reducible to linear factors, i.e. quadratic factors of the form $ax^2 + 2hxy + by^2$ where $h^2 < ab$. See *A.A.* p. 257.

Thus the locus $f(x, y) = 0$ consists of r lines through the origin. The quadratic factors merely give the origin. If $r = 0$, the locus is the origin only. If n is odd, $r \geqslant 1$ and there is at least one line in the locus.

Similarly, the lines whose coordinates satisfy an equation of the form $A_0 X^n + A_1 X^{n-1} Y + \ldots + A_n Y^n = 0$ constitute r ($\leqslant n$) sets of parallel lines.

EXERCISE 4D

In Nos. 1–6, use the method of 4·43 to find the value of λ for which the locus is degenerate, and give the separate equations of the lines.

1. $2x^2 + xy - y^2 + 11x + 2y + \lambda = 0$.

2. $\lambda x^2 + 5xy + y^2 + 5x + 2y + 1 = 0$.

3. $x^2 + 5xy + \lambda y^2 - 13x - 30y + 36 = 0$.

4. $2x^2 + \lambda xy - 3y^2 - x - 4y - 1 = 0$.

5. $\lambda xy - 35x + 6y - 15 = 0$.

6. $12x^2 - 13xy - 14y^2 + 38x - 81y + \lambda = 0$.

In Nos. 7–9, use the method of 4·44 to find the values of λ for which the envelope is degenerate, and state the coordinates of the separate points.

7. $3X^2 + 7XY + 4Y^2 + X + 2Y + \lambda = 0$.

8. $2X^2 + XY - Y^2 + \lambda X - 3Y - 2 = 0$.

9. $\lambda XY + 6X - 20Y - 10 = 0$.

In Nos. 10–15, find the equation of the pair of lines through the given point satisfying the given condition.

10. $(0,0)$, parallel to $x^2 - xy - 2y^2 + 3x + 2 = 0$.

11. $(1,1)$, parallel to $x^2 + xy - 2y^2 = 0$.

12. $(2, -3)$, parallel to $x^2 + xy - 6y^2 + 4x + 2y + 4 = 0$.

13. $(0,0)$, perpendicular to $x^2 + 3xy + 2y^2 - x + y - 6 = 0$.

14. $(2,3)$, perpendicular to $2x^2 + 3xy - 4y^2 = 0$.

15. $(-3, -1)$, perpendicular to $3x^2 - xy - y^2 + 4x - 1 = 0$.

In Nos. 16–18, find the value of λ, given that the locus or envelope is degenerate.

16. $2x^2 + 6xy - y^2 + 4x - 2y + \lambda = 0$.

17. $\lambda x^2 + 3xy - 2y^2 - 5y - \lambda = 0$.

18. $X^2 + 2\lambda XY + 5Y^2 + 6X + 2Y - 1 = 0$.

In Nos. 19–21, find the angle-bisectors of the line-pairs.

19. $20x^2 + 63xy + 36y^2 - 103x - 135y + 119 = 0$.

20. $2x^2 + xy - y^2 - 11x + 4y + 5 = 0$.

21. $x^2 + 4xy + y^2 - 6x - 3 = 0$.

22. Find the product of the lengths of the perpendiculars from (f, g) to $ax^2 + 2hxy + by^2 = 0$.

4·6. Common Points of Two Loci

4·61. The coordinates of the points common to two loci $f(x, y) = 0$, $g(x, y) = 0$ satisfy both $f(x, y) = 0$ and $g(x, y) = 0$ and are found by solving these simultaneous equations.

Any other equation deducible from $f(x, y) = 0$ and $g(x, y) = 0$ must also be satisfied by the coordinates of any common point. Such an equation therefore represents some locus which passes through all the common points of the two loci. This principle is of great importance. Several examples of its application are given here. Further developments are considered in Chapter 16.

4·62. $11x = 60y$ and $99x + 20y = 1100$ are two lines. The bisector of one of the angles between them is found in 3·52 to be
$$101(11x - 60y) = -61(99x + 20y - 1100).$$

The truth of this last equation follows from that of the other two. This corresponds to the fact that the bisector passes through the meet of the two lines.

4·63. $x + y = 3$ and $y - 3x + 5 = 0$ are two lines.

$x + y - 3 = \kappa(y - 3x + 5)$ represents a locus which passes through the meet of the two lines, because the coordinates of that point when substituted for x and y reduce both sides to zero.

This is true for all values of κ. If κ is numerical (not involving x or y) the locus represented by $x + y - 3 = \kappa(y - 3x + 5)$ is a line because the equation is of the first degree in x, y. It is therefore a line through the meet of $x + y = 3$, $y - 3x + 5 = 0$. If κ involves x or y, the equation represents some curve through the meet of the two lines.

4·64. $y = 3$ and $x^2 + y^2 = 25$ represent a line and circle.

The equation $x^2 + 3^2 = 25$, which can be derived from them, must represent some locus through the common points ($\pm 4, 3$) of the line and circle. Actually it reduces to $x^2 = 16$ and is the equation of a line-pair. This line-pair meets the circle in two points besides the common points.

4·65. $x^2 = 4$ and $y^2 = 9$ represent pairs of parallel lines meeting in the four points ($\pm 2, \pm 3$).

$x^2 + y^2 = 13$ passes through these four points. So do the loci $x^2 y^2 = 36$, $y^2 - x^2 = 5$, $(x^2 + y^2)^2 = 5x^2 + 7y^2 + 86$, etc.

4·66. EXAMPLE. Find the equation of the line joining $(3, -2)$ to the meet of $2x + 5y + 1 = 0$ and $4x - 7y - 11 = 0$.

The line $2x + 5y + 1 = \kappa(4x - 7y - 11)$ passes through the meet. It also passes through the point $(3, -2)$ if

$$6 - 10 + 1 = \kappa(12 + 14 - 11), \qquad \text{i.e. } \kappa = -\tfrac{1}{5}.$$

Hence the required equation is

$$5(2x + 5y + 1) + (4x - 7y - 11) = 0, \qquad \text{i.e. } 7x + 9y = 3.$$

4·67. EXAMPLE. Find the equation of the perpendicular to $4x + 5y + 6 = 0$ from the meet of $3x - 8y = 4$ and $7x + y = 1$.

The line $3x - 8y - 4 + \kappa(7x + y - 1) = 0$ passes through the meet. It is perpendicular to $4x + 5y + 6 = 0$ if

$$4(3 + 7\kappa) + 5(-8 + \kappa) = 0, \qquad \text{i.e. } \kappa = \tfrac{28}{33}.$$

Hence the required equation is

$$33(3x - 8y - 4) + 28(7x + y - 1) = 0, \qquad \text{i.e. } 295x - 236y = 160.$$

4·7. Join of a Point to the Meet of Two Lines

4·71. If $\kappa_1 : \kappa_2$ is an arbitrary constant and

$$X_1 x + Y_1 y + Z_1 = 0,$$
$$X_2 x + Y_2 y + Z_2 = 0$$

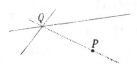

are two lines meeting at Q, then

$$\kappa_1(X_1 x + Y_1 y + Z_1) + \kappa_2(X_2 x + Y_2 y + Z_2) = 0$$

is an arbitrary line through Q. For the coordinates of Q satisfy $X_1 x + Y_1 y + Z_1 = 0$ and $X_2 x + Y_2 y + Z_2 = 0$ and therefore they satisfy $\kappa_1(X_1 x + Y_1 y + Z_1) + \kappa_2(X_2 x + Y_2 y + Z_2) = 0$. But this equation is of the first degree in x, y. Hence it represents some line through Q.

Also $\kappa_1 : \kappa_2$ can be chosen so that the equation represents a given line through Q. For let P be a point (f, g) on such a line.

Then $\kappa_1(X_1 x + Y_1 y + Z_1) + \kappa_2(X_2 x + Y_2 y + Z_2) = 0$

passes through P if

$$\kappa_1(X_1 f + Y_1 g + Z_1) + \kappa_2(X_2 f + Y_2 g + Z_2) = 0.$$

This gives a value of $\kappa_1 : \kappa_2$ for which the line passes through P as well as Q, and therefore coincides with PQ.

4·72. The constants Z_1, Z_2 are introduced in 4·71 so that the work may be applicable to lines through the origin. If the lines do not pass through the origin, we may put $Z_1 = Z_2 = 1$. The lines are then $[X_1, Y_1]$ and $[X_2, Y_2]$ and the arbitrary line through their meet is
$$\left[\frac{\kappa_1 X_1 + \kappa_2 X_2}{\kappa_1 + \kappa_2}, \frac{\kappa_1 Y_1 + \kappa_2 Y_2}{\kappa_1 + \kappa_2} \right].$$
This should be compared with the result in 1·51.

4·73. If $\alpha = 0$ and $\beta = 0$ are two intersecting lines, it is proved in 4·71 that an arbitrary line through their point of intersection is
$$\kappa_1 \alpha + \kappa_2 \beta = 0.$$

In practice it is often convenient to replace this equation by $\alpha = \kappa\beta$, in which κ takes the place of the arbitrary constant $-\kappa_2/\kappa_1$.

4·74. EXAMPLE. Find the equation of the line joining the meet of
$$2x + 3y = 4 \quad \text{and} \quad 3x + 7y = 2$$
to the meet of
$$2x + 3y + 1 = 0 \quad \text{and} \quad 3x + 7y = 5.$$

The equation is
$$2x + 3y - 4 = \kappa(3x + 7y - 2),$$

and is also $\qquad 2x + 3y + 1 = \lambda(3x + 7y - 5)$.

$$\therefore \quad \frac{2 - 3\kappa}{2 - 3\lambda} = \frac{3 - 7\kappa}{3 - 7\lambda} = \frac{-4 + 2\kappa}{+1 + 5\lambda}.$$

From the first equality, $\kappa = \lambda$. Hence from the second

$$-4 + 2\kappa = 1 + 5\kappa. \qquad \therefore \quad \kappa = -\tfrac{5}{3}.$$

Hence the equation is

$$3(2x + 3y - 4) + 5(3x + 7y - 2) = 0, \qquad \text{i.e. } 21x + 44y = 22.$$

4·75. EXAMPLE. The line $3x - 2y = 1$ meets the curve $x^2 = 5y + 4$ in two points P, Q. Find the equation of the line-pair OP, OQ.

Consider the equation

$$x^2 = 5y(3x - 2y) + 4(3x - 2y)^2.$$

It is satisfied by the coordinates of P and Q, since these coordinates satisfy $3x - 2y = 1$ and $x^2 = 5y + 4$. Hence it represents a locus passing through P and Q. Also the equation is
$$35x^2 - 33xy + 6y^2 = 0,$$

and this represents two lines through O. Hence it represents OP, OQ.

4·8. If the line $lx + my + n = 0$ meets the curve

$$px^2 + qxy + ry^2 + ux + vy + w = 0$$

in points P, Q, the equation of the line-pair OP, OQ can be written down by the method used in 4·75. Since the co-ordinates of P and Q satisfy

$$px^2 + qxy + ry^2 + ux + vy + w = 0 \quad \text{and} \quad -(lx + my)/n = 1,$$

they also satisfy

$$px^2 + qxy + ry^2 - (ux + vy)(lx + my)/n + w(lx + my)^2/n^2 = 0.$$

This represents a locus to which P and Q belong, and as it is of the form $ax^2 + 2hxy + by^2 = 0$, it must represent OP, OQ.

EXERCISE 4ᴇ

In Nos. 1–7, find the common points of the loci.

1. $y = x^2$, $2y = x+6$.   **2.** $y^2 = x$, $y+1 = 6x$.

3. $y^2 = x^3$, $3x = 2y+1$.

4. $(x-1)(y-2) = 0$, $(x-3)(y-4) = 0$.

5. $(x-1)(y-2) = 0$, $(x-2y+2)(2x+y-8) = 0$.

6. $x^3 + 2xy^2 = 2x$, $y^3 + x^2y = 2y$.

7. $y^2(x+y) = x(x+y)$, $x(x+y) = (x+y)$.

8. Discuss the common points of the loci $y^2 = 4ax$, $y = t(x+a)$ for different values of t.

In Nos. 9–21, find the line through the meet of the two lines satisfying the given condition.

9. $x+y = 7$, $2x-y = 4$, through $(1, 1)$.

10. $8x = 9y$, $x+y+2 = 0$, through $(2, 7)$.

11. $x+y+4 = 0$, $3x+4y = 11$, through $(4, 5)$.

12. $3x+5y = 7$, $x+y+3 = 0$, through $(0, 0)$.

13. $x+y = 0$, $2x+11y = 27$, parallel to $x+3y = 40$.

14. $2x+3y = 8$, $3x = y+1$, parallel to $17y = x$.

15. $x+y = 11$, $3x+2y = 7$, perpendicular to $2x-4y = 7$.

16. $11x = y+52$, $7x = 131-18y$, perpendicular to $7x+5y = 1$.

17. $7x+13y = 15$, $x+y = 126$, parallel to the join of $(8, -4)$, $(-7, 2)$.

18. $x/a+y/b = 1$, $ax+by = ab$, through the origin.

19. $a_1x+b_1y+c_1 = 0$, $a_2x+b_2y+c_2 = 0$, through (f, g).

20. $y = ax+b$, $y = cx+d$, parallel to OX.

21. $ax+by = 0$, $cx+dy+e = 0$, perpendicular to $y = mx$.

In Nos. 22, 23, find the join of the meets of the two pairs of lines.

22. $x-y = 7$, $3x+2y = 8$; $x-y = 5$, $3x+2y+3 = 0$.

23. $5x-4y+7 = 0$, $4x+5y+3 = 0$; $2y-x+2 = 0$, $2x-y+1 = 0$.

24. Prove that $cx+y = c+3$ passes through a fixed point for all values of c.

25. Find the fixed point through which $(2-3k)x+(3-4k)y = 5-7k$ passes for all values of k.

26. Find the envelope of the line $(2-\lambda)x+(\lambda+2)y = \lambda$ when λ varies.

27. Prove that the curve $(x^2 - 1)(1 - k) = y^2(2k - 1)$ passes through two fixed points for all values of k.

28. Through what fixed points does $kx(x^2 - 1) = y(y - 4)$ pass for all values of k?

29. Find the points common to the curves given by

$$2a(x^2 + 4y^2) + 8bxy = 5a + 3b,$$

where a and b vary.

30. Find the equations of the diagonals of the parallelogram formed by $ax + by = 2a$, $bx + ay = 4a$, $ax + by = 3a$, $bx + ay = 5a$.

In Nos. 31–35, find the equations of the lines joining the origin to the points of intersection of the given line and curve.

31. $y = x^2$, $2y = x + 3$.   **32.** $y^2 = x + 3$, $3x - y = 1$.

33. $2x^2 + 3y^2 = 6$, $x + y = 1$.   **34.** $2x^2 + 3y = 6$, $x + y = 1$.

35. $x^2 + xy + y^2 + 2x + 3y - 8 = 0$, $x + y + 2 = 0$.

4·9. Common Lines of Two Envelopes

4·91. The coordinates of the lines common to two envelopes $F(X, Y) = 0$ and $G(X, Y) = 0$ satisfy both $F(X, Y) = 0$ and $G(X, Y) = 0$ and are found by solving the simultaneous equations. Any other equation deducible from $F(X, Y) = 0$ and $G(X, Y) = 0$ must also be satisfied by the coordinates of any common line. Such an equation therefore represents some envelope which contains all the common lines of the two envelopes.

This is the dual of the principle illustrated in 4·62–4·67. Further illustrations are given here.

4·92. $X + Y + 1 = 0$ and $X + Y - 1 = 0$ are two points. $X + Y + 1 = \kappa(X + Y - 1)$ represents an envelope one of whose lines is the join of these two points, because the coordinates of that line when substituted for X and Y reduce both sides of the equation to zero. This is true for all values of κ. If κ is numerical, the envelope $X + Y + 1 = \kappa(X + Y - 1)$ is a point because the equation is of the first degree in X, Y. It is therefore a point on the join of the points

$$X + Y + 1 = 0, \quad X + Y - 1 = 0.$$

4·93. $X^2 = 1$ and $Y^2 = 4$ represent point-pairs. These are the corners of a rhombus and the common lines are the sides of the rhombus. These lines belong to all derived envelopes such as $X^2 + Y^2 = 5$, $Y^2 = 4X^2$, $(X^2 + Y^2)^2 = 5(9X^2 - Y^2)$.

4·94. $Y = 2$ and $Y^2 = X$ represent a point and a curve envelope shown in the figure. From $Y = 2$ and $Y^2 = X$ can be

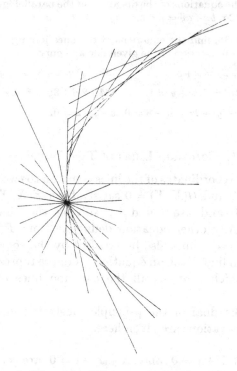

derived $X = 4$. Hence [4, 2] is a common line of the envelopes. It is shown in the figure as the tangent from the point $Y = 2$, i.e. $(0, -\frac{1}{2})$. The other tangent from this point is OY which has no envelope coordinates.

4·95. Example. Find the point where the line $3x + 4y = 6$ meets the join of $(4, -4)$ and $(8, -2)$.

The join is $x - 2y = 12$ and this meets $3x + 4y = 6$ at $(6, -3)$.

4·96. EXAMPLE. Find the point where the line $[X_1, Y_1]$ meets the join of (x_1, y_1) and (x_2, y_2).

1st method. Any point on the join is

$$\left(\frac{\kappa_1 x_1 + \kappa_2 x_2}{\kappa_1 + \kappa_2}, \; \frac{\kappa_1 y_1 + \kappa_2 y_2}{\kappa_1 + \kappa_2} \right)$$

and this lies on $[X_1, Y_1]$ if

$$(\kappa_1 x_1 + \kappa_2 x_2) X_1 + (\kappa_1 y_1 + \kappa_2 y_2) Y_2 + (\kappa_1 + \kappa_2) = 0,$$

i.e. $\qquad \kappa_1 (x_1 X_1 + y_1 Y_1 + 1) + \kappa_2 (x_2 X_1 + y_2 Y_1 + 1) = 0.$

This gives $\kappa_1 : \kappa_2$ and hence the coordinates of the point are determined.

2nd method. The equation of the point can be written down in the form

$$\begin{vmatrix} x_1 X + y_1 Y + 1 & x_2 X + y_2 Y + 1 \\ x_1 X_1 + y_1 Y_1 + 1 & x_2 X_1 + y_2 Y_1 + 1 \end{vmatrix} = 0.$$

For this equation is of the first degree in X, Y and therefore represents a point. It is satisfied by the coordinates of X_1, Y_1. Also it is satisfied by the coordinates of the join of (x_1, y_1), (x_2, y_2) because these satisfy both $x_1 X + y_1 Y + 1 = 0$ and $x_2 X + y_2 Y + 1 = 0$. Hence it is the equation of the meet of $[X_1, Y_1]$ with the join.

3rd method. The method of 4·95 can also be used for 4·96.

4·97. Meet of a Line with the Join of Two Points

This is the dual of 4·71. If $\kappa_1 : \kappa_2$ is an arbitrary constant, and
$$x_1 X + y_1 Y + 1 = 0 \quad \text{and} \quad x_2 X + y_2 Y + 1 = 0$$
are the two points, and if q is the line joining them, then

$$\kappa_1 (x_1 X + y_1 Y + 1) + \kappa_2 (x_2 X + y_2 Y + 1) = 0$$

is an arbitrary point on q.

For the coordinates of q satisfy $x_1 X + y_1 Y + 1 = 0$ and $x_2 X + y_2 Y + 1 = 0$ and therefore they satisfy

$$\kappa_1 (x_1 X + y_1 Y + 1) + \kappa_2 (x_2 X + y_2 Y + 1) = 0.$$

But this equation is of the first degree in X, Y. Hence it represents some point on q.

Also $\kappa_1:\kappa_2$ can be chosen so that the equation represents a given point on q. For let p be a line $[F, G]$ through such a point. Then $\kappa_1(x_1 X + y_1 Y + 1) + \kappa_2(x_2 X + y_2 Y + 1) = 0$ lies on p if $\kappa_1(x_1 F + y_1 G + 1) + \kappa_2(x_2 F + y_2 G + 1) = 0$. This gives a value of $\kappa_1:\kappa_2$ for which the point lies on p as well as on q and is therefore the point pq.

4·98. It is proved in 4·97 that if $A = 0$ and $B = 0$ are two points, then an arbitrary point on their join is $\kappa_1 A + \kappa_2 B = 0$.

In practice it is often convenient to replace this equation by $A = \kappa B$, where κ takes the place of the arbitrary constant $-\kappa_2/\kappa_1$.

4·99. EXAMPLE. Find the condition for two of the lines $x^3 - pxy^2 + qy^3 = 0$ to be at right angles.

The equation of any two perpendicular lines through the origin is
$$x^2 + axy - y^2 = 0.$$

Hence, if the equation represents lines two of which are at right angles, $x^3 - pxy^2 + qy^3 \equiv (x^2 + axy - y^2)(x - qy)$. Equating coefficients of $x^2 y$, xy^2,
$$0 = a - q, \qquad p = aq + 1.$$
$$\therefore \; p - 1 = q^2.$$

Conversely, when this condition holds $x^2 + qxy - y^2$ is a factor of $x^3 - pxy^2 + qy^3$, the other factor being $x - qy$. Hence $p - 1 = q^2$ is a necessary and sufficient condition.

EXERCISE 4F

In Nos. 1–6, find the common lines of the envelopes.

1. $X^2 + Y^2 = 17,\ 2X + Y = 6$.   **2.** $4X^2 = 1,\ 9Y^2 = 1$.

3. $(X+1)(Y+2) = 0,\ (X-1)(2Y-1) = 0$. **4.** $aY^2 = X,\ aX = 1$.

5. $c^2(2X^2 + 3Y^2) = 1 \pm 2cX$.   **6.** $XY = X,\ XY^2 = X^3$.

7. Find the equation of the point in which the join of $2X = 2Y + 1$ and $X + Y = 1$ meets $[-1, 1]$.

8. Find the equation of the point in which the join of $(\frac{1}{2}, -\frac{7}{8})$, $(-\frac{3}{23}, -\frac{2}{23})$ meets $4x - 3y + 1 = 0$.

9. Use the three methods of 4·96 to find the point in which the join of $(1, -1)$ and $(2, -5)$ meets $2x - y = 9$.

10. Find the equation of the point in which the join of

$$p_1 X + q_1 Y + r_1 = 0 \quad \text{and} \quad p_2 X + q_2 Y + r_2 = 0$$

meets $[A, B]$.

11. Prove that $Y - kX = 2k + 3$ lies on a fixed line for all values of k.

12. Find the coordinates of the line on which

$$kX + 3Y + 2 = 2X + 2kY + k$$

lies for all values of k.

13. Find the locus of $4X + (2\lambda - 1) Y + 2 - \lambda = 0$ when λ varies.

14. Prove that the curve $X^2 - (1 - \lambda) Y^2 = \lambda$ has four fixed tangents for all values of λ and give their equations.

EXERCISE 4G

1. What is represented by

$$(ax + by + c) \{(px + qy + r)^2 + (lx + my + n)^2\} = 0$$

and by the corresponding envelope equation?

2. An equilateral triangle has centre the origin and one side $x + y = 1$. Find the other sides and show that the equation of the three perpendiculars from the vertices to the opposite sides is

$$x^2(x + 3y) = y^2(y + 3x).$$

3. Find the equation of the bisectors of the angles between lines through $(1, -4)$ parallel to the lines

$$15x^2 + 34xy + y^2 + 60x + 68y + 60 = 0.$$

4. Show that the lines $ax + by = c$, $(ax + by)^2 = 3(ay - bx)^2$ form an equilateral triangle.

5. Find the area of the rectangle formed by the lines

$$x^2 + 2xy - y^2 = 0, \quad x^2 + 2xy - y^2 - 4y - 2 = 0.$$

6. Find the equation of the remaining sides of the rectangle centre the origin two of whose sides are $x^2 - y^2 + 3x + y + 2 = 0$.

7. Find the distance between the parallel lines

$$9x^2 + 24xy + 16y^2 + 12x + 16y - 5 = 0.$$

8. State the conditions for $ax^2 + 2hxy + by^2 + 2gx + 2fy + c = 0$ to represent two parallel lines and assuming that they are satisfied show that the distance between the lines is $2\sqrt{\{(g^2 - ac)/(a^2 + ab)\}}$.

9. $\{\omega = \frac{1}{3}\pi\}$. Find the angle between the lines $2(x^2 + y^2) = 5xy$.

10. $\{\omega = \tfrac{1}{3}\pi\}$. Find the equation of the bisectors of the angles between the lines $5x^2 + 14xy - y^2 = 0$.

11. $\{\omega\}$. Find the condition that the lines $ax^2 + 2hxy + by^2 = 0$ should be equally inclined to the axis of x.

12. $\{\omega\}$. Find the condition that the lines $ax^2 + 2hxy + by^2 = 0$ should be at right angles.

13. What is the general equation of a line-pair having

$$x^2 + kxy - y^2 = 0$$

for angle-bisectors?

14. Find the area of the triangle formed by the lines

$$ax^2 + 2hxy + by^2 = 0, \quad Xx + Yy + 1 = 0.$$

15. Prove that the length intercepted by $ax^2 + 2hxy + by^2 = 0$ on $lx + my + n = 0$ is $2n \sqrt{\{(l^2 + m^2)(h^2 - ab)\}}/(am^2 - 2hlm + bl^2)$.

16. Find the circumcentre and centroid of the triangle formed by $ax^2 + 2hxy + by^2 = 0$ and $lx + my = 1$.

17. Find the foot of the perpendicular from (f, g) to $lx + my + n = 0$. Prove that the locus of a point such that the distance between the feet of the perpendiculars from it to $ax^2 + 2hxy + by^2 = 0$ is $2k$ is

$$(x^2 + y^2)(h^2 - ab) = k^2\{(a - b)^2 + 4h^2\}.$$

18. If $ax^2 + 2hxy + by^2 + 2gx + 2fy + c = 0$ represents lines meeting at P, prove that $OP^2 = (A + B)/C$.

19. Prove that any line-pair, centre the origin, making equal angles with $lx + my + n = 0$ is given by

$$\{(l^2 - m^2)x^2 + 2lmxy\} + \lambda\{(l^2 - m^2)y^2 - 2lmxy\} = 0,$$

and explain geometrically why the equation can also be written

$$(lx + my)^2 + \mu(mx - ly)^2 = 0.$$

20. Obtain the condition that one of the lines $a_1 x^2 + 2h_1 xy + b_1 y^2 = 0$ should coincide with one of the lines $a_2 x^2 + 2h_2 xy + b_2 y^2 = 0$.

21. Obtain the condition that one of the lines $a_1 x^2 + 2h_1 xy + b_1 y^2 = 0$ should be perpendicular to one of the lines $a_2 x^2 + 2h_2 xy + b_2 y^2 = 0$.

22. Find the condition that two of the lines $x^3 + 3Hxy^2 + Gy^3 = 0$ should coincide.

23. Find the conditions for each of the lines

$$ax^3 - 3bx^2 y - 3cxy^2 + dy^3 = 0$$

to bisect an angle between the other two.

24. Find the condition for two of the lines

$$ax^4 + bx^3y + cx^2y^2 + dxy^3 + ey^4 = 0$$

to be at right angles.

25. Write down the equation of the lines joining the origin to the points of intersection of $x^3 + y^3 = 3axy$ and $x - y = b$ and find the values of b for which the line is a tangent to the curve.

26. Prove that the product of the lengths of the perpendiculars from (h, k) to the n lines $a_0x^n + a_1x^{n-1}y + \ldots + a_ny^n = 0$ is

$$\pm \frac{a_0h^n + a_1h^{n-1}k + \ldots + a_nk^n}{\sqrt{\{(a_0 - a_2 + \ldots)^2 + (a_1 - a_3 + \ldots)^2\}}}.$$

Chapter 5

THE CIRCLE

5·1. Equation of Circle

5·11. The form of the equation of a circle referred to rectangular axes is suggested in 2·41.

The circle centre (p, q) and radius r is the locus of a point whose distance from (p, q) is r and its equation is therefore

$$(x-p)^2 + (y-q)^2 = r^2.$$

This is of the form

$$x^2 + y^2 + 2gx + 2fy + c = 0.$$

Conversely $x^2 + y^2 + 2gx + 2fy + c = 0$ can be written

$$(x+g)^2 + (y+f)^2 = g^2 + f^2 - c.$$

If $g^2 + f^2 > c$, this equation represents the circle centre $(-g, -f)$ and radius $\sqrt{(g^2 + f^2 - c)}$.

If $g^2 + f^2 = c$, the equation is satisfied only by $x = -g$, $y = -f$. It represents the point $(-g, -f)$ and may be said to represent the circle centre $(-g, -f)$ and radius 0.

If $g^2 + f^2 < c$, the equation is not satisfied by any values of x, y.

5·12. The circle centre the origin and radius a is

$$x^2 + y^2 = a^2.$$

Any point P on this circle has polar coordinates (a, t) where $t = \angle XOP$. Hence $x = a\cos t$, $y = a\sin t$ are parametric equations of the circle.

By 3·65 the tangent at the point t is

$$x\cos\alpha + y\sin\alpha = a.$$

Parametric envelope equations are therefore

$$-aX = \cos t, \quad -aY = \sin t.$$

Elimination of t gives the envelope equation $X^2 + Y^2 = 1/a^2$.

5·13. Putting m for $\tan \frac{1}{2}t$ we obtain the alternative parametric equations

$$x : y : a = 1 - m^2 : 2m : 1 + m^2$$

of the circle $x^2 + y^2 = a^2$.

Also the tangent $x \cos \alpha + y \sin \alpha = a$ becomes

$$x(1 - m^2) + 2ym = a(1 + m^2).$$

Many of the properties of the circle are proved in elementary geometry. It is not proposed to prove them by analytical methods in this book.

5·14. EXAMPLE. Find the equation of the circle through

$$(1, 0) \quad (3, 4) \quad (-1, 5).$$

1st method. Let the equation be $x^2 + y^2 + 2gx + 2fy + c = 0$.

Then
$$1 + 2g + c = 0,$$
$$25 + 6g + 8f + c = 0,$$
$$26 - 2g + 10f + c = 0,$$

therefore $\quad g = -\frac{5}{9}, \quad f = -\frac{49}{18}, \quad c = \frac{1}{9}$

and the equation is

$$9(x^2 + y^2) - 10x - 49y + 1 = 0.$$

2nd method. The equation can be written down in the form

$$\begin{vmatrix} x^2 + y^2 & x & y & 1 \\ 1^2 + 0^2 & 1 & 0 & 1 \\ 3^2 + 4^2 & 3 & 4 & 1 \\ 1^2 + 5^2 & -1 & 5 & 1 \end{vmatrix} = 0$$

by eliminating g, f, c from the equations used in the first method. Alternatively, the expansion of the determinant by its top row shows that the equation represents a circle; and the other rows have been chosen so that the equation is satisfied by the coordinates $(1, 0)$, $(3, 4)$, $(-1, 5)$.

5·15. EXAMPLE. Find the equation of the circle on (x_1, y_1) (x_2, y_2) as diameter.

The equation $(x - x_1)(x - x_2) = 0$ represents the pair of lines

P_1Q_1, P_2Q_2 parallel to OY, and $(y-y_1)(y-y_2) = 0$ represents the pair of lines P_1Q_2, P_2Q_1 parallel to OX. The equations are both satisfied by the coordinates of P_1, P_2, Q_1, Q_2. Therefore

$$(x-x_1)(x-x_2) + (y-y_1)(y-y_2) = 0$$

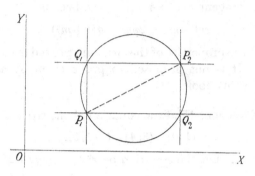

is also satisfied by the coordinates of those points. Hence it represents a curve through the points. But it is of the form $x^2 + y^2 + 2gx + 2fy + c = 0$. Therefore it represents a circle. This must be the circle through P_1, P_2, Q_1, Q_2, which is the circle on P_1P_2 as diameter.

5·16. EXAMPLE. PQ is the chord $lx + my = n$ of the circle $x^2 + y^2 = a^2$. Find the equation of the circle on PQ as diameter.

Eliminating y from the equations,

$$m^2x^2 + (lx - n)^2 = m^2a^2.$$

Similarly, eliminating x,

$$l^2y^2 + (my - n)^2 = l^2a^2.$$

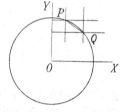

The form of these equations shows that they represent pairs of lines parallel to the axes. Since they are derived from $lx + my = n$ and $x^2 + y^2 = a^2$, they are satisfied by the coordinates of P and Q. Hence the lines must be the parallels to the axes through P, Q.

Adding the equations,

$$(l^2 + m^2)(x^2 + y^2) - 2lnx - 2mny + 2n^2 = (l^2 + m^2)a^2,$$

and this must be satisfied by the coordinates of four points of intersection of the line-pairs, i.e. the corners of a rectangle with PQ as a diagonal. But the form of the equation shows that it is the equation of a circle. Hence it must be the equation of the circle on PQ as diameter.

EXERCISE 5A

In Nos. 1–5, find the equation of the circle with the given centre and radius.

1. $(2, 3)$, 4.   2. $(-1, 0)$, 2.   3. $(4, -5)$, 7.

4. $(-1\frac{1}{2}, -2)$, $2\frac{1}{2}$.   5. $(a, -b)$, $a + b$.

In Nos. 6–12, find the centre and radius of the circle.

6. $x^2 + y^2 + 2x - 6y = 15$.   7. $x^2 + y^2 + 2x - 8 = 0$.

8. $x^2 + y^2 - 3x + 4y + 5 = 0$.   9. $3x^2 + 3y^2 + 12x + y = 2$.

10. $5x^2 + 5y^2 + 6y - 1 = 0$.   11. $x^2 + y^2 = 2a(x + a)$.

12. $a(x^2 + y^2) = bx + cy$.

In Nos. 13–16, show in a diagram the position of the circle.

13. $x^2 + y^2 - 2x - 4y = 4$.

14. $x^2 + y^2 - 2ax + 2by = 0$, $(a > 0, b > 0)$.

15. $x^2 + y^2 + ay = 0$, $(a > 0)$.

16. $x(x - f) + y(y + g) = 0$, $(f > 0, g > 0)$.

In Nos. 17–31, find the equation of the circle through the given points satisfying the given condition.

17. $(1, 2)$, centre $(-4, 5)$.   18. $(0, 0)$, centre $(19, -37)$.

19. $(12, c + 5)$, centre $(0, c)$.   20. $(3, 5)$, $(7, -1)$, $(0, 0)$.

21. $(8, 4)$, $(-5, 3)$, $(7, 6)$.   22. $(1, 2)$, $(2, 3)$, $(-3, -5)$.

23. (a, b), $(a, -b)$, $(a + b, a - b)$.

24. $(a + 36, b + 77)$, $(a - 84, b + 13)$, $(a - 40, b - 75)$.

25. $(1, 2)$, $(3, 4)$, centre on OX.

26. $(-1, 1)$, $(3, -3)$, centre on $[-1, 6]$.

27. $(1, -2)$, $(3, -4)$, touching OX.

28. $(-2, 3)$, touching OX and OY.

29. $(0, 0)$, cutting off intercepts $+3$, -4 from OX, OY.

30. $(a, 0)$, (b, c), centre on OY.

31. $(b, 0)$, $(-b, 0)$, radius a, $(a > b)$.

In Nos. 32, 33, prove that the points are concyclic.

32. $(0, 1)$, $(2, 3)$, $(3, 5)$, $(-7, 0)$.

33. $(a-5, b+3)$, $(a-1, b+7)$, $(a+3, b+9)$, $(a-7, b-11)$.

34. Find the condition that the points $(a, 0)$, $(0, b)$, (c, d), $(0, 0)$ should be concyclic.

In Nos. 35–39, find the equation of the circle on the given line as diameter.

35. $(3, 5)$ $(4, -7)$.   **36.** $(a, 0)$ $(0, b)$.

37. Chord $x + 3y = 35$ of $x^2 + y^2 = 125$.

38. Chord $x - y + 1 = 0$ of $x^2 + y^2 - 10x - 12y + 51 = 0$.

39. Chord $lx + my = 1$ of $x^2 + y^2 + 2gx + 2fy + c = 0$.

5·2. Tangent

5·21. The tangent at $(a \cos t,\ a \sin t)$ to $x^2 + y^2 = a^2$ is found in 5·12. In general the tangent to a circle at a given point is found by using that fact that it is perpendicular to the radius. Let $P_1(x_1, y_1)$ be a point on the circle

$$x^2 + y^2 + 2gx + 2fy + c = 0.$$

The centre of this circle is $(-g, -f)$ and the tangent at P_1 is the line through (x_1, y_1) perpendicular to the join of (x_1, y_1) and $(-g, -f)$. Its equation is therefore

$$x(x_1 + g) + y(y_1 + f) = x_1(x_1 + g) + y_1(y_1 + f).$$

Since P_1 lies on the circle,

$$x_1^2 + y_1^2 + 2gx_1 + 2fy_1 + c = 0$$

and the equation of the tangent may be written

$$xx_1 + yy_1 + g(x + x_1) + f(y + y_1) + c = 0.$$

5·22. EXAMPLE. Find the tangents to $x^2 + y^2 = 10x$ that are parallel to $3x + 4y = 0$.

The equation of the circle is $(x - 5)^2 + y^2 = 5^2$. Therefore the centre is $(5, 0)$ and the radius is 5.

An arbitrary line parallel to $3x + 4y = 0$ is $3x + 4y = k$. The length of the perpendicular from $(5, 0)$ to this line is $\pm \frac{1}{5}(15 - k)$. Equating this to the radius gives

$$15 - k = 25 \quad \text{or} \quad 15 - k = -25,$$

$$k = -10 \quad \text{or} \quad k = 40.$$

Thus the tangents are $3x + 4y + 10 = 0$ and $3x + 4y - 40 = 0$.

5·23. EXAMPLE. Find the envelope equation of the circle

$$x^2 + y^2 + 2gx + 2fy + c = 0.$$

The required equation is the condition of tangency for $Xx + Yy + 1 = 0$ and it is found by expressing that the length of the perpendicular from the centre $(-g, -f)$ to $Xx + Yy + 1 = 0$ is equal to the radius $\sqrt{(g^2 + f^2 - c)}$. Hence

$$\pm \frac{-Xg - Yf + 1}{\sqrt{(X^2 + Y^2)}} = \sqrt{(g^2 + f^2 - c)}.$$

$$\therefore \quad (Xg + Yf - 1)^2 = (g^2 + f^2 - c)(X^2 + Y^2).$$

EXERCISE 5B

In Nos. 1–3, write down the equation of the tangent.

    1. At $(3, 4)$ to the circle centre $(0, 0)$, radius 5.

    2. At $(4, -1)$ to the circle centre $(-1, 11)$, radius 13.

    3. At (x_1, y_1) to the circle centre (f, g), radius h.

In Nos. 4–6, use the method of 5·21 to find the equation of the tangent at the given point to the circle.

    4. $(5, -2)$, $x^2 + y^2 = 29$.   **5.** (x_1, y_1), $x^2 + y^2 = a^2$.

    6. $(-1, 4)$, $x^2 + y^2 + 3x + 2y = 22$.

    7. Obtain the result of 5·21 by calculating dy/dx at (x_1, y_1).

In Nos. 8–12, write down the equation of the tangent at the given point to the circle.

    8. $(3, -2)$, $x^2 + y^2 = 13$.   **9.** $(1/\sqrt{3}, 2/\sqrt{3})$, $3x^2 + 3y^2 = 5$.

    10. $(0, 1)$, $x^2 + y^2 + 5x + 11y = 12$.

    11. $(7, -9)$, $5x^2 + 5y^2 - 7x - 9y = 682$.

    12. (x_1, y_1), $k(x^2 + y^2) + 2ux + 2vy + w = 0$.

In Nos. 13–17, find the equation of the circle.

13. Centre $(-5, -4)$, touching OX.

14. Centre $(2, 3)$, touching $3x + 4y = 23$.

15. Centre $(0, 0)$, touching $ax + by + c = 0$.

16. Centre $(0, 0)$, touching the circle centre $(3, -4)$ and radius 7.

17. Centre the origin, touching $x^2 + y^2 + 10x + 24y + 69 = 0$.

18. Find the condition that $x/a + y/b = 1$ should be a tangent to $x^2 + y^2 = c^2$.

19. If $y + 2x\sqrt{2} = k$ is a tangent to $x^2 + y^2 + 6y = 0$, what is the value of k?

20. Prove that $y = 3x + a\sqrt{10}$ touches $x^2 + y^2 = a^2$ and find the point of contact.

21. Find the condition that $lx + my = 1$ should be a tangent to $(x - f)^2 + (y - g)^2 = h^2$.

In Nos. 22–25, find the tangents to the given circle satisfying the given condition.

22. $x^2 + y^2 = 1$, parallel to $y = x\sqrt{3}$.

23. $x^2 + y^2 = 4$, perpendicular to $3x + 4y + 5 = 0$.

24. $x^2 + y^2 + 4x - 2y + 1 = 0$, parallel to OY.

25. $x^2 + y^2 = a^2$, perpendicular to $lx + my + n = 0$.

26. Find the equation of the chord of $x^2 + y^2 = 81$ with mid-point $(-2, 3)$.

27. Find the equation of the chord of $x^2 + y^2 - 6x - 16 = 0$ with mid-point $(0, -1)$.

28. Find the length of the chord $y = 3x + 2$ of $x^2 + y^2 - 4x + 4y - 7 = 0$.

29. Find the length of the chord $x/a + y/b = 1$ of $x^2 + y^2 = c^2$.

5·3. Power of a Point wo a Circle

5·31. The *power* of a point P wo a circle centre C and radius r is defined to be $PC^2 - r^2$.

When P is outside the circle, the power is equal to the square of the tangent from P to the circle. It is also equal to $PQ \cdot PQ'$, where PQQ' is any secant through P (figure a).

When P is inside the circle, the power

$$= (PC + CA')(PC - AC) \quad \text{(figure b)}$$

$$= -AP \cdot PA' = -QP \cdot PQ',$$

and is negative.

The power of a point on the circle is zero.

When the radius of the circle is zero, the power of P is the square of the distance PC.

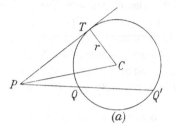
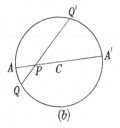

$$(a) \qquad\qquad\qquad (b)$$

The power of (x_1, y_1) wo the circle $x^2 + y^2 + 2gx + 2fy + c = 0$ whose centre is $(-g, -f)$ and radius $\sqrt{(g^2 + f^2 - c)}$ is

$$(x_1 + g)^2 + (y_1 + f)^2 - (g^2 + f^2 - c) = x_1^2 + y_1^2 + 2gx_1 + 2fy_1 + c.$$

Thus it is the expression formed by substituting x_1 for x and y_1 for y in the left side of the equation of the circle. This assumes that the equation has been written in the standard form with the coefficients of x^2 and y^2 equal to unity.

5·32. When (x_1, y_1) is outside the circle

$$x^2 + y^2 + 2gx + 2fy + c = 0,$$

the formula $\qquad x_1^2 + y_1^2 + 2gx_1 + 2fy_1 + c$

is also the formula for the square of the tangent from (x_1, y_1) to the circle.

5·35. *Power of Two Circles.* This is defined as $C_1 C_2{}^2 - r_1{}^2 - r_2{}^2$, where C_1, C_2 are the centres and r_1, r_2 are the radii. When $r_2 = 0$ it reduces to the power of the "point circle" C_2 wo the circle centre C_1 and radius r_1. When $r_1 = r_2 = 0$ it reduces to $C_1 C_2{}^2$.

The power of the circles

$$x^2 + y^2 + 2g_1 x + 2f_1 y + c_1 = 0,$$

$$x^2 + y^2 + 2g_2 x + 2f_2 y + c_2 = 0$$

is $\qquad (g_1 - g_2)^2 + (f_1 - f_2)^2 - (g_1{}^2 + f_1{}^2 - c_1) - (g_2{}^2 + f_2{}^2 - c_2)$

$$= c_1 + c_2 - 2g_1 g_2 - 2f_1 f_2.$$

5·4. Angle between Circles. Orthogonal Circles

An *angle between two circles* is an angle between their tangents at a point of intersection. Orthogonal circles are those whose tangents at a point of intersection are at right angles. It follows that the tangent to each at either point of intersection passes through the centre of the other. Hence $d^2 = r_1{}^2 + r_2{}^2$, i.e. the power of orthogonal circles is zero, and the condition for two circles to cut orthogonally is $c_1 + c_2 - 2g_1g_2 - 2f_1f_2 = 0$.

5·45. An *angle between any two curves* at a point of intersection is defined to be an angle between the tangents to the curves at the point.

EXERCISE 5c

In Nos. 1–4, find the power of:

1. $(3, 4)$ wo $x^2 + y^2 + 3x - 5y + 2 = 0$.
2. $(1, -1)$ wo $x^2 + y^2 = 4x + 2y$.
3. $(6, -7)$ wo $3x^2 + 3y^2 = 7x + 6y + 12$.
4. (a, b) wo $(x - a)^2 + (y - b)^2 = c^2$.

In Nos. 5–10, find the lengths of the tangents from:

5. $(7, 4)$ to $x^2 + y^2 = 5$.   6. $(3, -8)$ to $4x^2 + 4y^2 = 7$.
7. $(0, 5)$ to $x^2 + y^2 + 2x + 3y = 4$.
8. $(5, -6)$ to $2x^2 + 2y^2 + x = 15$.
9. $(0, 0)$ to $(x - a)^2 + (y - b)^2 = c^2$.
10. (f, g) to $ax^2 + ay^2 + 2ux + 2vy + w = 0$.

In Nos. 11, 12, find the product of the segments into which a chord of the circle is divided at the point.

11. $x^2 + y^2 = 49$, $(2, 3)$.   12. $x^2 + y^2 + 4x + 6y = 87$, $(-1, 4)$.

In Nos. 13, 14, find the power of the circles.

13. $x^2 + y^2 = 1$, $x^2 + y^2 + 2x + 14y = 14$.
14. $x^2 + y^2 + 2y = 3$, $x^2 + y^2 = 2x - 4y - 5$.

In Nos. 15, 16, find the angle of intersection of the circles.

15. $x^2 + y^2 - 14x + 4y + 28 = 0$, $x^2 + y^2 + 4y = 5$.
16. $2x^2 + 2y^2 + 3x - 5y + 4 = 0$, $9x^2 + 9y^2 - 21x + 33y - 70 = 0$.

17. Prove that the circles $x^2 + y^2 = ax$, $x^2 + y^2 = by$ cut orthogonally at $((ab^2/(a^2 + b^2),\ a^2b/(a^2 + b^2))$.

18. If a circle cuts

$$s_1 \equiv x^2 + y^2 + 2g_1 x + 2f_1 y + c_1 = 0$$

and

$$s_2 \equiv x^2 + y^2 + 2g_2 x + 2f_2 y + c_2 = 0$$

orthogonally, prove that it also cuts $\kappa_1 s_1 + \kappa_2 s_2 = 0$ orthogonally.

19. Find the circles which cut

$$x^2 + y^2 + 2a_1 x + b = 0 \quad \text{and} \quad x^2 + y^2 + 2a_2 x + b = 0$$

orthogonally.

5·5. Radical Axis

5·51. The locus of a point which has equal powers wo two circles is a line called the *radical axis* of the circles.

If the circles are

$$s \equiv x^2 + y^2 + 2gx + 2fy + c = 0,$$

$$s' \equiv x^2 + y^2 + 2g'x + 2f'y + c' = 0,$$

the powers of (x, y) are s and s'. They are equal if and only if

$$s = s',$$

i.e. $$2(g - g')x + 2(f - f')y + (c - c') = 0.$$

Hence the locus is a line unless $g = g'$ and $f = f'$.

When $g = g'$ and $f = f'$, the circles are concentric and there is no locus.

5·52. The radical axis of the circles $s = 0$, $s' = 0$ is $s = s'$ provided that the equations of the circles are written, as in 5·51, with unit coefficients of x^2 and y^2.

The form of the equation $s = s'$ shows that the radical axis passes through the common points of the circles when these exist: for the coordinates of these points satisfy $s = 0$ and $s' = 0$ and therefore they satisfy $s = s'$. That these points belong to the locus is also an immediate consequence of the definition.

The tangents from any point of the radical axis which is outside the circles are equal. But if the radical axis had been defined by the equality of tangents instead of by the equality of the powers, the radical axis of intersecting circles would have been only part of the line.

5·53. The radical axis
$$2(g-g')x+2(f-f')y+(c-c') = 0$$
is perpendicular to the join of the centres of the circles.

If the line of centres is chosen as x-axis, $f = f' = 0$. The equations of the circles are then
$$x^2+y^2+2gx+c = 0, \qquad x^2+y^2+2g'x+c' = 0$$
and the radical axis is
$$2(g-g')x+(c-c') = 0.$$

If, at the same time, the radical axis is chosen as y-axis, $c = c'$. The circles are then
$$x^2+y^2+2gx+c = 0, \qquad x^2+y^2+2g'x+c = 0.$$
These have centres $(-g, 0)$, $(-g', 0)$, and radii $\sqrt{(g^2-c)}$, $\sqrt{(g'^2-c)}$. Hence the radical axis of circles with centres A, B and radii a, b is the perpendicular to AB at the point O in AB such that
$$AO^2 - OB^2 = a^2 - b^2.$$

5·54. *Radical Centre.* The radical axes of three circles
$$s \equiv x^2+y^2+2g_1x+2f_1y+c_1 = 0,$$
$$s' \equiv x^2+y^2+2g_2x+2f_2y+c_2 = 0,$$
$$s'' \equiv x^2+y^2+2g_3x+2f_3y+c_3 = 0,$$
taken in pairs, are $s' = s''$, $s'' = s$, $s = s'$; and unless these coincide, they meet in a point given by $s = s' = s''$. This is called the *radical centre* of the three circles.

5·6. Coaxal Circles

5·61. The powers of $P(x, y)$ wo the circles
$$s \equiv x^2+y^2+2gx+2fy+c = 0,$$
$$s' \equiv x^2+y^2+2g'x+2f'y+c' = 0$$
are s and s'. These are equal when P lies on the radical axis $(s = s')$ of the circles.

Consider the circle $\qquad s = \kappa s'$,

i.e. $\quad (1-\kappa)(x^2+y^2)+2(g-\kappa g')x+2(f-\kappa f')y+(c-\kappa c') = 0.$

The power of P wo this circle is

$$x^2 + y^2 + 2\frac{g - \kappa g'}{1 - \kappa} x + 2\frac{f - \kappa f'}{1 - \kappa} y + \frac{c - \kappa c'}{1 - \kappa},$$

and this is equal to $(s - \kappa s')/(1 - \kappa)$. But if s is equal to s' each of them is equal to $(s - \kappa s')/(1 - \kappa)$. Hence if P lies on the radical axis of $s = 0$ and $s' = 0$, it has the same power wo all the circles $s = \kappa s'$. Hence every pair of circles of the system $s = \kappa s'$ has the same radical axis ($s = s'$). Such a system is called a *coaxal system*. There are three types of coaxal system.

5·62. If $s = 0$ and $s' = 0$ are intersecting circles, every circle of the coaxal system $s = \kappa s'$ defined by them passes through their points of intersection. For the coordinates of these points when substituted in $s = \kappa s'$ reduce both sides to zero. Thus the coaxal system consists of circles which pass through two fixed points. This is called an *intersecting* system of coaxal circles.

With the central axis and radical axis as OX and OY, the equations of two circles are, by 5·53,

$$s \equiv x^2 + y^2 + 2gx + c = 0,$$
$$s' \equiv x^2 + y^2 + 2g'x + c = 0.$$

The points of intersection are $(0, \sqrt{-c})$ and $(0, -\sqrt{-c})$. Hence c must be negative for an intersecting system.

5·63. When c is positive the circles $s = 0$, $s' = 0$ do not meet. And no two circles of the system $s = \kappa s'$ meet. The circles are said to form a *non-intersecting* coaxal system.

Any circle $s = \kappa s'$ of the system has an equation

$$x^2 + y^2 + 2\lambda x + c = 0, \quad (c > 0)$$

and this may be written

$$(x + \lambda)^2 + y^2 = \lambda^2 - c.$$

The circles given by $\lambda = \sqrt{c}$, $\lambda = -\sqrt{c}$ have zero radius and centres $(\sqrt{c}, 0)$, $(-\sqrt{c}, 0)$. These are called the *limiting points* of the coaxal system.

5·64. The circles given by $c = 0$ form an *intermediate* type. They all touch the y-axis at the origin.

5·65. A coaxal system is determined by two of its members. If it is a non-intersecting system, either or both of the determining members may be a limiting point.

Many of the properties of coaxal circles are conveniently proved by the methods of pure geometry. (*M.G.* Chapter XI.)

5·66. *Circle of Apollonius.* The equation $s = \kappa s'$ expresses that the power of P wo the circle $s = 0$ is κ times the power wo $s' = 0$. Thus:

The locus of a point whose powers wo two given circles are in a constant ratio is a circle coaxal with the given circles.

In particular, replacing the circles by limiting points:

The locus of a point whose distances from two given points are in a constant ratio is a circle of the coaxal system which has the given points as limiting points.

A circle formed in this way from two given points is called a *Circle of Apollonius.*

5·67. *Orthogonal Coaxal Systems.* The power of

$$(0, \mu) \text{ wo } x^2 + y^2 + 2\lambda x + c = 0$$

is $\mu^2 + c$. The circle centre $(0, \mu)$ and radius $\sqrt{(\mu^2 + c)}$ is

$$x^2 + y^2 - 2\mu y - c = 0.$$

Such circles, given by different values of μ, form a coaxal system. If c is a positive constant, $x^2 + y^2 + 2\lambda x + c = 0$ defines a non-intersecting coaxal system, with centres on OX, having OY as radical axis. And $x^2 + y^2 - 2\mu y - c = 0$ defines an intersecting system with centres on OY, having OX as radical axis. The limiting points of the first system (5·63) are the points of intersection of the second (5·62). The condition in 5·4 shows that any two circles, one from each system, intersect orthogonally.

EXERCISE 5D

In Nos. 1–4, find the radical axis of the circles.

1. $x^2 + y^2 + 2x + 3y = 4$, $x^2 + y^2 + x + 2y = 5$.

2. $x^2 + y^2 + 3x - y = 0$, $2x^2 + 2y^2 + 3y - 4 = 0$.

3. $3x^2 + 3y^2 + 2x + y - 4 = 0$, $2x^2 + 2y^2 + 10x - 7y - 100 = 0$.

4. $x^2 + y^2 = r^2$, $k(x^2 + y^2) + 2gx + 2fy + c = 0$.

5. Find the equation of the common chord of $x^2 + y^2 = 6x + 7$ and $x^2 + y^2 = 4x + 12$.

6. Find the radical centre of $x^2 + y^2 = 4$, $x^2 + y^2 + 7x - 3y - 5 = 0$, $x^2 + y^2 + 4x - y - 1 = 0$.

7. Circles of radii 3, 7 have centres A, B at distance k apart. Find, for $k = 8, 12, 10, 2$, the ratio in which the radical axis divides AB.

8. Find the points of intersection of
$$2(x^2 + y^2) + x + y = 31, \quad 3(x^2 + y^2) + x + y = 44.$$

9. Find the circle through (3, 5) coaxal with
$$x^2 + y^2 + 3x + 5y + 7 = 0 \quad \text{and} \quad x^2 + y^2 + x - y - 1 = 0.$$

10. Find the radius of the circle through (1, 1) coaxal with
$$2x^2 + 2y^2 + 3x = 0 \quad \text{and} \quad (x - 1)^2 + y^2 = 0.$$

11. Find the centre and radius of the least circle coaxal with
$$x^2 + y^2 + 2x - 4 = 0 \quad \text{and} \quad x^2 + y^2 - 3x - 4 = 0.$$

12. Find the length of the common chord of the circles
$$x^2 + y^2 + 12x + 88y + 676 = 0 \quad \text{and} \quad x^2 + y^2 - 14x - 80y - 4280 = 0.$$

In Nos. 13–16, find the limiting points.

13. $x^2 + y^2 = 1$, $x^2 + y^2 + 6x + 8 = 0$.

14. $x^2 + y^2 = 16$, $x^2 + y^2 - 6x + 8y + 24 = 0$.

15. $\lambda(x^2 + y^2 - 6x + 4y + 12) = \mu(x^2 + y^2 - 12x - 4y + 36)$.

16. $a(x^2 + y^2) + \lambda y + b = 0$, where a and b are positive constants.

17. Find the circles coaxal with
$$x^2 + y^2 + 52x = 12 \quad \text{and} \quad 2x^2 + 2y^2 + 13x = 178$$
which touch $5x - 12y = 130$.

18. A coaxal system includes the circle $x^2 + y^2 = 2x$ and has (4, 4) as a limiting point. Find the other limiting point.

19. If a, b, c, r are constants and k varies, prove that the circles $x^2 + y^2 - r^2 = k(ax + by + c)$ are coaxal.

20. If t varies, prove that the circles centre $(t, 0)$ and radius $\sqrt{(t^2-4)}$ are coaxal.

21. Find the radius of the circle centre $(h, 0)$ belonging to the coaxal system with limiting points $(\pm k, 0)$.

22. Find the general equation of a circle of the coaxal system with limiting points $(-4, -3)$ and $(2, 5)$.

23. If the circles centres $(a_1, 0)$ $(a_2, 0)$ $(a_3, 0)$ and radii r_1, r_2, r_3 are coaxal, prove that

$$r_1{}^2(a_2-a_3)+r_2{}^2(a_3-a_1)+r_3{}^2(a_1-a_2)+(a_2-a_3)(a_3-a_1)(a_1-a_2)=0.$$

24. Find the locus of a point whose distances from $(a, 0)$ and $(-a, 0)$ are in the ratio $\lambda : 1$.

25. Find the locus of a point whose distances from (x_1, y_1) and (x_2, y_2) are in the ratio $k_2 : k_1$.

5·7. Polar Equation

5·71. Let $P_1(r_1, \theta_1)$ be the centre and let c be the radius of a circle. Take any point $P(r, \theta)$ on the circle and put $\mathbf{r} = \mathbf{OP}$, $\mathbf{r}_1 = \mathbf{OP}_1$. Then

$$P_1P^2 = (\mathbf{r}-\mathbf{r}_1)^2 = \mathbf{r}^2+\mathbf{r}_1{}^2-2\mathbf{r}.\mathbf{r}_1.$$

$$\therefore\quad c^2 = r^2+r_1{}^2-2rr_1\cos(\theta-\theta_1).$$

This is the **polar** equation of the circle.

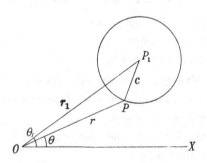

5·72. EXAMPLE. Find the polar equations of the circles of radius a which touch the initial line at the pole.

Let B be the point diametrically opposite to the pole and

let $P(r, \theta)$ be an arbitrary point on the circle. Then

$$OP = OB \cos POB.$$

$$\therefore \quad r = 2a \cos (\tfrac{1}{2}\pi - \theta) \text{ in figure } (a)$$

$$= 2a \sin \theta,$$

and $\quad\quad r = 2a \cos (-\tfrac{1}{2}\pi - \theta) \text{ in figure } (b)$

$$= -2a \sin \theta.$$

Thus the circles are $r = \pm 2a \sin \theta$.

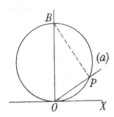 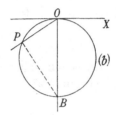

5·8. Inversion

5·81. The *inverse* of a point P wo the circle centre O and radius k is the point P' in OP, on the same side of O as P, such that $OP . OP' = k^2$. Every point except O has an inverse, and this inverse is unique. The inverse of the inverse is the original point.

When P describes a curve locus f, its inverse P' describes another curve f', and this is called the inverse of f wo the circle. Inversion is sometimes said to take place wo the point O as an abbreviation for wo the circle centre O.

5·82. Let $P(x, y)$ and $P'(x', y')$ have polar coordinates r, θ and r', θ'. Then if P' is the inverse of P wo the circle centre O and radius k,

$$rr' = k^2 \quad \text{and} \quad \theta' = \theta,$$

$$x' = r' \cos \theta' = k^2 \cos \theta / r = k^2 x / r^2 = k^2 x / (x^2 + y^2).$$

Similarly $\quad\quad y' = k^2 y / (x^2 + y^2).$

Hence the inverse of (x, y) is given by $x' : y' : k^2 = x : y : x^2 + y^2$.

5·83. To find the inverse of a given curve it is often convenient to use polar coordinates with O as pole. If $f(r,\theta) = 0$ is the polar equation of the given curve and $P'(r',\theta')$ is the inverse of an arbitrary point $P(r,\theta)$ on the given curve, then $rr' = k^2$ and $\theta' = \theta$.

But $f(r,\theta) = 0$. $\therefore$ $f(k^2/r',\theta') = 0$.

Hence P' lies on the curve whose polar equation is found from the given polar equation by changing r into k^2/r.

The inversion of lines and circles is investigated in books on pure geometry. (*M.G.* Chapter x.) See also Exercise 5E, Nos. 1, 2.

5·84. EXAMPLE. Find the inverse of $y = x^2$ wo the circle centre the origin and radius k.

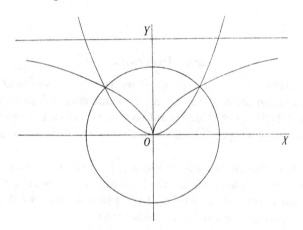

The polar equation of the given curve is

$$r \sin \theta = r^2 \cos^2 \theta,$$

i.e. $\sin \theta = r \cos^2 \theta.$

Hence the polar equation of the inverse is

$$r \sin \theta = k^2 \cos^2 \theta.$$

This curve may be sketched by giving values to θ.

The cartesian equation is

$$y = k^2 x^2/r^2,$$

i.e. $$(x^2 + y^2)\,y = k^2 x^2.$$

5·85. EXAMPLE. Show that a curve and its inverse cut OPP' at supplementary angles at the inverse points P and P'.

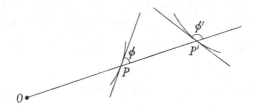

From $rr' = k^2$, by differentiation,

$$\frac{1}{r}\frac{dr}{d\theta} + \frac{1}{r'}\frac{dr'}{d\theta} = 0.$$

$$\therefore \quad \cot\phi + \cot\phi' = 0, \quad (E.C. \text{ vol. } \text{II, p. 335})$$

$$\therefore \quad \phi \text{ and } \phi' \text{ are supplementary.}$$

It follows that an angle between two curves at a point of intersection is equal to an angle between the inverse curves at the corresponding point of intersection.

5·86. Inversion is an example of a transformation by which, from a given curve or figure, a new one can be derived. Known properties of one figure may then lead to properties of the other. For example, if ABC is a line, $AB + BC = AC$, and this leads by inversion to Ptolemy's property

$$AB.CD + AD.BC = AC.BD$$

of a cyclic quadrilateral.

Other methods of deriving new curves are suggested by two of the equations in Exercise 2c, No. 4.

The graph of $r = cf(\theta)$ is a curve *similar* to that of $r = f(\theta)$.

The graph of $r = f(\theta) + c$ is called a *conchoid* of the graph of $r = f(\theta)$. See Exercise 5E, Nos. 16, 17.

Another curve can be derived from a given curve by taking the locus of the foot of the perpendicular to a tangent of the given curve from a fixed point O. This is called the *pedal* of the given curve wo the point O. See Exercise 5E, No. 18.

A more important transformation can be obtained by taking the inverse wo the circle centre O and radius k of the foot of the perpendicular from O to the tangent, thus combining the operation of inversion and pedals. This transformation will be considered from a different standpoint in a later chapter.

EXERCISE 5E

1. Write down the polar equation of the circle centre $(-1, \frac{1}{2}\pi)$ and radius 2.

2. Write down the polar equation of the circle of radius a passing through the pole and having its centre on the initial line. Prove that its inverse wo $r = k$ is a line at a distance $k^2/2a$ from the pole.

3. Prove that the inverse wo $r = k$ of the circle centre (r_1, θ_1) and radius c is a circle unless $c^2 = r_1{}^2$ and find the radius of this inverse circle.

4. O is a fixed point and P a variable point on a fixed circle of radius a through O. Find the locus of the point P' in OP such that $OP' = kOP$, where k is a constant.

5. O is a fixed point, and P is a variable point on the fixed circle centre $(b, 0)$ and radius a. Find the locus of the point P' in OP such that $OP' = kOP$, where k is a constant.

In Nos. 6–8, find the inverses wo $r = k$ of the line.

6. $x = 0.$ **7.** $y = c.$ **8.** $a = r \cos(\theta - \alpha).$

9. Find the inverse of O wo $x^2 + y^2 + 2gx + 2fy + c = 0$.

10. Find the degree of the inverse of

$$ax^2 + 2hxy + by^2 + 2gx + 2fy + c = 0$$

wo $x^2 + y^2 = 1$ when $c \neq 0$. What happens when $a = b$ and $h = 0$?

11. Find the inverse of $(t^2, t^3 + \frac{1}{2})$ wo $x^2 + y^2 = 1$. Sketch the curve $x = t^2$, $y = t^3 + \frac{1}{2}$ and its inverse.

12. Find the inverse of $r = \operatorname{cosec}^2 \frac{1}{2}\theta$ wo $r = 1$ and sketch the two curves.

13. Find the inverse wo $r = a$ of $r = a(1 + \cos\theta)$ and of its double tangent $r \cos\theta = -\frac{1}{4}a$.

14. Find the inverse wo $r = k$ of $y = ax^2 + bxy + cy^2$.

15. Show that $x(x^2+y^2) = ax^2 - by^2$ is its own inverse wo

$$x^2+y^2 = 2ax.$$

16. Sketch the conchoid of the line $r = a\sec\theta$ and show that its cartesian equation is $(x^2+y^2)(x-a)^2 = b^2x^2$.

17. Show that the conchoid of a circle through the pole is

$$r = 2a\cos(\theta-\gamma)+c$$

and sketch this curve in the special cases $\gamma = 0, c = 2a$ and $\gamma = 0, c = a$.

18. Show that the pedal of the circle $r = 2a\cos\theta$ wo the pole is the curve $r = a(1+\cos\theta)$.

EXERCISE 5F

1. Find the line-pair joining the origin to the points of intersection of $x^2+y^2-7x+5y-1 = 0$ and $3x-2y = 5$.

2. Find the line-pair joining the origin to the points of intersection of $x^2+y^2+2gx+2fy+c = 0$ and $Xx+Yy+1 = 0$. By expressing that the lines coincide, find the condition for $Xx+Yy+1 = 0$ to touch the circle.

3. Find the condition for $(ct_1, c/t_1)(ct_2, c/t_2)(ct_3, c/t_3)(ct_4, c/t_4)$ to be concyclic.

4. Find the equation of the tangents from $(b, 0)$ to $x^2+y^2 = a^2$.

5. Find the condition for the circles $x^2+y^2+2g_1x+2f_1y+c_1 = 0$ and $x^2+y^2+2g_2x+2f_2y+c_2 = 0$ to touch.

6. Find the equations of the common tangents of $x^2+y^2 = 225$ and $(x-20)^2+y^2 = 9$.

7. Find the lengths (between the points of contact) of the common tangents of $(x-24)^2+(y+67)^2 = 55^2$ and $(x+20)^2+(y-50)^2 = 20^2$.

8. Find the angle of intersection of the circles through $(0, 2)$ and $(1, 1)$ which touch OX.

9. Find the length of the common chord of the circles

$$(x-f_1)^2+(y-g_1)^2 = h_1{}^2 \quad \text{and} \quad (x-f_2)^2+(y-g_2)^2 = h_2{}^2.$$

10. $\{\omega = \tfrac{1}{3}\pi\}$. Find the equation of the circle centre $(2, 3)$ and radius 3.

11. $\{\omega\}$. Prove that the circle centre (u, v) and radius r is

$$(x-u)^2+(y-v)^2+2(x-u)(y-v)\cos\omega = r^2.$$

12. By using No. 11 and the method of 5·11, find the condition for $x^2+2xy\cos\omega+y^2+2gx+2fy+c = 0$ to represent a circle. If it is satisfied, find the centre and radius.

13. $\{\omega\}$. Find the circle of radius k touching OY at O.

14. $\{\omega\}$. Find the circle through $(a, 0)$, $(0, b)$, and the origin.

15. $\{\omega\}$. Find the tangent to $x^2 + y^2 + 2xy \cos \omega = k^2$ at (x_1, y_1).

16. $\{\omega\}$. Find the tangents to $x^2 + y^2 + 2xy \cos \omega = 1$ parallel to $y = 2x$.

17. Find the common chord and line of centres of the circles $r = a \cos(\theta - \alpha)$ and $r = b \cos(\theta - \beta)$.

18. A variable line through O meets the circles $r = a \cos(\theta - \alpha)$ and $r = b \cos(\theta - \beta)$ in P and Q. Find the locus of the mid-point of PQ.

19. Prove that the limiting points of a coaxal system are inverse wo each circle of the system.

20. P is a variable point on a fixed circle of a coaxal system of which L is a limiting point. Prove that PL^2 varies as the distance of P from the radical axis.

21. If c is constant and μ varies, prove that $x^2 + y^2 = 2\mu(x - c)$ represents coaxal circles with limiting points $(0, 0)$ and $(2c, 0)$. Show that the inverses of these circles wo the circle centre O and radius $2c$ all have centre $(2c, 0)$.

22. Find the locus of the foot of the perpendicular from $(b, 0)$ to a variable tangent to $x^2 + y^2 = a^2$.

23. Find the locus of the foot of the perpendicular from the origin to a variable tangent to $x^2 + y^2 + 2gx + 2fy + c = 0$.

24. A is the fixed point $(a, 0)$ outside the fixed circle $x^2 + y^2 = c^2$ and P is a variable point on the circle. Find the locus of the point of intersection of the tangent at P and the perpendicular at A to AP.

25. A, B are $(a, 0)$, $(0, b)$ and AP, BP are drawn so that $\angle OAP = \theta$ and $\angle OBP = \pi - \theta$. Find the locus of P when θ varies.

26. Find the square (τ) of the length of the direct common tangent of the circles
$$s \equiv (x - a_1)^2 + y^2 - r_1^2 = 0, \quad r_1 > 0,$$
$$s' \equiv (x - a_2)^2 + y^2 - r_2^2 = 0, \quad r_2 > 0.$$
Interpret the equation $s^{\frac{1}{2}} + s'^{\frac{1}{2}} = \tau^{\frac{1}{2}}$.

27. Prove that
$$\{x(r_1 - r_2) - (a_1 r_1 - a_2 r_2)\}^2 = \{(a_1 - a_2)^2 - (r_1 - r_2)^2\} y^2$$
is the equation of the direct common tangents of the circles in No. 26.

28. Find the equations of the two pairs of common tangents to the circles
$$(x - a_1)^2 + (y - b_1)^2 = r_1^2 \quad \text{and} \quad (x - a_2)^2 + (y - b_2)^2 = r_2^2.$$

29. Prove that the four common tangents to the circles

$$x^2 + y^2 - 2(a+b)x + c = 0 \quad \text{and} \quad x^2 + y^2 - 2(a-b)x + c = 0$$

are the lines $[X, Y]$ given by

$$X = aY^2, \quad acX^2 + X(a^2 - b^2 + c) + a = 0$$

and that their equation is

$$(y^2 - 4ax)\{cy^2 + (x^2 + c)(c + a^2 - b^2)\} + \{a(c + x^2) + x(c + a^2 - b^2)\}^2 = 0.$$

30. Write down the equation of the sphere centre (a, b, c) and radius r.

31. Find the centre and radius of the sphere whose equation is

$$x^2 + y^2 + z^2 + 2ux + 2vy + 2wz + d = 0, \quad (u^2 + v^2 + w^2 > d).$$

MISCELLANEOUS EXERCISE A

[These are arranged in sets of seven]

1. Find the area of the triangle $(13, 15)$ $(27, 31)$ $(14, 17)$.

2. A, B are points on OX, OY and AB is of constant length $3c$. P, Q are points on AB such that $AP = PQ = QB$. Find the loci of P, Q.

3. Find the angles of the triangle formed by

$$x + y = 2, \quad 4y = 3x + 11, \quad x = 7y - 12.$$

4. Find the polar equation of the line through $(-k, -\tfrac{1}{3}\pi)$ perpendicular to the initial line.

5. Find the coordinates of the line on which the point

$$(3k + 1)(X - Y) = 2X + Y + 1$$

lies for all values of k.

6. Find the angle between the lines joining the origin to the points of intersection of

$$y = 2x + 3 \quad \text{and} \quad x^2 + 2y^2 + 12x + 9 = 0.$$

7. Find the equations of the tangents to $x^2 + y^2 = a^2$ through $(a + b, 0)$.

8. $\{\omega = \tfrac{1}{3}\pi\}$. Find the distance between $(3, -5)$ and $(7, 2)$.

9. Prove that the condition for $lx + my + n = 0$ to cut all three sides of the triangle formed by $l_r x + m_r y + n_r = 0$ $(r = 1, 2, 3)$ externally is that the expressions

$$\begin{vmatrix} l & m & n \\ l_2 & m_2 & n_2 \\ l_3 & m_3 & n_3 \end{vmatrix} \begin{vmatrix} l_2 & m_2 \\ l_3 & m_3 \end{vmatrix}, \quad \begin{vmatrix} l & m & n \\ l_3 & m_3 & n_3 \\ l_1 & m_1 & n_1 \end{vmatrix} \begin{vmatrix} l_3 & m_3 \\ l_1 & m_1 \end{vmatrix}, \quad \begin{vmatrix} l & m & n \\ l_1 & m_1 & n_1 \\ l_2 & m_2 & n_2 \end{vmatrix} \begin{vmatrix} l_1 & m_1 \\ l_2 & m_2 \end{vmatrix}$$

should have the same sign.

10. Find the angle-bisectors of $a(x^2 + y^2) = bx(x + 2y)$.

· **11.** Given that $4X^2 - 6XY + Y^2 + 24X - 18Y + \lambda = 0$ represents a point-pair, find the value of λ.

12. Find the equation of the line joining $(11, -3)$ to the meet of $13x + y = 12$ and $4x - y = 17$.

13. Find the equations of the diameters of $x^2 + y^2 + 4x - 10y - 7 = 0$ that meet the circle on the y-axis.

14. Find the locus of the point at which the circles $(x - a)^2 + y^2 = b^2$ and $(x + a)^2 + y^2 = c^2$ subtend equal angles.

15. Find the points distant $13\frac{2}{3}$ from $(-\frac{2}{3}, 15)$ having coordinates x and y in the ratio $7:5$.

16. Find the area of the rhombus formed by $13x + 9y = 0$, $x + 3y = 0$, $13x + 9y = 150$, and $x + 3y = 30$.

17. Sketch the graph given by the polar parametric equations $r = 1 + t^2$, $\theta = t^2$ and find the cartesian equation of the curve.

18. Find the equation of the bisector of that angle between $16x - 63y + 11 = 0$ and $12x + 5y = 3$ in which the origin lies.

19. Find the line-pair joining the origin to the points of intersection of $3x + y = 2$ and $5x^2 + 3xy - 2y^2 + 7x - 2y + 1 = 0$.

20. Prove that the circles

$$x^2 + y^2 - 4x - 6y = 36 \quad \text{and} \quad x^2 + y^2 - 10x + 2y + 22 = 0$$

touch internally.

21. Find the line of centres of the coaxal system given by

$$x^2 + y^2 + 2ax + 2by + c + 2\lambda(ax - by + 1) = 0$$

when λ varies. Also find in a similar form the equation of an arbitrary circle of the orthogonal system.

22. $P_1 P_2$ is bisected at Q_1, $Q_1 P_3$ is divided at Q_2 so that $2Q_1 Q_2 = Q_2 P_3$, $Q_2 P_4$ is divided at Q_3 so that $3Q_2 Q_3 = Q_3 P_4$, and so on. Find the coordinates of Q_{n-1} in terms of those of $P_1, P_2, ..., P_n$.

23. Find the four points that are equidistant from

$$14x - 27y + 109 = 0, \quad 22x + 21y + 53 = 0, \quad \text{and} \quad 30x + 5y = 131.$$

24. Express the equation $4x - 5y + 6 = 0$ in the form

$$x \cos \alpha + y \cos \beta = p,$$

where $\alpha + \beta = \frac{2}{3}\pi$.

25. Find the pair of lines through $(-4, 5)$ parallel to

$$x^2 - 7xy + y^2 - 11x + y - 1 = 0.$$

26. Find the equation of the point in which the join of $5X + Y = 1$ and $X + 5Y = 1$ meets the join of $3X - Y = 3$ and $X - 3Y = 3$.

27. To what point must the origin be moved so that

$$ax^2 + 2hxy + by^2 + 2gx + 2fy + c = 0$$

may become $ax^2 + 2hxy + by^2 + c' = 0$ and what is the value of c'?

28. If the square of the tangent from P to $x^2 + y^2 + 3x + 4y = 0$ varies as the distance of P from OY, prove that the locus of P is a circle of radius not less than 2.

29. Find the centroid of 4 at $(3, 1)$, 3 at $(0, 0)$, 2 at $(-3, -1)$, -5 at $(1, -7)$.

30. $ABCD$ is a square and A and C move on fixed perpendicular lines. Prove that B moves on a fixed line.

31. Find the bisectors of the angles between $y - k = (x - h) \tan \alpha$ and $y - k = (x - h) \tan \beta$.

32. Find the angle between the two sets of parallel lines given by $2X^2 + 5XY + Y^2 = 0$.

33. Find the condition that the lines joining $(c, 0)$ to the points of intersection of $x^2 + y^2 = a^2$ and $x \cos \alpha + y \sin \alpha = p$ should be perpendicular.

34. Find the centre and radius of the circle through $(0, 7)$, $(2, 1)$ and $(4, -1)$.

35. Find the tangent at $(3, 2)$ to the circle
$$x^2 + y^2 + 2xy \cos \omega + x - 8y - 12 \cos \omega = 0.$$

36. Find the angle between the vectors $\mathbf{i} + \mathbf{j} + \mathbf{k}$ and $\mathbf{i} + \mathbf{j} - \mathbf{k}$.

37. Find the ecentre opposite $(1, 1)$ of the triangle $(1, 1)$ $(40, 81)$ $(31, 41)$.

38. Find the lines through $(2a, 0)$ whose perpendicular distance from $(a, -2a)$ is $\frac{1}{2}a \sqrt{2}$.

39. Find the angle between the lines $3x^2 - 2xy - y^2 + 12x + 4y = 0$.

40. If the chord $x + y = b$ of $x^2 + y^2 - 2ax - 4a^2 = 0$ subtends a right angle at the origin, prove that $b(b - a) = 4a^2$.

41. Find the length of the chord $3x + 8y = 0$ of the circle
$$x^2 + y^2 - 6x + 8y = 0.$$

42. If the origin is one limiting point of a coaxal system of circles to which $x^2 + y^2 + 2ax + 2by + c = 0$ belongs, find the other limiting point.

43. Verify that the triangle $(5, 1 - \sqrt{3})$ $(3, 1 + \sqrt{3})$ $(7, 1 + \sqrt{3})$ is equilateral.

44. $\{\omega\}$. Find the locus of a point P such that the line joining the feet of the perpendiculars from P to the axes is of constant length c.

45. Find the orthocentre of the triangle formed by
$$x(u + 1) - yu + u(u + 1) = 0, \quad x(v + 1) - yv + v(v + 1) = 0, \quad y = 0.$$

46. What is represented by $x^4 + 2x^2y^2 + y^4 = 1$?

47. Prove that the lines $ax^2 + 2hxy + by^2 = 0$ are equally inclined to $p(x^2 + y^2) + q(ax^2 + 2hxy + by^2) = 0$.

48. If the origin is changed to $(2, -1)$ and the axes are then rotated through an angle $\frac{1}{4}\pi$, find the new equation of the circle $x^2 + y^2 = 4$.

49. Write down the equation of the lines joining the origin to the points of intersection of $x^2 + y^2 = 2cx$ and $x/c - y/d + 1 = 0$. Hence find the values of d for which the line is a tangent to the circle.

50. Find the values of x', y', α if $x^2 + 2xy + 2y^2 - 4x - 6y + 3 = 0$ can be transformed into $ax^2 + by^2 + c = 0$ by change of origin to (x', y') and rotation of axes through an angle α.

51. Find the lines through $(-2, 3)$ inclined at $60°$ to $x + y\sqrt{3} = 1$.

52. Find the incentre of the triangle $(1, -5)$ $(6, 7)$ $(10, 7)$.

53. Find the line joining the meet of

$$2x + y + 1 = 0 \quad \text{and} \quad 3x - 5y - 7 = 0$$

to the meet of $\quad 4x - y - 5 = 0 \quad$ and $\quad x + 3y + 11 = 0$.

54. $\{\tan \omega = 2\tfrac{2}{5}\}$. Find the angle-bisectors of $5x^2 + 4xy - 2y^2 = 0$.

55. If $4x + 7y = k$ is a tangent to $x^2 + y^2 = 10y$, find the value of k.

56. Find the locus of the mid-points of chords of a circle centre A that subtend a right angle at a fixed point B.

57. Prove that $(3, 3)$ $(5, 9)$ $(-1, 7)$ $(-3, 1)$ is a rhombus.

58. $\{\omega\}$. Find the angle between

$$ax - by + c = 0 \quad \text{and} \quad a(y + x\cos\omega) + b(x + y\cos\omega) + c = 0.$$

59. Find the bisector of the obtuse angle between

$$57x - 176y + 1 = 0 \quad \text{and} \quad 35x + 12y = 0.$$

60. Find the pair of lines through $(1, 2)$ parallel to

$$2x^2 + 17xy - y^2 = 0.$$

61. Find the coordinates of the line through (a, b) and the meet of $[X_1, Y_1]$ and $[X_2, Y_2]$.

62. By moving the origin and rotating the axes, transform

$$2x^2 + 4xy + y^2 - 8x + 2y - 3 = 0$$

into the form $ax^2 + by^2 + c = 0$.

63. If $s_r \equiv x^2 + y^2 + 2g_r x + 2f_r y + c_r$, interpret the equation

$$\kappa_1 s_1 + \kappa_2 s_2 + \kappa_3 s_3 = 0,$$

and show that in general it is possible to find one circle orthogonal to three given circles.

64. Find the ratio in which $(-5, -8)$ divides the line joining $(-1, -2)$ to $(5, 7)$.

65. What kind of surface is represented by an equation of the form $z = f(x^2 + y^2)$?

66. Find the area of the parallelogram bounded by

$$4(x-5) = 3(y-6), \quad 3(x-5) = 4(y-6),$$
$$3(x-5) = 4(y-6)+7, \quad 4(x-5) = 3(y-6)+7.$$

67. $\{\omega = \tfrac{1}{4}\pi\}$. Find the angle between the lines $8x^2 + 4xy - y^2 = 0$.

68. Prove that two of the lines joining the origin to the points of intersection of $x^3 + y^3 = 3xy$ and $2x - 4y + 3 = 0$ are at right angles.

69. Find the equation of the circle through $(4, -9)$ and $(-2, -3)$ having its centre on $11x - 28y = 9$.

70. Coplanar circles have collinear centres A, B, C and radii a, b, c such that $a^2BC + b^2CA + c^2AB + BC.CA.AB = 0$. Prove that the circles are coaxal.

71. Forces P, Q, $P+Q$ act along the sides of the triangle of reference $A_1A_2A_3$. Find the areal equation of the line of action of their resultant.

Prove that if $P:Q$ is constant, the resultant passes through a fixed point and give its coordinates.

72. Find the line joining the feet of the perpendiculars from the origin to $2x + 3y = 7$ and $4x + 5y = 9$.

73. Find the distance between the parallel lines $y = mx + c$ and $y = mx + d$.

74. Find, for $\lambda = 2$ and for $\lambda = 1$, the equation of the point in which the join of $3X - Y - 1 = 0$ and $2X + \lambda Y + 1 = 0$ meets $[-1, 1]$.

75. Find the equations of the diagonals of the trapezium formed by $x = 0$, $y = 7$, $x = 2y - 5$, $x = 2y$.

76. Find the length of the chord $ax + by = 0$ of the circle

$$x^2 + y^2 + 2gx + 2fy + c = 0.$$

77. If A, B, C are fixed points and $PA^2 + PB^2 + PC^2$ is constant, prove that the locus of P is a circle whose centre is the centroid of A, B, C.

78. A, B lie on OX, OY; AB is of constant length $2a$, and M is the mid-point of AB. Find the locus of a point P such that MP is perpendicular to AB and of constant length b.

79. Determine k so that one of the bisectors of the angles between $[l, m]$ and $[l, n]$ is $[l, k]$.

80. Find the line through $(4, -5)$ having the part of it cut off between OX and OY' divided at that point in the ratio $2:3$.

81. Find the angle-bisectors of $4x^2 + 5xy - y^2 + 8x + 5y + 4 = 0$.

82. If $ax^2 + 2hxy + by^2 + 2gx + 2fy + c = 0$ represents two lines, give the equation of the lines through (x_1, y_1) perpendicular to them.

83. Find the equation of the tangent at $(a+c\cos t, b+c\sin t)$ to the circle $(x-a)^2+(y-b)^2 = c^2$.

84. Find the angle of intersection of the circles through $(3, -2)$ which touch $3y = 4x$ and $y = 0$.

85. Prove that the points $(m_1{}^3, m_1{}^2), (m_2{}^3, m_2{}^2), (m_3{}^2, m_3{}^2)$ are collinear if $m_2 m_3 + m_3 m_1 + m_1 m_2 = 0$.

86. If the vertices A, B of the triangle ABC are fixed and $\cot A + k \cot B$ is constant, prove that the locus of C is a line.

87. What lines through $(2, -1)$ have coordinates such that $4X^2 = 9Y^2$?

88. Given that $3x^2 - 4xy - \lambda y^2 + \lambda x + 12y - 13 = 0$ is a line-pair, find the value of λ.

89. Find the perpendicular to $x/a + y/b = 1$ from the meet of $l_1 x + m_1 y = 1$ and $l_2 x + m_2 y = 1$.

90. Find the equation of the circle on the chord $x - y = 1$ of $2x^2 + 2y^2 - 2x - 6y = 25$ as diameter.

91. $\{\omega\}$. Find the square of the tangent from (x_1, y_1) to the circle $x^2 + 2xy \cos \omega + y^2 + 2gx + 2fy + c = 0$.

92. Find the point dividing the join of $(r_1, \theta_1), (r_2, \theta_2)$ in the ratio of $\kappa_2 : \kappa_1$.

93. Find the length of the perpendicular from the origin to the join of $(a\cos\alpha, a\sin\alpha)$ to $(a\cos\beta, a\sin\beta)$.

94. Find the medians of the triangle $(10, -1)$ $(4, 7)$ $(2, 5)$ and verify that they are concurrent.

95. Find the line-pair joining $(2, 3)$ to the points of intersection of $2x + 3y = 1$ and $x^2 + y^2 = 1$.

96. Find the angle between the lines
$$(x^2 + y^2)\cos^2\alpha = (x\cos\beta + y\sin\beta)^2.$$

97. $\{\omega = \tfrac{1}{3}\pi\}$. If OX is rotated through $\tfrac{1}{3}\pi$ and OY through $\tfrac{2}{3}\pi$, find the new equation of the curve $xy = k^2$.

Also answer the same question for rotations $\tfrac{1}{6}\pi$ and $\tfrac{1}{3}\pi$.

98. Find the centre and radius of the inverse of the circle
$$x^2 + y^2 + 2gx + 2fy + c = 0 \quad \text{wo} \quad x^2 + y^2 = k^2.$$

99. $\{\omega = \tfrac{5}{6}\pi\}$. Find the area of the triangle $(7, 3)$ $(4, 2)$ $(-5, 6)$.

100. Find the angle between
$$a^2 y - b^2 x = ab(x - y) \quad \text{and} \quad a^2 x + b^2 y = ab(x + y) + c.$$

101. Prove that the origin lies in the acute angle between

$$l_1 x + m_1 y + n_1 = 0 \quad \text{and} \quad l_2 x + m_2 y + n_2 = 0$$

if $n_1 n_2 (l_1 l_2 + m_1 m_2)$ is negative.

102. Find the pair of perpendiculars from $(4, 7)$ to the lines

$$x^2 - xy - y^2 - x + 3y - 1 = 0.$$

103. Find the point through which

$$(3 - 5k) x + (4 + 7k) y - (11 + 9k) = 0$$

passes for all values of k.

104. Find the length of the common chord of the circles

$$x^2 + y^2 = 4 \quad \text{and} \quad x^2 + y^2 - 24x - 10y + 25 = 0.$$

105. Find the limiting points of the coaxal system determined by

$$x^2 + y^2 - 6x - 6y + 4 = 0 \quad \text{and} \quad x^2 + y^2 - 2x - 4y + 3 = 0.$$

Also find the circles of the system which touch $x + y = 5$.

Chapter 6

PARAMETRIC EQUATIONS

6·1. Equations of a Curve

The representation of a curve by parametric equations is explained in 2·11, 2·23, and some examples are given in Exercises 2 A, 2 B. In the present chapter parametric methods are illustrated by applications to certain curves which are themselves of some importance. In 12·9 it is shown that every curve which has a cartesian equation of the second degree also has rational parametric equations of the form

$$x:y:1 = a_1t^2 + 2b_1t + c_1 : a_2t^2 + 2b_2t + c_2 : a_3t^2 + 2b_3t + c_3.$$

When a curve is given by

$$x = f(t), \ y = g(t) \quad \text{or by} \quad x:y:c = f(t):g(t):h(t),$$

the point of the curve given by $t = t_1$ is called the *point* t_1. The chord joining the points t_1 and t_2 is called the *chord* $t_1 t_2$.

6·2. The Parabola

6·21. A *parabola* is the locus of a point whose distance from a line varies as the square of its distance from a perpendicular line. If the lines are taken as axes of coordinates, the equation of the parabola is $x^2 = ky$ or $y^2 = kx$.

All parabolas are similar. For if the value of k in $y^2 = kx$ is changed to λk, this is equivalent to the substitution of x/λ for x and y/λ for y. It only changes the scale of the graph.

Writing
$$\frac{y}{k} = \frac{x}{y} = t,$$

$$y = kt \quad \text{and} \quad x = kt^2.$$

Thus parametric equations of $y^2 = kx$ are

$$x = kt^2, \qquad y = kt,$$

or
$$x:y:k = t^2:t:1.$$

The parabola $y^2 = kx$ is symmetrical about the line $y = 0$, which is called the *axis* of the parabola.

6·22. *Intersections of* $ux + vy + wk = 0$ *and* $y^2 = kx$. The point t of the parabola $x : y : k = t^2 : t : 1$ lies on $ux + vy + wk = 0$ if
$$ut^2 + vt + w = 0.$$

This is a quadratic for t whose roots are the parameters of the points of intersection of the line and curve.

6·23. *Chord* $t_1 t_2$. By 6·22 $ux + vy + wk = 0$ meets the parabola where $ut^2 + vt + w = 0$. If $ux + vy + wk = 0$ is to be the chord $t_1 t_2$, the roots of $ut^2 + vt + w = 0$ must be t_1 and t_2 and therefore this quadratic equation must be equivalent to
$$t^2 - t(t_1 + t_2) + t_1 t_2 = 0.$$
Hence $$u : v : w = 1 : -(t_1 + t_2) : t_1 t_2$$
and the equation of the chord is
$$x - (t_1 + t_2) y + kt_1 t_2 = 0.$$

Alternatively, from 1·72 the chord $t_1 t_2$ is
$$\begin{vmatrix} x & y & 1 \\ kt_1^2 & kt_1 & 1 \\ kt_2^2 & kt_2 & 1 \end{vmatrix} = 0, \quad \text{i.e.} \quad \begin{vmatrix} x & y & k \\ t_1^2 & t_1 & 1 \\ t_2^2 & t_2 & 1 \end{vmatrix} = 0, \quad (1)$$

i.e. $$x(t_1 - t_2) - y(t_1^2 - t_2^2) + kt_1 t_2 (t_1 - t_2) = 0, \quad (2)$$
or, as $t_1 \neq t_2$, $$x - y(t_1 + t_2) + kt_1 t_2 = 0. \quad (3)$$

Equations (1), (2) express that (x, y), (kt_1^2, kt_1), (kt_2^2, kt_2) are collinear. If $t_1 = t_2$, collinearity holds for all values of x, y; and the equations do in fact then reduce to $0 = 0$.

But in (3) it is assumed that $t_1 \neq t_2$, and therefore this equation represents the line joining two distinct points t_1, t_2.

6·24. If a chord of the parabola joins the points on the curve whose parameters are the roots of $ut^2 + vt + w = 0$, then the equation of the chord can be written down immediately as $ux + vy + wk = 0$.

6·25. *Tangent at* t_1. This is the limit when $t_2 \to t_1$ of the chord $t_1 t_2$. Therefore, from 6·23 (3), its equation is
$$x - 2t_1 y + kt_1^2 = 0.$$

This can actually be derived from 6·23 (3) simply by substituting t_1 for t_2; for in fact the limit when $t_2 \to t_1$ of

$$x - (t_1 + t_2)\, y + kt_1 t_2$$

is the same as the value of that expression when $t_2 = t_1$.

But the equation 6·23 (3) is obtained on the assumption that $t_1 \neq t_2$ and it represents the chord joining two distinct points. It is not the chord joining two coincident points: such a chord would be indeterminate like 6·23 (2) with $t_1 = t_2$.

6·26. *Normal at t_1.* This is the line through t_1 perpendicular to the tangent at t_1. Its equation is therefore

$$2t_1 x + y = \ldots$$

and it must be satisfied by $x = kt_1{}^2$, $y = kt_1$.

Hence it is $\qquad 2t_1 x + y = 2kt_1{}^3 + kt_1$.

6·27. *Envelope Equation.* The tangent $x - 2ty + kt^2 = 0$ is the line $[X, Y]$ if

$$X : Y : 1/k = 1 : -2t : t^2 \tag{1}$$

and X, Y satisfy $kY^2 = 4X$.

Hence $kY^2 = 4X$ is the equation of the parabola regarded as an envelope, and (1) gives parametric envelope equations.

6·28. Some properties of the parabola are given in Exercise 6A, Nos. 16–24, where the curve is represented by $y^2 = 4ax$ or $x : y : a = t^2 : 2t : 1$. The origin A is called the *vertex* and the point $S(a, 0)$ is called the *focus* of the parabola. The line $x + a = 0$ is called the *directrix*.

EXERCISE 6A

[It is recommended that solutions of Nos. 16–24 should be preserved for reference in Chapter 13. The notation of 6·28 is used]

In Nos. 1–3, state what is represented by the equations.

1. $x = \dfrac{a}{t^2}, \; y = \dfrac{a}{t}$.   **2.** $x = at^4, \; y = at^2$.   **3.** $x = \sqrt{t}, \; y = t$.

In Nos. 4–8, give the points of intersection of the line and curve.

4. $x - 5y + 6 = 0$ and $x = t^2,\ y = t$.

5. $4x - 14y = 5$ and $x:y:1 = t^2:1:2t$.

6. $11x - 6y + 1 = 0$ and $x:y:1 = t:t^2+1:t^3$.

7. $5x - y = 2k$ and $x:y:k = t:t^2:1+t^3$.

8. $x/a - y/b = 1$ and $x = a(1-t^2)/(1+t^2),\ y = 2bt/(1+t^2)$.

In Nos. 9–14, give the equation of the chord $t_1 t_2$ and of the tangent and normal at t.

9. $x:y:a = t^2:2t:1$.   **10.** $x:y:k = t^2:1:t$.

11. $x:y:a = t+1:t^2:t-1$.   **12.** $x:y:a = t^2-1:t^2+1:t$.

13. $x = at^2,\ y = at^3$.   **14.** $x = kt,\ y = kt^{-2}$.

15. Find the condition for the lines

$$x = a_1 t + b_1,\ y = c_1 t + d_1 \quad \text{and} \quad x = a_2 t + b_2,\ y = c_2 t + d_2$$

to be (i) parallel, (ii) perpendicular.

16. Draw an accurate graph of $x:y:a = t^2:2t:1$ taking $a = \frac{1}{2}$ inch. From an arbitrary point P whose parameter is t draw PN perpendicular to the axis, and let the tangent and normal at P meet the axis in T and G.

17. Prove that $TA = AN$ and $NG = 2a$.

18. Find SP in terms of t and deduce that the parabola is the locus of a point which is equidistant from the fixed point $(a, 0)$ and the fixed line $x + a = 0$.

19. If SY is drawn perpendicular to PT, prove that Y is midway between S and the foot of the perpendicular from P to the directrix, and that Y lies on the tangent at the vertex.

20. Prove that ASY, YST, TSP are similar triangles and that $\angle SPT = \angle TPK$.

21. Prove that rays of light emitted from the focus and reflected at the curve become parallel to the axis.

22. If the chord $t_1 t_2$ passes through the focus, prove that the tangents at t_1, t_2 are at right angles and that they meet at a point F on the directrix. Give the coordinates of F.

23. In No. 22, prove that FS is perpendicular to the chord.

24. State the coordinates of the mid-point of the chord $t_1 t_2$, and the condition for the chord to be in a fixed direction. Find the locus of the mid-points of parallel chords of a parabola.

6·3. Note on Orthogonal Projection

6·31. The *orthogonal projection* of a point P on a plane is the foot of the perpendicular from P to the plane.

Consider the various points P of any figure F in a plane α, and their orthogonal projections P' on a plane α' which makes with α an angle θ such that $0 < \theta < \tfrac{1}{2}\pi$. These points P' compose a figure F' which is called the orthogonal projection of F on α'.

The orthogonal projection of a line is a line. The orthogonal projection of a segment PQ of a line is an equal segment $P'Q'$ if and only if PQ is parallel to the line of intersection of α and α'. If PQ is perpendicular to that line, $P'Q' = PQ\cos\theta$. See Exercise 6 B, No. 1.

The orthogonal projection of a circle is not a circle, but is a curve which is investigated in 6·4.

6·32. Let P be any point in the plane α whose coordinates are x, y referred to rectangular axes OX, OY of which the former lies along the line of intersection of α, α'. In α' take rectangular axes so that OX' coincides with OX, and let the orthogonal projection P' of P have coordinates x', y' referred to these axes. Then

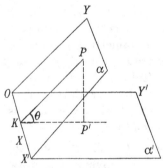

$$x' = x, \qquad y' = y\cos\theta.$$

In other words (x, y) projects into $(x, y\cos\theta)$.

If P describes a curve $x = f(t)$, $y = g(t)$, then P' describes the curve

$$x' = f(t), \qquad y' = g(t)\cos\theta.$$

If P describes a curve $f(x, y) = 0$, then, since $x' = x$, $y' = y\cos\theta$ and $f(x, y) = 0$, it follows that $f(x', y'\sec\theta) = 0$. This is the equation of the projection of the original curve.

A fuller treatment of orthogonal projection is given in *P.G.* Chapter II.

EXERCISE 6 B

1. Two planes α and α' inclined to one another at an angle θ meet in a line l. PQ is a segment of a line in α which makes an angle ϕ with l, and $P'Q'$ is its orthogonal projection on α'. Prove that

$$P'Q' = PQ \sqrt{(1 - \sin^2\theta \sin^2\phi)}.$$

2. With the notation of 6·32, what point has (h, k) for orthogonal projection?

3. Find the orthogonal projections of $y^2 = kx$ on a plane through OX and on a plane through OY.

4. Show how to project $y^2 = 2ax$ orthogonally into (i) $y'^2 = ax'$, (ii) $y'^2 = 4ax'$.

5. If $\cos\theta = b/a$ and OX is the vanishing line, what is the orthogonal projection of $x^2 + y^2 = a^2$?

6. If $\cos\theta = b/a$ and OY is the vanishing line, what is the orthogonal projection of $x^2/a^2 + y^2/b^2 = 1$?

In Nos. 7–13, sketch the graph.

7. $x^2 + 4y^2 = 4.$ **8.** $4x^2 + 9y^2 = 36.$ **9.** $x^2 - \tfrac{1}{4}y^2 = 1.$

10. $x^2 - y^2 = 1.$ **11.** $x^2 - 2y^2 = 1.$ **12.** $xy = 1.$

13. $xy = -1.$

6·4. The Ellipse

6·41. The curve obtained by the orthogonal projection of a circle is called an ellipse.

Parametric equations of a circle centre the origin and radius a are $x = a\cos\phi$, $y = a\sin\phi$ (5·12).

Taking α' as the plane through the x-axis making an angle $\cos^{-1}(b/a)$ with the plane of $x = a\cos\phi$, $y = a\sin\phi$, the orthogonal projection is given by

$$x' = a\cos\phi, \quad y' = b\sin\phi.$$

It is convenient to draw the ellipse and circle in one diagram as in the figure. The dotted circle is equal to that from which the ellipse is obtained, but is drawn in the plane of the ellipse.

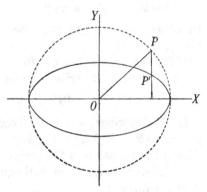

Using x, y instead of x', y',

the ellipse is given by the parametric equations $x = a\cos\phi$, $y = b\sin\phi$.

Elimination of ϕ gives

$$\frac{x^2}{a^2} + \frac{y^2}{b^2} = 1, \qquad (b < a).$$

The circle $x = a\cos\phi$, $y = a\sin\phi$ or $x^2 + y^2 = a^2$ is called the *auxiliary circle* of the ellipse.

By a property of orthogonal projection, the area of the ellipse $x^2/a^2 + y^2/b^2 = 1$ is b/a times that of the auxiliary circle. Therefore the area is πab. (*P.G.* p. 22.)

6·42. *Eccentric Angle.* The angle ϕ, $= \angle XOP$, in 6·41 is called the *eccentric angle* of the point P' on the ellipse. As ϕ increases from 0 to 2π, P describes the circle and $P'(a\cos\phi, b\sin\phi)$ describes the ellipse. Two values of ϕ which differ by a multiple of 2π give the same point of the ellipse.

In the orthogonal projection, the chord joining two points ϕ, ϕ' of the circle projects into the chord of the ellipse joining points whose eccentric angles are ϕ, ϕ'. Also the tangent at P to the circle projects into the tangent at the corresponding point to the ellipse. See Exercise 6c, Nos. 1, 2.

6·43. Rational algebraic parametric equations of

$$x^2/a^2 + y^2/b^2 = 1$$

may be obtained as follows.

Since
$$\frac{y^2}{b^2} = 1 - \frac{x^2}{a^2} = \left(1 - \frac{x}{a}\right)\left(1 + \frac{x}{a}\right),$$

$$1 - \frac{x}{a} : \frac{y}{b} = \frac{y}{b} : 1 + \frac{x}{a} = t : 1, \text{ say.}$$

$$\therefore \quad 1 - \frac{x}{a} : \frac{y}{b} : 1 + \frac{x}{a} = t^2 : t : 1,$$

$$\therefore \quad \frac{x}{a} : \frac{y}{b} : 1 = 1 - t^2 : 2t : 1 + t^2.$$

These equations may also be derived from $x = a\cos\phi$, $y = b\sin\phi$ by putting $\tan\tfrac{1}{2}\phi = t$.

6·44. Some properties of the ellipse are given in Exercise 6c, Nos. 12–15. The points $A(a, 0)$ and $A'(-a, 0)$ are called the *vertices* of the ellipse. AA' is called the *major axis*.

The line joining $B(0, b)$ and $B'(0, -b)$ is called the *minor axis*. Every chord through the origin is bisected there, and this point is called the *centre* (C).

EXERCISE 6c

[It is recommended that the solutions of Nos. 1–5 and 12–15 should be preserved for reference in Chapter 14. The notation of 6·44 is used]

1. Obtain in the form $x \cos \alpha + y \sin \alpha = p$ the equations of the chord $\phi_1 \phi_2$ and the tangent at ϕ_1 to the circle $x = a \cos \phi$, $y = a \sin \phi$.

2. Use orthogonal projection to deduce from No. 1 the equations of the chord $\phi_1 \phi_2$ and the tangent at ϕ_1 to the ellipse $x = a \cos \phi$, $y = b \sin \phi$.

3. Find the point of intersection of the tangents at $\phi = \alpha \pm \beta$ to the circle $x = a \cos \phi$, $y = a \sin \phi$ and deduce the point of intersection of the tangents at $\phi = \alpha \pm \beta$ to the ellipse $x = a \cos \phi$, $y = b \sin \phi$.

4. Interpret the equation

$$\begin{vmatrix} x/a & y/b & 1 \\ \cos \phi_1 & \sin \phi_1 & 1 \\ \cos \phi_2 & \sin \phi_2 & 1 \end{vmatrix} = 0$$

and simplify it when $\phi_1 \neq \phi_2$. Deduce the equation of the tangent to the ellipse $x^2/a^2 + y^2/b^2 = 1$ at the point $(a \cos \phi, b \sin \phi)$.

5. Find the equation of the normal to $x = a \cos \phi$, $y = b \sin \phi$ at the point ϕ_1.

6. Determine the points of intersection of the line $lx + my + n = 0$ and the ellipse $x/a : y/b : 1 = 1 - t^2 : 2t : 1 + t^2$.

7. Find the equation of the chord $t_1 t_2$ of the ellipse in No. 6.

8. Find the equation of the tangent at the point t_1 to the ellipse in No. 6. How many tangents to the ellipse may pass through a given point (f, g)?

9. Show that the point (x, y) given by

$$x : b + y : b - y = a(t^2 - 1) : b(t + 1)^2 : b(t - 1)^2$$

lies on the ellipse $x^2/a^2 + y^2/b^2 = 1$.

10. Show that the point (x, y) given by

$$x : 1 + y : 1 - y = t^2(t + 1) : (t + 1)^2 : t^4$$

lies on the circle $x^2 + y^2 = 1$.

11. What is the greatest number of normals to an ellipse that may be drawn through a given point?

12. Draw an accurate graph of $x = a\cos\phi$, $y = b\sin\phi$ taking $a = 1\frac{1}{2}$ in, $b = \frac{1}{2}$ in. From an arbitrary point P on the curve draw PN, PN' perpendicular to the axes and let the tangent and normal at P meet the axes in T, T' and G, G'.

13. Prove that $CN.CT = a^2$ and $CN'.CT' = b^2$.

14. Prove that $CG = (1 - b^2/a^2)\,CN$ and $CG' = (1 - a^2/b^2)\,CN'$.

15. Prove that TG' is perpendicular to $T'G$.

6·5. The Hyperbola

6·51. Curves such as those in Exercise 6 B, Nos. 9, 10, 11, given by equations of the form $\dfrac{x^2}{a^2} - \dfrac{y^2}{b^2} = 1$ are called *hyperbolas*.

The curve $\dfrac{x^2}{a^2} - \dfrac{y^2}{b^2} = 1$ is symmetrical about both axes, and, since $\dfrac{x^2}{a^2} = \dfrac{y^2}{b^2} + 1 \geqslant 1$, no part of the curve lies between the lines $x = \pm a$. The figures illustrate the cases $a > b$, $a = b$, $a < b$.

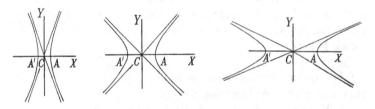

6·52. Just as $x^2/a^2 + y^2/b^2 = 1$ is satisfied by $x = a\cos\phi$, $y = b\sin\phi$, so $x^2/a^2 - y^2/b^2 = 1$ is satisfied by $x = a\operatorname{ch}\phi$, $y = b\operatorname{sh}\phi$. But $\operatorname{ch}\phi$ is always positive, and therefore the parametric equations $x = a\operatorname{ch}\phi$, $y = b\operatorname{sh}\phi$ represent only one branch of the curve.

6·53. Alternative equations may be found by expressing $\operatorname{ch}\phi$, $\operatorname{sh}\phi$ in terms of $\operatorname{th}\frac{1}{2}\phi$ ($= m$). These are

$$x/a : y/b : 1 = 1 + m^2 : 2m : 1 - m^2.$$

They represent the whole curve except the point $(-a, 0)$ if m

is unrestricted. But if $m = \operatorname{th}\tfrac{1}{2}\phi$, it can only take values between -1 and $+1$.

6·54. $x^2/a^2 - y^2/b^2 = 1$ may also be represented by

$$x = a \sec \phi, \qquad y = b \tan \phi.$$

6·55. The equation $x^2/a^2 - y^2/b^2 = 1$ may be written

$$\left(\frac{x}{a}+\frac{y}{b}\right)\left(\frac{x}{a}-\frac{y}{b}\right) = 1.$$

Put $\dfrac{x}{a}+\dfrac{y}{b} = t$. Then $\dfrac{x}{a}-\dfrac{y}{b} = \dfrac{1}{t}$.

$$\therefore \quad x = \tfrac{1}{2}a\left(t+\frac{1}{t}\right), \qquad y = \tfrac{1}{2}b\left(t-\frac{1}{t}\right).$$

These equations represent the whole of the curve. In the figure

$t = +1$ gives the point $A\,(a, 0)$,

$t = -1$ gives the point $A'(-a, 0)$.

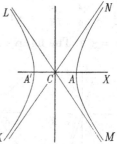

As t increases from $-\infty$ to $+\infty$, the portions KA', $A'L$, MA, AN of the curve are given by $t < -1$, $-1 < t < 0$, $0 < t < 1$, $1 < t$.

6·56. Some properties of the hyperbola are given in Exercise 6 D, Nos. 7–10. The points $A(a, 0)$ and $A'(-a, 0)$ are called the *vertices* of the hyperbola. AA' is called the *transverse* axis. The line $x = 0$ does not meet the curve. It is called the *conjugate* axis. The origin is called the *centre* (C). Any line through it is $y = mx$. This meets the curve $x^2/a^2 - y^2/b^2 = 1$ where $x^2(1/a^2 - m^2/b^2) = 1$.

$$\therefore \quad x = \pm ab/\surd(b^2 - a^2m^2).$$

Hence there are two intersections if and only if $m^2 < b^2/a^2$. The lines $y = \pm bx/a$ are called *asymptotes*. Lines through the centre meet the curve if they lie in those angles between the asymptotes which contain the transverse axis.

EXERCISE 6D

[It is recommended that solutions of Nos. 1, 4, 7–10 should be preserved for reference in Chapter 15. The notation of 6·56 is used]

In Nos. 1–4, find the equation of the chord of the hyperbola and deduce the equations of the tangent and normal.

1. $x/a : y/b : 1 = 1+t^2 : 2t : 1-t^2.$ **2.** $x = a\,\mathrm{ch}\,\phi, y = b\,\mathrm{sh}\,\phi.$

3. $x = a\sec\theta, y = b\tan\theta.$

4. $x/a - y/b : x/a + y/b : 1 = t^2 : 1 : t.$

5. Prove that the equations

$$x - y : x + y : k = t^2 : 1 : t$$

and $$x : y : k = 1+m^2 : 2m : 1 - m^2$$

represent the same curve and find the relation between t and m.

6. Find the product of the lengths of the perpendiculars from $(\pm a\,\mathrm{ch}\,\phi, b\,\mathrm{sh}\,\phi)$ to $x/a \pm y/b = 0$. Hence show that when the asymptotes are taken as oblique axes, the equation of the hyperbola is $xy = \frac{1}{4}(a^2 + b^2).$

7. Draw graphs of the hyperbola $x = \frac{1}{2}a\left(t + \dfrac{1}{t}\right), y = \frac{1}{2}b\left(t - \dfrac{1}{t}\right)$, when a, b are (i) 1 in., $\frac{1}{2}$ in., (ii) $\frac{1}{2}$ in., 1 in. From an arbitrary point P on the curve draw PN, PN' perpendicular to the axes and let the tangent and normal at P meet the axes in T, T' and G, G'.

8. Prove that $CN \cdot CT = a^2$ and $CN' \cdot CT' = -b^2.$

9. Prove that $CG = (1 + b^2/a^2)\,CN$ and $CG' = (1 + a^2/b^2)\,CN'.$

10. The tangent at A meets an asymptote at D. From CX, CS is cut off equal to CD and SY is drawn perpendicular to CD. Find the coordinates of S and Y.

6·6. $xy = k^2$

6·61. The product of the perpendiculars from a point (x_1, y_1) of the hyperbola $x^2/a^2 - y^2/b^2 = 1$ to the asymptotes

$$x/a + y/b = 0 \quad \text{and} \quad x/a - y/b = 0$$

is $$\dfrac{x_1/a + y_1/b}{\sqrt{(1/a^2 + 1/b^2)}}\dfrac{x_1/a - y_1/b}{\sqrt{(1/a^2 + 1/b^2)}}, \text{ i.e. } \dfrac{1}{1/a^2 + 1/b^2}.$$

If the asymptotes are taken as new axes of coordinates, the lengths of the perpendiculars are $x'\sin\omega$ and $y'\sin\omega$, where ω

is the angle $2\tan^{-1}\dfrac{b}{a}$ between the asymptotes. Hence the new equation of the curve is

$$x'y' = \frac{a^2b^2\cosec^2\omega}{a^2+b^2} = \tfrac{1}{4}(a^2+b^2).$$

Properties of the hyperbola are conveniently investigated from the equation $xy = k^2$ or the corresponding parametric equations.

When a hyperbola has its asymptotes at right angles, it is called a *rectangular hyperbola*. The equation of a rectangular hyperbola is $xy = k^2$ with rectangular axes.

From $xy = k^2$, $\dfrac{x}{k} = \dfrac{k}{y} = t$, say.

$$\therefore \quad x = kt, \qquad y = k/t,$$

or $$x:y:k = t^2 : 1 : t.$$

These equations represent the curve completely. No point of the curve is given by $t = 0$.

6·62. *Intersections of* $ux+vy+wk = 0$ *and* $xy = k^2$. The point t of the hyperbola lies on the line if

$$ut^2 + v + wt = 0.$$

Therefore the parameters of the points of intersection are the roots of this equation.

6·63. *The chord* t_1t_2. By 6·62, $ux+vy+wk = 0$ meets the curve where $ut^2+v+wt = 0$. If $ux+vy+wk = 0$ is to be the chord t_1t_2, the roots of this quadratic in t must be t_1, t_2, and therefore the quadratic must be the same as

$$t^2 - t(t_1+t_2) + t_1t_2 = 0.$$

$$\therefore \quad u:v:w = 1 : t_1t_2 : -(t_1+t_2)$$

and the equation of the chord is

$$x + t_1t_2 y = (t_1+t_2)k.$$

6·64. The chord joining the points whose parameters are the roots of $ut^2 + wt + v = 0$ is $ux + vy + wk = 0$.

6·65. The tangent to $x : y : k = t^2 : 1 : t$ at the point t_1 is the limit when $t_2 \to t_1$ of the chord $t_1 t_2$. Therefore by 6·63 the tangent is
$$x + t_1^2 y = 2t_1 k. \tag{1}$$
The tangent $x + t^2 y = 2tk$ is the line $[X, Y]$ if
$$X : Y : 1/k = 1 : t^2 : -2t.$$
Thus X, Y satisfy $4k^2 X Y = 1.$ (2)

This is the equation of the hyperbola regarded as an envelope.

6·66. EXAMPLE. Prove that a tangent to a hyperbola forms with the asymptotes a triangle of constant area.

1st method. The tangent 6·65 (1) meets the asymptotes $x = 0$, $y = 0$ in $(0, 2k/t_1)$, $(2kt_1, 0)$. Hence, in the figure, $OT = 2kt_1$ and $OT' = 2k/t_1$. The area OTT'
$$= \tfrac{1}{2} OT . OT' \sin \omega = \tfrac{1}{2} (2kt_1)(2k/t_1) \sin \omega = 2k^2 \sin \omega.$$

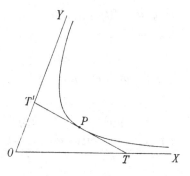

2nd method. The area OTT'
$$= \tfrac{1}{2} OT . OT' \sin \omega = \tfrac{1}{2} (-1/X)(-1/Y) \sin \omega = 2k^2 \sin \omega,$$
by 6·65 (2).

EXERCISE 6E

1. Prove that the curves given by $xy = k^2$ for different values of k are similar.

2. State the relation between t_1 and t_2 if the chord $t_1 t_2$ of
$$x : y : k = t^2 : 1 : t$$
(i) passes through the centre, (ii) has a fixed direction.

3. Prove that the tangents at t_1, t_2 to $x : y : k = t^2 : 1 : t$ meet on the line joining the centre to the mid-point of the chord $t_1 t_2$.

4. Prove that the part of a tangent to a hyperbola cut off between the asymptotes is bisected at the points of contact.

5. State the condition of perpendicularity for the chords $t_1 t_2$ and $t_3 t_4$ of $x : y : k = t^2 : 1 : t$.

6. Prove that the orthocentre of a triangle inscribed in a rectangular hyperbola lies on the curve.

7. Find the equation of the normal at t_1 to $x : y : k = t^2 : 1 : t$.

8. Find the chord $t_1 t_2$, the tangent at t_1, and the envelope equation of $x : y : a = t : t^3 : 1$.

6·7. Tangents and Envelope Equations

6·71. The equation of the chord t, $t + \epsilon$ of the curve
$$x : y : a = f(t) : g(t) : h(t)$$

is
$$\begin{vmatrix} x & y & a \\ f(t) & g(t) & h(t) \\ f(t+\epsilon) & g(t+\epsilon) & h(t+\epsilon) \end{vmatrix} = 0$$

or
$$\begin{vmatrix} x & y & a \\ f(t) & g(t) & h(t) \\ \dfrac{f(t+\epsilon)-f(t)}{\epsilon} & \dfrac{g(t+\epsilon)-g(t)}{\epsilon} & \dfrac{h(t+\epsilon)-h(t)}{\epsilon} \end{vmatrix} = 0.$$

The limit of this when $\epsilon \to 0$ is
$$\begin{vmatrix} x & y & a \\ f(t) & g(t) & h(t) \\ f'(t) & g'(t) & h'(t) \end{vmatrix} = 0.$$

This is the equation of the tangent at the point t. Hence the envelope coordinates X, Y of the tangent are such that

$$X:-Y:1 = \begin{vmatrix} g(t) & h(t) \\ g'(t) & h'(t) \end{vmatrix} : \begin{vmatrix} f(t) & h(t) \\ f'(t) & h'(t) \end{vmatrix} : \begin{vmatrix} f(t) & g(t) \\ f'(t) & g'(t) \end{vmatrix}.$$

These are parametric envelope equations of the curve. The determinants are those formed from

$$\begin{bmatrix} f(t) & g(t) & h(t) \\ f'(t) & g'(t) & h'(t) \end{bmatrix}$$

by omitting the columns in turn.

The use of parameters is further illustrated in 6·72, 6·73, and in 6·8. When the curve of 6·72 is sketched, it is found to have two branches which cross at the origin. The origin is called a *double point* of the curve. In 6·73 the origin is called a *cusp*. Such points are discussed more fully in Chapter 7, and further details about these points and about points of inflexion may be found in $E.C.$ vol. I, p. 71 and vol. II, pp. 427–430.

6·72. EXAMPLE. Find the condition of collinearity of the distinct points t_1, t_2, t_3 of the curve $x = t^2 - 1$, $y = t - 1/t$.

The points lie on the line $x + uy + v = 0$ if t_1, t_2, t_3 satisfy

$$t(t^2 - 1) + u(t^2 - 1) + vt = 0. \tag{1}$$

Hence, as t_1, t_2, t_3 are all different, they are the roots of the cubic equation (1), and therefore

$$t_1 + t_2 + t_3 = -u, \quad t_2 t_3 + t_3 t_1 + t_1 t_2 = v - 1, \quad t_1 t_2 t_3 = u.$$

Therefore $\qquad t_1 + t_2 + t_3 + t_1 t_2 t_3 = 0$

is a necessary condition. It is also sufficient; for when it holds and u, v are given the values $-t_1 - t_2 - t_3$, $1 + t_2 t_3 + t_3 t_1 + t_1 t_2$, the third equation is satisfied; and these values of u, v determine the line of collinearity.

The cartesian equation is $xy^2 = x^2 - y^2$.

6·73. EXAMPLE. Find where the tangent to the curve $x : y : a = t^3 : t^2 : 1$ at the point t meets the curve again.

1st method. A tangent to $x:y:a = t^3:t^2:1$ is

$$\begin{vmatrix} x & y & a \\ t^3 & t^2 & 1 \\ 3t^2 & 2t & 0 \end{vmatrix} = 0$$

or, if $t \neq 0$, $2x - 3ty + at^3 = 0.$

[This result holds also if $t = 0$. For the chord joining O to the point t is $x = ty$ and its limit when $t \to 0$ is $x = 0$.]

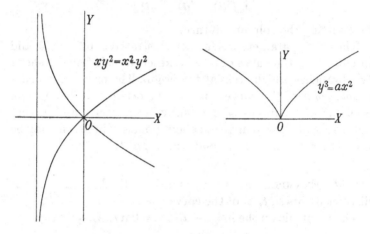

The tangent meets the curve at the point t_1 if

$$2t_1{}^3 - 3tt_1{}^2 + t^3 = 0$$

i.e. $$(t_1 - t)^2 (2t_1 + t) = 0.$$

This expresses that two of the points of intersection of the curve with the tangent are at the point of contact and that the remaining point of intersection is at $t_1 = -\tfrac{1}{2}t$, i.e. the point $(-\tfrac{1}{8}at^3, \tfrac{1}{4}at^2)$.

2nd method. The three points t_1, t_2, t_3 of the curve lie on the line $x + uy + va = 0$ if t_1, t_2, t_3 satisfy $t^3 + ut^2 + v = 0$

$$t_1 + t_2 + t_3 = -u, \quad t_2 t_3 + t_3 t_1 + t_1 t_2 = 0, \quad t_1 t_2 t_3 = -v.$$

Hence by an argument like that used in 6·72 the points are collinear if

$$t_2 t_3 + t_3 t_1 + t_1 t_2 = 0.$$

Putting $t_1 = t$ and making $t_2 \to t$,

$$2tt_3 + t^2 = 0.$$

$$\therefore \quad t_3 = -\tfrac{1}{2}t.$$

Thus $-\tfrac{1}{2}t$ is the parameter of the point in which the tangent meets the curve again.

The cartesian equation is $y^3 = ax^2$. The curve is symmetrical about the line $x = 0$, and if a is positive it lies in the first and second quadrants. The point O is a cusp.

6·8. Condition of Collinearity

6·81. The second solution in 6·73 illustrates the application of the condition of collinearity of three points of a cubic curve to the determination of the point in which a tangent meets the curve again. Other uses of the condition will now be illustrated.

Although the condition of collinearity is usually found as in the above examples on the assumption that t_1, t_2, t_3 are unequal, it takes the same form when $t_2 \to t_1$ as when $t_2 = t_1$ and takes the same form when $t_2 \to t_1$ and $t_3 \to t_1$ as when $t_1 = t_2 = t_3$.

6·82. If t_1, t_2, t_3 satisfy the condition of collinearity when they are all equal to t, this means that the tangent at t meets the curve in three coincident points at t instead of the usual two coincident points. This can happen in different ways. For $x = t$, $y = t^3$ the condition is $t_1 + t_2 + t_3 = 0$, and if $t_1 = t_2 = t_3 = t$, then $t = 0$; the point $(0, 0)$ given by $t = 0$ is a point of inflexion. In 6·73 the condition is $t_2 t_3 + t_3 t_1 + t_1 t_2 = 0$, and if $t_1 = t_2 = t_3 = t$, then $t = 0$; the point $(0, 0)$ given by $t = 0$ is a cusp.

6·83. If there are values of t_1, t_2 for which the condition of collinearity is an identity in t_3, then the points t_1, t_2 are collinear with an arbitrary point t_3 of the curve. Hence the points t_1, t_2 coincide.

The condition in 6·72 is an identity if $t_1 + t_2 = 0$, $t_1 t_2 = -1$, i.e. if t_1, t_2 are $+1, -1$, and these values give the double point. See also 6·84. But in 6·73 the condition is an identity when $t_1 = t_2 = 0$ and this gives the cusp.

6·84. Example. Find the double point and point of inflexion of
$$x:y:a = 2t^2+1:t^3-3t^2+2t:2t-1.$$

The line $ux+y+va = 0$ meets the curve in points t given by
$$u(2t^2+1)+t^3-3t^2+2t+v(2t-1) = 0.$$

The roots of this cubic equation satisfy
$$t_1+t_2+t_3 = 3-2u,$$
$$t_2t_3+t_3t_1+t_1t_2 = 2+2v,$$
$$t_1t_2t_3 = v-u.$$

Hence the condition for the points t_1, t_2, t_3 to be collinear is
$$2t_1t_2t_3 = (t_2t_3+t_3t_1+t_1t_2)-2+(t_1+t_2+t_3)-3.$$

This is an identity in t_3 if
$$2t_1t_2 = t_1+t_2+1,$$
$$t_1t_2 = -t_1-t_2+5.$$
$$\therefore\quad t_1+t_2 = 3,\ t_1t_2 = 2,\qquad \therefore\quad t_1, t_2 \text{ are 2 and 1.}$$

The double point is $x:y:a = 3:0:1$, i.e. it is $(3a, 0)$.

Putting $t_1 = t_2 = t_3$,
$$2t^3-3t^2-3t+5 = 0.$$

This has one root $t \simeq -1\cdot26$, which shows that there is a point of inflexion near $(-1\cdot2a, 2\cdot6a)$.

6·85. Example. Prove that the curve
$$x:y:a = t^2-1:t^3-t:t^3$$

has no point of inflexion.

The line $ux+vy+(1-v)a = 0$ meets the curve in points t given by
$$u(t^2-1)+v(t^3-t)+(1-v)t^3 = 0.$$

The roots of this equation satisfy
$$t_1+t_2+t_3 = -u,\quad t_2t_3+t_3t_1+t_1t_2 = -v,\quad t_1t_2t_3 = u.$$

Hence the condition for the points t_1, t_2, t_3 to be collinear is

$$t_1 + t_2 + t_3 + t_1 t_2 t_3 = 0.$$

The parameters of any inflexions are found by putting t_1, t_2, t_3 all equal to t. This gives

$$3t + t^3 = 0. \qquad \therefore \quad t = 0.$$

But this value of t gives no point of the curve.

6·86. EXAMPLE. Find the condition of collinearity for the points t_1, t_2, t_3 of $\quad x : y : a = t^4 - 1 : t : t^2.$

What conclusions can be drawn from the condition? Show that the parametric equations are equivalent to $y^2(y^2 + ax) = a^4$ and sketch the curve.

The line $x + uy + va = 0$ meets the curve in points t_1, t_2, t_3, t_4 given by

$$t^4 - 1 + ut + vt^2 = 0.$$

$$\therefore \quad \Sigma t = 0, \quad \Sigma t_1 t_2 = v, \quad \Sigma t_1 t_2 t_3 = -u, \quad t_1 t_2 t_3 t_4 = -1.$$

Therefore if the two conditions

$$t_1 + t_2 + t_3 + t_4 = 0 \quad \text{and} \quad t_1 t_2 t_3 t_4 = -1$$

hold, the four points lie on the line $x + uy + va = 0$ given by $u = -\Sigma t_1 t_2 t_3$, $v = \Sigma t_1 t_2$.

When t_1, t_2, t_3 are collinear, they are collinear with some point t_4, and the two conditions hold. The condition of collinearity for t_1, t_2, t_3 is found by elimination of t_4 from the two conditions. This gives $t_1 t_2 t_3 (t_1 + t_2 + t_3) = 1$.

The tangent at t_1 meets the curve again in points t given by

$$t_1^2 t(2t_1 + t) = 1.$$

$$\therefore \quad t = -t_1 \pm \sqrt{(t_1^2 + t_1^{-2})}.$$

Any inflexions are given by $3t_1^4 = 1$, i.e. $t = \pm 1/\sqrt[4]{3}$. These are the points $(-2a/\sqrt{3}, \pm a \sqrt[4]{3})$.

The condition of collinearity cannot be an identity in t_3 and there is no double point (8·77). Since $t = a/y$ and $x/a = t^2 - t^{-2}$,

$$\frac{x}{a} = \frac{a^2}{y^2} - \frac{y^2}{a^2}.$$

$$\therefore \quad y^2(y^2 + ax) = a^4 \quad \text{or} \quad x = a^3/y^2 - y^2/a.$$

By adding the values of x in the graphs of $x = a^3/y^2$ and $x = -y^2/a$, the form of the curve is found to be that shown in the figure.

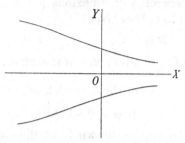

6·9. A Method of Finding Parametric Equations

6·91. Consider the circle $x^2 + y^2 = ax$ through the origin A. An arbitrary line through A meets the circle in one point P besides A. Let this line be $y = tx$; the points of intersection are given by $x^2(1 + t^2) = ax$, and are the origin and the point $x = a/(1 + t^2)$, $y = tx = at/(1 + t^2)$. Also every point P on the circle is given by $y = tx$ for some value of t. Thus the co-ordinates of an arbitrary point of the circle have been found in terms of the parameter a, and

$$x : y : a = 1 : t : 1 + t^2$$

are parametric equations of the circle.

6·92. If the method of 6·91 is applied to any curve through the origin with a cartesian equation of degree 2, it gives rational algebraic parametric equations. If it is applied to a curve of higher degree, it will in general fail to do so. For example the line $y = tx$ meets $x^3 + y^3 = x$ where $x = \pm 1/\sqrt{(1 + t^3)}$ and so the parametric equations given in this way are not rational. Again for $x^5 + x^3 + y^3 = x$, the solution with $y = tx$ leads to $2x^2 = -1 - t^3 \pm \sqrt{(5 + 2t^3 + t^6)}$. Irrational parametric representations of this kind are of little value. Trigonometrical parametric equations can often be made algebraic and rational by substitutions like $\tan \tfrac{1}{2}\theta = t$.

6·93. In general if the equation of the curve is of degree n, the solution with $y = tx$ leads to an equation of degree $n - 1$ for x after the root $x = 0$ has been removed. In particular for a cubic curve the method leads to a quadratic for x. But if the cubic curve has a double point or cusp at the origin, this point will count as two among the points of intersection of $y = tx$ and the curve; there will then be two roots $x = 0$ and the coordinate of P will be given by a simple equation. Thus rational parametric equations will be found.

6·94. EXAMPLE. Find parametric equations of $y^2 = x^2(1 + x)$. The origin is a double point. Put $y = tx$; then

$$t^2 x^2 = x^2(1 + x).$$

$$\therefore \quad x = 0, 0, t^2 - 1,$$

and the parametric equations are

$$x = t^2 - 1, \quad y = t^3 - t.$$

6·95. For a curve whose equation is of degree 2, the line AP may be drawn through any point $A(f, g)$ of the curve. $y = tx$ is replaced by $y - g = t(x - f)$, and this gives parametric equations. The same method gives parametric equations of a cubic curve with a double point or cusp at (f, g).

It will be found that in general a cubic curve with no double point or cusp does not possess rational parametric equations. More generally it is found that the existence of a rational parametric representation implies that the curve has the maximum number of double points or singular points appropriate to the degree of its cartesian equation (8·78).

EXERCISE 6F

In Nos. 1–6, find the equation of the tangent at t and give parametric envelope equations.

1. $x = at^2 + b$, $y = ct + d$.
2. $x = t/(t^2 - 1)$, $y = 1/(t + 1)$.
3. $x : y : a = t^2 : t^3 : 1 + t^2$.
4. $x : y : 2a = t^2 - t^4 : 2t^3 : (1 + t^2)^2$.
5. $x : y : 1 = 2 + t^2 : 2t + t^3 : 1 + t^2$.
6. $x : y : a = t - \sin t : 1 - \cos t : 1$.

In Nos. 7, 8, give the equation of the chord $t_1 t_2$.

7. $x = at^2, y = a/t$. **8.** $x = 4t^2 + 1, y = 3t + 2$.

9. Find the condition of collinearity for the points t_1, t_2, t_3 of $x:y:a = 1 + 2t^2 : t + 2t^3 : 1 + t^2$ and use it to find the points of inflexion. Is there a double point? Sketch the curve.

10. Find the condition of collinearity for the points t_1, t_2, t_3 of $x:y:3a = t:t^2:1+t^3$ (Folium of Descartes). If a chord subtends a right angle at the origin, prove that it passes through the point $(\frac{3}{2}a, \frac{3}{2}a)$.

11. Find the point of inflexion of

$$x:y:1 = 5(1+t)^2 : 3t(1+t)^2 : 5t(1-3t+6t^2).$$

12. Show that the points t_1, t_2, t_3 of the curve

$$x:y:1 = (t-3)^3 : (t-3)(2t-3) : t(t-2)$$

are collinear if $2(t_1+t_2+t_3) - (t_2t_3+t_3t_1+t_1t_2) = 6 - \frac{5}{9}t_1t_2t_3$. Hence find the double point.

13. If the points t_1, t_2, t_3 of $x:y:2a = t^2-t^4 : 2t^3 : (1+t^2)^2$ are collinear, prove that $t_1t_2t_3(\Sigma t - 3t_1t_2t_3) = (\Sigma t_1t_2)(1+\Sigma t_1t_2)$. Find the relation between the parameters of the ends of chords which pass through the origin.

14. Tangents to $y^2 = x^3$ at P, Q, R meet the curve again in P', Q', R'. Prove that if P, Q, R are collinear, so also are P', Q', R'.

15. Find where the tangent to $x:y:a = 1:t^3:t$ at the point t meets the curve again.

16. Find the parameters of the contacts of the tangents from $(-\frac{16}{33}a, \frac{4}{33}a)$ to the curve $x:y:a = t:t^2:t^3+1$.

17. Obtain parametric equations of $3x^2 + 2xy = y^3$.

18. Give parametric equations of $x(1+y^2) = 1$ and find where the tangent at (x_1, y_1) meets the curve again.

19. Find parametric equations of $(x^2+y^2)(9x-7y) = x^2-y^2$ and find the point of inflexion.

20. Three points θ_1, θ_2, θ_3 of the curve $x = \sin 3\theta$, $y = \sin \theta$ are collinear. Find the relation between θ_1, θ_2, θ_3.

21. Show that from any point Q on the curve $y^2 = x^2(1+x)$ two lines QP_1, QP_2 can be drawn to touch the curve at P_1, P_2, and that when Q varies the envelope of P_1P_2 is $y^2+8(x+1)(x+2) = 0$.

22. Sketch the part of the curve $x = t$, $y = 2+t^3$ given by small values of t, and the corresponding part of its inverse wo $x^2+y^2 = 1$.

23. Find the inverse of $(y-2)^2 = x^3$ wo $x^2+y^2 = 1$. Sketch the parts of these curves near $(0, 2)$ and $(0, \frac{1}{2})$.

Chapter 7

THE GENERAL ALGEBRAIC CURVE

7·1. The General Curve

7·11. The general algebraic equation of degree n in x, y is

$$a + (bx + cy) + (dx^2 + 2exy + fy^2) + u_3 + u_4 + \ldots + u_n = 0,$$

where u_r is a homogeneous polynomial of degree r in x, y. The curve represented by it is called a curve of *order* n. The number of constants in the equation

$$= \{1 + 2 + 3 + \ldots + (n + 1)\} - 1$$
$$= \tfrac{1}{2}(n + 1)(n + 2) - 1 = \tfrac{1}{2}n(n + 3).$$

Thus there are $\infty^{n(n+3)/2}$ curves of order n in a plane.

7·12. In general one curve of order n can be drawn to pass through $\tfrac{1}{2}n(n + 3)$ points. For the equations which express that (x_1, y_1), (x_2, y_2), ... satisfy the equation in 7·11 are linear equations from which the independent constants can be obtained. Some exceptions will be considered in 7·5.

7·2. Intersection of Line and Curve

7·21. Any line may be represented by parametric equations $x = At + C$, $y = Bt + D$. See 2·23, p. 28.

This line meets the general curve given in 7·11 at points whose parameters t are such that

$$a + b(At + C) + c(Bt + D) + \ldots + w(Bt + D)^n = 0.$$

At the present stage this is an equation of real algebra. It is of degree n in t and therefore has n roots or fewer. Hence a line meets the general curve of order n in n points or fewer.

It may happen that the equation reduces to an identity; that is it may be true for all values of t. In this exceptional case every point of the line belongs to the curve. The curve is then degenerate and consists of the line and a curve of order $n-1$. A line meets a non-degenerate curve of order n in n points or fewer.

7·3. Neighbourhood of a Point on a Curve

7·31. At a point P_1 of a non-degenerate algebraic curve there is usually a unique tangent. This line is the limit of the line $P_1 P_2$ when P_2 is another point on the curve which tends to P_1.

Take the point P_1 to be the origin. The constant a in the equation of 7·11 is then zero. Any line through P_1 is

$$x = At, \quad y = Bt, \tag{1}$$

and this meets the curve in points t given by

$$t(bA + cB) + t^2(\ldots) + \ldots + wB^n t^n = 0. \tag{2}$$

One root of this equation is $t = 0$. If (1) represents $P_1 P_2$, another root is the parameter of P_2. When P_2 tends to P_1 this second root tends to zero; and if (1) represents the tangent at P_1, which is the limit of $P_1 P_2$, (2) will have a repeated root $t = 0$. This happens when $bA + cB = 0$. Hence the tangent is given by

$$x = At, \quad y = Bt, \quad \text{where} \quad bA + cB = 0.$$

Thus the tangent is $bx + cy = 0$ unless b and c are both zero.

7·32. EXAMPLE. Find the tangent at the origin to

$$x^2 + y^3 = x + y.$$

$x = at, y = bt$ meets the curve in points given by

$$a^2 t^2 + b^3 t^3 = (a + b) t.$$

$$\therefore \quad t = 0 \quad \text{or} \quad b^3 t^2 + a^2 t - (a + b) = 0.$$

One point of intersection is the origin. If $a + b = 0$, there is another root $t = 0$ and $x = at, y = bt$ is the tangent. Hence the tangent is $x + y = 0$.

7·33. *Coincident Intersections.* There is a convention in algebra by which an equation $(x - a)^2 = 0$ is said to have two equal roots $x = a$. This does not alter the fact that the equation is true for just one value of x. But it is a convenient form of language to adopt.

The convention is extended. An equation $(x-a)^n \phi(x) = 0$, where $\phi(x)$ is a polynomial not zero for $x = a$, is said to have n equal roots $x = a$.

In view of these algebraic conventions it is suitable to make corresponding conventions in geometry and to say for example that the line $x + y = 0$ meets the curve in 7·32 in two coincident points in virtue of the occurrence of the factors x^2 and y^2 when the equations are solved for x and y.

The solution of the equations of a curve and a chord $P_1 P_2$ leads to equations for x, y or for a parameter t containing the factors

$$(x-x_1)(x-x_2), \quad (y-y_1)(y-y_2) \quad \text{or} \quad (t-t_1)(t-t_2).$$

The equation of the tangent can be found by taking the limit of the equation of $P_1 P_2$, and if the equation of the tangent is solved with the equation of the curve, it leads to equations for x, y or for t containing the factors $(x-x_1)^2$, $(y-y_1)^2$ or $(t-t_1)^2$.

Therefore we agree to say that the tangent meets the curve in two coincident points P_1. This convention does not alter the fact that a curve and its tangent have just one common point at P_1. It is important to avoid the mistake of supposing that

$$\lim_{P_2 \to P_1} (\text{the line joining } P_1 \text{ to } P_2),$$

which is the definition of the tangent at P_1, can be replaced by

the line joining two coincident points P_1.

Nevertheless, in analytical geometry, a tangent is often found by means of a repeated factor instead of by a direct appeal to the definition involving the calculation of a limit. Thus the conventional language has the merit of suggesting the process which is commonly used in finding tangents.

More generally two loci are said to have n coincident points of intersection at (a, b) if the solution of their equations leads to $(x-a)^n = 0$ and $(y-b)^n = 0$ either directly or after obtaining $(t-t_1)^n = 0$.

7·34. EXAMPLE. Find the tangent at the origin to $x = y^3 + x^5$. $x = at$, $y = bt$ meets the curve in points given by

$$at = b^3t^3 + a^5t^5.$$

$$\therefore \quad t = 0 \quad \text{or} \quad a^5t^4 + b^3t^2 - a = 0.$$

One point of intersection is the origin. If $a = 0$, there are two more roots $t = 0$, and $x = at$, $y = bt$ is the tangent. Hence the tangent is $x = 0$. It meets the curve in three coincident points at the origin: it is an inflexional tangent.

7·35. If in the general equation $a = 0$ and b, c are not both zero, the origin is called an *ordinary* or *simple* point of the curve. Any line other than the tangent at O through an ordinary point O of a curve has just one of its points of intersection with the curve at O.

7·36. When $b = c = 0$ as well as $a = 0$, every line through the origin has the origin for two of its points of intersection with the curve. The parameters of the intersections are given by

$$t^2(dA^2 + 2eAB + fB^2) + t^3(\ldots) = 0.$$

It follows that the line $x = At$, $y = Bt$ has three intersections with the curve at the origin if

$$dA^2 + 2eAB + fB^2 = 0.$$

If $e^2 > fd$, there are two values of $A : B$ which satisfy this equation. The corresponding lines have three intersections with the curve at O. O is then called a *double point*. The coordinates x, y of a point on either tangent at O satisfy

$$x : A = y : B$$

and $\quad dA^2 + 2eAB + fB^2 = 0.$

Hence they satisfy

$$dx^2 + 2exy + fy^2 = 0,$$

which is therefore the equation of the pair of tangents at the origin.

The limit definition fails to give a tangent at a double point P_1. There is in fact no such line as *the* tangent at a double point. The limit definition can be modified so as to give the tangent to each branch of the curve separately by confining the point P_2 which tends to P_1 to one branch at a time. But the tangents are usually found by the method of repeated roots.

If $e^2 < fd$, the equation of the parameters gives $t = 0$ or

$$dA^2 + 2eAB + fB^2 + (pt + qt^2 + \ldots) = 0,$$

and $dA^2 + 2eAB + fB^2$ has the same sign for all values of A, B. But if $|t|$ is sufficiently small,

$$|pt + qt^2 + \ldots| < |dA^2 + 2eAB + fB^2|.$$

Hence the equation cannot be true for such small values of t. Therefore there are no points of the curve in the immediate neighbourhood of the origin for any line $x = At$, $y = Bt$. The origin is then called an *isolated point* of the curve. See Exercise 7 A, No. 8.

If $e^2 = fd$, but d, e, f are not all zero, the equation

$$dA^2 + 2eAB + fB^2 = 0$$

can be written

$$(dA + eB)^2 = 0 \quad \text{or} \quad (eA + fB)^2 = 0$$

and it is true for one value of $A : B$. For this value, the line $x = At$, $y = Bt$ has three intersections with the curve at P. The form of the curve near P depends upon the remaining terms $gx^3 + \ldots$ of the equation. A simple example in which the origin is a *cusp* is $x^2 = y^3$. But there is not always a cusp when $e^2 = fd$. See 7·38.

7·37. When $d = e = f = 0$ as well as $b = c = 0$ and $a = 0$, the general equation reduces to

$$gx^3 + 3hx^2y + 3ixy^2 + jy^3 + u_4 + u_5 + \ldots = 0.$$

In general any line through the origin meets the curve three times there. If the cubic $gx^3 + 3hx^2y + 3ixy^2 + jy^3 = 0$ has three distinct roots, then there are three tangents at the origin which is called an ordinary *triple point*. Each of the tangents has four intersections with the curve at the origin.

Other forms of triple point arise when the cubic does not possess three distinct roots (7·39).

Multiple points of order higher than 3 arise when there are no terms of degree less than 4 in the equation of the curve.

7·38. EXAMPLE. Discuss the points of intersection of the curve $y^2 = 2x^2y + 3x^4 + x^5$ with an arbitrary line through the origin. What is the form of the curve near the origin?

The line $x = At$, $y = Bt$ meets the curve in points given by

$$t^2(B^2 - 2A^2Bt - 3A^4t^2 - A^5t^3) = 0.$$

The origin counts as two points of intersection. If $B = 0$, it counts as three. The origin is a double point with coincident tangents $y = 0$. The equation may be written

$$(y - x^2)^2 = 4x^4 + x^5.$$

$$\therefore \quad y \simeq 3x^2 \quad \text{or} \quad y \simeq -x^2.$$

7·39. EXAMPLE. Discuss, as in 7·38, the curve $x^4 + y^5 = x^2y$. The line $x = At$, $y = Bt$ meets the curve in points given by

$$t^4(A^4 + B^5t) = A^2Bt^3, \quad \text{i.e. } t^3(B^5t^2 + A^4t - A^2B) = 0.$$

The origin counts as three points of intersection. If $B = 0$, it counts as four, and if $A = 0$, it counts as five. Thus OX meets the curve four times at O and OY meets it five times. These lines have no other points of intersection with the curve. They are the tangents at the origin.

If $AB \neq 0$, there are two points of intersection given by the roots of $B^5t^2 + A^4t - A^2B = 0$. Since these roots have opposite signs, the points of intersection are in opposite quadrants.

If (x, y) is a point near the origin on the branch that touches OX, y is small compared to x; y^5 is small compared to x^4; thus $x^4 \simeq x^2y$; therefore the form of this branch of the curve near the origin is approximately that of $x^2 = y$. If (x, y) is on the other branch, x is small compared to y. Suppose that x is of order y^n; then $n > 1$ and the three terms are of orders y^{4n}, y^5, y^{2n+1}.

Hence $n = 2$ and the term x^4 is negligible compared to the others. Therefore the form of this branch is approximately that of $y^5 = x^2 y$, i.e. $x = \pm y^2$. The origin is a triple point, and by an obvious convention we say that it has two of its tangents coincident with OY.

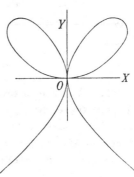

If x^4 and y^5 are of the same order for large values of x and y, $x^2 y$ is negligible compared to them and so $x^4 + y^5 = 0$ is an approximation to the form of the curve at a great distance from the origin. The curve is closed in the first and second quadrants and is symmetrical about OY.

Other examples of the forms of curves near the origin will be found in *E.C.* vol. II, p. 427.

EXERCISE 7A

1. Find the points of intersection of the line $x = at$, $y = bt$ and the curve $x = y^3$.

2. Find the point of intersection other than the origin of $y^2(x + 2y + 4) = x^2$ and $x = ky$. What happens when $k = -2$?

3. Find the points of intersection of $y = kx$ and $(x^2 + y^2)^2 = (x^2 - y^2)$, distinguishing between the cases $k^2 < 1$, $k^2 = 1$, $k^2 > 1$.

4. Show that $x = at$, $y = bt$ meets $x^2 - 4y^2 = x + y$ in one point besides the origin unless $a = -b$, $a = 2b$, $a = -2b$. Discuss these cases.

In Nos. 5–14, find the points of intersection of $x = at$, $y = bt$ with the curve and give the equations of the tangents at the origin.

5. $x^3 = x - y$.

6. $y^3 = x(x - 2y)$.

7. $4x^2 + y^4 = y^2$.

8. $y^2 = x^2(x - 1)$.

9. $(x + y)^3 = (x - y)^2$.

10. $y(y^2 - x^2) = x^4$.

11. $y^4 - 5x^2 y^2 + 4x^4 = y^5$.

12. $x^3 + y^3 = x^2 y^2$.

13. $x^5 + 2x^3 y = y^3$.

14. $x^5 + y^4 = x^2 y$.

15. State the coordinates of nine particular points which lie on the cubic $ax(x^2 - 1) = by(y^2 - 1)$ for all values of the constants a, b. Choose a, b so that (i) the cubic is degenerate, (ii) the cubic passes through $(2, 3)$.

16. Find a, b, c, d so that $x^3 = y(ax^2 + bxy + cy^2 + dy)$ passes through $(0, 2)$, $(1, 1)$, $(2, -1)$, $(3, -2)$. What is its form near the origin?

In Nos. 17–20, discuss the form of the curve near the origin.

17. $y^2 = 2yx^2 + x^4 + x^5$.   **18.** $y(y^2 - 7x^4) + 6x^6(1 + x) = 0$.

19. $x(x^2 - y)^2 = y^5$.   **20.** $x^7 + y^7 = xy^5 + x^3y^2$.

7·4. Conics

7·41. A curve of order 2 is called a conic section or *conic* because these curves were first obtained as plane sections of (circular) cones. This method of approach is explained in a later chapter.

The general equation of a conic is written

$$ax^2 + 2hxy + by^2 + 2gx + 2fy + c = 0.$$

Conics are of three main types and these are illustrated in 6·2, 6·4, 6·5, 6·6, where the equations are particularly simple on account of the special choice of axes. Properties of the conics are considered in greater detail in Chapters 13, 14, 15. Here and in Chapter 12 some properties common to all conics are proved.

7·42. A non-degenerate conic meets an arbitrary line in two points or fewer. For the point t of the line l whose parametric equations are

$$x = At + C, \qquad y = Bt + D$$

lies on the conic given by the general equation if

$$a(At + C)^2 + 2h(At + C)(Bt + D) + b(Bt + D)^2$$
$$+ 2g(At + C) + 2f(Bt + D) + c = 0,$$

and this is a quadratic for t. Exceptionally it may be an identity in t. Every point of the line then belongs to the conic, which consists of l and another line.

7·43. At an ordinary point of an algebraic curve there is an unique tangent. See 7·31. This applies to a conic. If the curve in 7·41 is a conic and $a = b = c = 0$, the equation is

$$dx^2 + 2exy + fy^2 = 0,$$

and by 4·3 this represents a line-pair, coincident lines, or one point. Thus only a degenerate conic can possess a double point. At every point of a non-degenerate conic there is a unique tangent.

7·44. EXAMPLE. Prove that chords of a conic which subtend a right angle at a fixed point on the curve are concurrent.

Let the tangent and normal to the conic at the fixed point be taken as axes of x and y. Then by 7·31 the equation of the conic is

$$cy + dx^2 + exy + fy^2 = 0.$$

Suppose that a chord PQ which subtends a right angle at the origin is

$$lx + my = 1.$$

Then by 4·8 the equation of OP, OQ is

$$cy(lx + my) + dx^2 + exy + fy^2 = 0.$$

Since these lines are at right angles,

$$cm + d + f = 0.$$

Thus the chord $lx + my = 1$ passes through the fixed point $(0, -c/(d+f))$.

This fixed point is called the *Frégier point* of O. It lies on the normal at O. If $d + f = 0$, no chord subtends a right angle at O. This is proved by the condition $cm + d + f = 0$ obtained above.

7·45. The Conic through Five Points

The six constants a, b, c, f, g, h in the equation of 7·41 are not independent. Only the five independent ratios $a : b : c : f : g : h$ are relevant. There are ∞^5 conics in a plane.

In general a unique conic can be drawn to pass through five given points (x_r, y_r), $r = 1$ to 5. The ratios of the constants in its equation are given by the five simple equations

$$ax_r^2 + 2hx_r y_r + by_r^2 + 2gx_r + 2fy_r + c = 0, \quad r = 1 \text{ to } 5$$

and the equation of the conic is

$$\begin{vmatrix} x^2 & xy & y^2 & x & y & 1 \\ x_1^2 & x_1y_1 & y_1^2 & x_1 & y_1 & 1 \\ x_2^2 & x_2y_2 & y_2^2 & x_2 & y_2 & 1 \\ x_3^2 & x_3y_3 & y_3^2 & x_3 & y_3 & 1 \\ x_4^2 & x_4y_4 & y_4^2 & x_4 & y_4 & 1 \\ x_5^2 & x_5y_5 & y_5^2 & x_5 & y_5 & 1 \end{vmatrix} = 0.$$

A more convenient way of obtaining the equation of the conic through five points is given in Chapter 16.

7·46. If three of the five given points lie on a line l, that line must be part of the conic. By 7·42 a line can only meet a non-degenerate conic in two points. If four of the five points lie on a line l, the conic through the five points is not unique. It consists of l and an arbitrary line through the fifth point. In this case the five equations for $a : b : c : f : g : h$ are not independent; any one of the equations corresponding to the collinear points is deducible from the other three. If the five given points are collinear, there are ∞^2 conics through them, each conic consisting of the line of collinearity and another line.

7·5. Curves of Order n

7·51. Curves of order 3, 4, 5, ... are called cubics, quartics, quintics, ... By 7·12 one curve of order n can be drawn to pass through $\frac{1}{2}n(n+3)$ points. It may happen that there is not a unique curve of order n through the $\frac{1}{2}n(n+3)$ points, or there may be a unique curve which is degenerate.

7·52. For example, the quartic through 14 points of which five are collinear consists of the line of collinearity and a cubic; the quartic through 14 points, of which five lie on one line and five others lie on another line, consists of the two lines and an arbitrary conic through the remaining four points; the quartic through 14 points of which six are collinear is not unique.

7·53. An exception of another kind may be given. The number of points required to determine a cubic curve is nine, and nine points usually determine a unique cubic. But, in general, two cubics intersect in nine points. Consider for example the cubics

$$C \equiv x(x^2 - 1) - 2y(y^2 - 1) \quad \text{and} \quad C' \equiv x(x - 1) - 3y(y^2 - 1)$$

which meet in the points $(0, 0)$, $(0, 1)$, $(0, -1)$, $(1, 0)$, $(1, 1)$, $(1, -1)$, $(-1, 0)$, $(-1, 1)$, $(-1, -1)$. Also the cubics given by $C = kC'$ pass through these same nine points. Thus if the nine given points happen to be the points of intersection of two cubics, there will be ∞^1 cubics through the points. Such a set of points, therefore, does not determine a unique cubic.

This kind of exception does not arise with conics, because two conics meet in four points and five points are needed to determine a conic. But such exceptions arise with curves of higher order. See Exercise 7B, No. 18. Hence it is necessary to include the phrase "in general" in 7·12 and 7·45.

7·6. Intersections of Two Curves

7·61. To find the points of intersection of two curves it is convenient to use rational parametric equations of one of them and the ordinary cartesian equation of the other. The special conics considered in Chapter 6 are all given by parametric equations which are special cases of

$$x : y : 1 = a_1 t^2 + 2b_1 t + c_1 : a_2 t^2 + 2b_2 t + c_2 : a_3 t^2 + 2b_3 t + c_3,$$

and it is shown in 6·9 that every non-degenerate conic has equations of this form. Hence the points t in which a non-degenerate conic meets the general curve of 7·11 are found from

$$a(a_3 t^2 + 2b_3 t + c_3)^n + b(a_1 t^2 + 2b_1 t + c_1)(a_3 t^2 + 2b_3 t + c_3)^{n-1}$$
$$+ \dots = 0,$$

which is an equation of degree $2n$. In particular the points in which the conic meets the general cubic

$$a + bx + cy + dx^2 + 2exy + fy^2 + gx^3 + 3hx^2y + 3ixy^2 + jy^3 = 0$$

are given by

$$a(a_3 t^2 + 2b_3 t + c_3)^3 + b(a_3 t^2 + 2b_3 t + c_3)^2 (a_1 t^2 + 2b_1 t + c_1) + \ldots = 0$$

which is of the form

$$At^6 + Bt^5 + Ct^4 + Dt^3 + Et^2 + Ft + G = 0.$$

Thus the conic meets the cubic in not more than six points unless the coefficients in this equation are all zero. In this exceptional case every point of the conic lies on the cubic; the cubic is then degenerate, consisting of the conic and a line. Similarly, subject to the exceptions due to degeneracy, a conic meets a curve of order n in not more than $2n$ points.

7·62. More generally two non-degenerate curves of orders m, n have not more than mn points of intersection. But parametric equations are not always available for the proof of this statement. It can be proved by showing that the elimination of x (or y) from the equations

$$a + (bx + cy) + \ldots + \lambda y^m = 0,$$
$$a' + (b'x + c'y) + \ldots + \lambda' y^n = 0$$

of the curves leads to an equation of degree not greater than mn in y (or x).

7·63. EXAMPLE. Show that a non-degenerate curve of order 5 cannot have more than six double points.

If A, B, C, D, E, F, G are seven double points of such a curve, it is possible to draw a cubic curve to pass through these points and through two other points of the quintic. A, B, C, D, E, F, G count for two each amongst the points of intersection of the cubic and quintic. Hence there will be 16 points of intersection. But a cubic and a non-degenerate quintic cannot meet in more than 15 points. Thus the seven double points cannot exist.

7·7. Curves of Class m

7·71. Lines $[X, Y]$ such that

$$A + (BX + CY) + (DX^2 + 2EXY + FY^2)$$
$$+ U_3 + U_4 + \ldots + U_m = 0,$$

where U_r is a homogeneous polynomial of degree r in X, Y, form an envelope which is called a curve envelope of *class m*.

There are $\frac{1}{2}m(m+3)$ independent constants in the equation and therefore in general one curve envelope of class m can be found which contains $\frac{1}{2}m(m+3)$ given lines.

Any point except the origin may be represented by parametric envelope equations $X = at+c$, $Y = bt+d$. If the point is (x_1, y_1), the values of a, b may be $y_1, -x_1$, and c, d are then numbers such that $cx_1 + dy_1 = -1$. It can be proved as in 7·2 that this envelope point has in general m lines or fewer in common with the curve envelope of class m. Exceptionally it can happen that all lines through the point belong to the envelope; the curve envelope is then degenerate and consists of the point and a curve envelope of class $m-1$.

7·72. *Duality.* There exists a unique tangent at an ordinary point of a curve locus. There are exceptional points such as double points at which there are two tangents. Dually there exists a unique contact on an ordinary line of a curve envelope (4·12). There are exceptional lines, such as double tangents on which there are two contacts.

The same algebra which proves that curves of orders m, n have in general mn common points also proves that curve envelopes of classes m, n have in general mn common lines.

7·73. EXAMPLE. Find the contact of $[1, 1]$ with $Y = X^3$.

1st method. The line $[t, t^3]$ belongs to the envelope. It meets $[1, 1]$ where

$$xt + yt^3 + 1 = 0 = x + y + 1.$$

$$\therefore \quad x = -(t^2 + t + 1)/(t^2 + t), \quad y = 1/(t^2 + t).$$

The limit of this point when $t \to 1$ is $(-\frac{3}{2}, \frac{1}{2})$.

2nd method. Any point on the line $[1, 1]$ is $(-m, m-1)$. The line $[X, Y]$ goes through this point if

$$-mX + (m-1)Y + 1 = 0 \quad \text{and} \quad Y = X^3.$$

$$\therefore \quad (m-1)X^3 - mX + 1 = 0.$$

The root $X = 1$ is a repeated root if it satisfies

$$3(m-1)X^2 - m = 0,$$

i.e. if $m = \frac{3}{2}$, giving $(-\frac{3}{2}, \frac{1}{2})$.

EXERCISE 7B

1. Find the conic through $(0, 2)$, $(0, 4)$, $(1, 0)$, $(3, 0)$, $(2, 1)$.

2. Find the conic through $(5, 2)$, $(3, -4)$, $(2, -7)$, $(1, 2)$, $(3, 5)$.

3. Find the curve of class 2 touching $[-1, 2]$, $[-1, 4]$, $[0, 1]$, $[1, 1]$, $[1, 4]$.

4. Prove that the conic through $(1, 2)$, $(2, 4)$, $(5, 0)$, $(-1, 1)$, $(3, 6)$ is degenerate.

5. Find the conic through $(-1, 1)$, $(0, 2)$, $(3, 3)$ touching the axis of x at the origin.

6. Show that ∞^1 conics pass through $(1, -2)$, $(2, 1)$, $(4, 7)$, $(-2, 2)$, $(-1, -8)$, and find that which also passes through the origin.

In Nos. 7, 8, find the common points of the conic and cubic.

7. $x^2 - 21xy + 84y^2 = 64a^2$, $x:y:a = t^3:t:1$.

8. $x:y:1 = 2t:1+t^2:1-t^2$, $(x-y)(x+y)^2 - (x^2-y^2) + (x+y) - 1 = 0$.

9. If the normal at the origin to $cy + dx^2 + exy + fy^2 = 0$ bisects the angle between the chords OP, OQ, prove that PQ passes through a fixed point on the tangent at O.

10. Find the Frégier point of the origin wo $ax^2 + 2hxy + by^2 = 2x$.

11. Find the Frégier point of (kt^2, kt) wo the curve $y^2 = kx$.

12. Verify that the elimination of y between

$$a_1 x^2 + 2h_1 xy + b_1 y^2 + 2g_1 x + 2f_1 y + c_1 = 0$$

and

$$a_2 x^2 + 2h_2 xy + b_2 y^2 + 2g_2 x + 2f_2 y + c_2 = 0$$

gives a quartic for x.

13. Find the cubic through $(0, 0)$, $(1, 0)$, $(-1, 0)$, $(1, 1)$, $(1, -1)$, $(-1, 1)$, $(-1, -1)$, $(-10, 1)$, $(10, -1)$.

14. What cubic curve passes through nine given points of which four are on a line l? What happens if three of the remaining five points are collinear?

15. Prove that a non-degenerate cubic cannot have two double points.

16. Prove that a non-degenerate curve of class 4 cannot have four double tangents.

17. State the points of intersection of the degenerate cubics $x^3 - 3x^2 + 2x = 0$ and $(y-x)^3 = y-x$ and find a cubic which passes through these nine points and through $(3, 5)$.

18. A unique quintic can in general be drawn through 20 given points. What happens when (i) six of the points lie on a line l, (ii) 11 of the points lie on a conic s, (iii) 16 of the points lie on a cubic c, (iv) six of the points lie on a line l and 11 of the others lie on a conic s?

19. Touching how many lines is it, in general, possible to draw a curve of class 4? How many of the lines may be concurrent without making the curve degenerate? How many of the lines may touch a curve of class 2?

20. Through how many points of space is it possible, in general, to draw a surface whose equation is of degree n in x, y, z?

21. Through how many points of space is it possible in general to draw a cone of order n with vertex at a given point?

Chapter 8

ABSTRACT GEOMETRY

8·1. Generalisation in Algebra and Geometry

8·11. The process of generalisation is a characteristic feature of mathematics. In algebra the concept of "number" is generalised by the successive replacement of the natural numbers $1, 2, 3, \ldots$ by rational numbers, by real numbers, and then by complex numbers. See $A.T.$ Chapter VIII.

Logically the new algebras which are created are independent of one another. The class of complex numbers does not include the class of real numbers as a sub-class. But it includes certain numbers which have an exact correspondence with the real numbers and enjoy similar properties. In virtue of this correspondence it is possible to use complex algebra to prove results of real algebra.

In the present chapter some generalisations of geometry are introduced. Logically these are independent of one another and of elementary geometry. The process consists of the creation of new geometries rather than the extension of the original ones. This is analogous to the development of algebra. Also, just as complex algebra can be applied to problems of real algebra, so the new geometries can be applied to problems of elementary geometry.

8·12. The solutions of the simultaneous equations

$$x^2 + y^2 = 10, \qquad 2x + y = 1$$

are $\{x = -1, y = +3\}$ and $\{x = 1\tfrac{4}{5}, y = -2\tfrac{3}{5}\}$.

This means geometrically that the circle $x^2 + y^2 = 10$ and the line $2x + y = 1$ have two points of intersection.

8·13. The elimination of y from the equations

$$x^2 + y^2 = 10, \qquad 3x + y = 10$$

leads to $(x-3)^2 = 0$. Hence these equations have only one solution $\{x = 3, y = 1\}$.

This means geometrically that the circle and line have only one common point. The line is a tangent to the circle.

With the usual convention, the tangent is said to meet the circle in two coincident points (7·33).

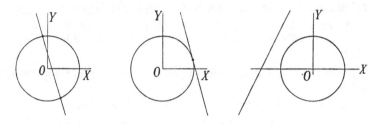

8·14. Elimination of x from the equations

$$x^2 + y^2 = 10, \qquad y - 2x = 10$$

leads to $(x+4)^2 = -2$ and this equation is never true. Thus the simultaneous equations have no solution. Geometrically this means that the line and circle have no common point.

The results of 8·12, 8·13, 8·14 are illustrated in the figures.

8·15. In 8·12 there are two points of intersection; in 8·13 there is one, and we may say that there are two coincident points of intersection; in 8·14 there is no point of intersection.

Although special numerical examples have been chosen, similar reasoning may be applied more generally. See Exercise 8 A, No. 2. It may also be applied to loci other than lines and circles.

The mathematician in his search for generality would like to be able to say that there are two points of intersection of a line and circle in every case. But the plain fact is that there are no such points in an example like 8·14. If the mathematician is dissatisfied with this state of affairs, he must make a new geometry for himself, and that is precisely what he does. In the new geometry which he creates, new meanings are given to point, line, circle, intersect, etc., and yet, as we shall see, it is possible to apply the new geometry to the solution of problems of the old.

EXERCISE 8 A

1. Prove that the circle $x^2+y^2+2x-2y=2$ meets the lines $x+2y=2$, $24x+7y=33$, $5x+3y=13$ in 2, 2 coincident, and 0 points.

2. Prove that the line $lx+my=1$ and the circle $x^2+y^2=r^2$ have 2, 2 coincident, 0 points of intersection according as $r^2(l^2+m^2) >, =, < 1$.

3. Prove that $y=k$ meets $y=x^3-3x$ in three points if $|k|<2$, in three points of which two are coincident if $|k|=2$, and in 1 point if $|k|>2$.

In Nos. 4–8, find the number of points of intersection for different values of k.

4. $x=k$, $x^2/a^2+y^2/b^2=1$.   **5.** $x=k$, $x^2/a^2-y^2/b^2=1$.

6. $y=2x+k$, $9x^2+4y^2=36$.   **7.** $6x-y=k^2$, $x^3=y$.

8. $x=k-1$, $y^4=2y^2+x$.   **9.** $y^2=x-k$, $y^2=x^3$.

10. Show that certain lines meet $(x^2+y^2)^2=x^2-y^2$ in four points, others in two points, and others not at all.

8·2. It is convenient to assign numbers to the various geometries that are introduced in this chapter. These numbers are useful mainly in the preliminary exposition. At a later stage, it is frequently not specified which geometry is in use, because it is obvious from the context, but even then it is sometimes desirable to avoid confusion by being explicit in this matter. Similar confusion is apt to arise between real and complex algebra.

Geometry number 1 or G_1 is the geometry of Euclid, and G_2 is the geometry of graphs due to Descartes in the simple form of a preliminary course of coordinate geometry.

We now introduce our first abstract geometry G_3. This is the same as the geometry called "common cartesian geometry" by Hardy in *P.M.* Appendix III.

8·3. Common Cartesian Geometry, G_3

In abstract geometry the entities point, line, ... are discarded so far as their original meanings are concerned. The fundamental elements of G_3 are ordered pairs of real numbers (x, y). The significance of the word ordered is that (y, x) is distinct from (x, y) unless $x = y$.

The nomenclature of G_1 and G_2 is retained and the elements (x, y) of G_3 are called points. Thus the first definition of G_3 is:

An ordered pair of real numbers is called a *point*. In G_2 points are represented by coordinates and the convention is made that (x, y) means the point whose coordinates are x, y. In G_3 (x, y) still denotes a point, but the point *is* the pair of numbers. Points are no longer represented by coordinates, because the points *are* the coordinates.

The aggregate of points (x_1, y_1) which are such that $\{x = x_1, y = y_1\}$ satisfies an equation $f(x, y) = 0$ is called a *locus*, and $f(x, y) = 0$ is called the equation of the locus. Briefly we refer to the locus $f(x, y) = 0$.

A locus $ax + by + c = 0$, where a, b, c are numbers and a, b are not both zero, is called a *line*. The line is also called $[a/c, b/c]$ when $c \neq 0$. The convention about $[a, b]$ in 0·2 thus becomes unnecessary.

The ideas involved in such phrases as "lie on", "meet" or "intersect", "pass through", etc. are introduced into G_3 by such definitions as the following:

If $f(x_1, y_1) = 0$, the point (x_1, y_1) is said to *lie on* the locus $f(x, y) = 0$, or the locus is said to *pass through* the point.

Two loci $f(x, y) = 0$ and $g(x, y) = 0$ are said to *meet* or *intersect* at (x_1, y_1) if $f(x_1, y_1) = 0 = g(x_1, y_1)$.

In general the language of G_1, G_2 is adopted for G_3. In G_3 a *circle* means a locus $x^2 + y^2 + 2gx + 2fy + c = 0$; the square of the *distance* $P_1 P_2$ is defined by $d^2 = (x_1 - x_2)^2 + (y_1 - y_2)^2$; the *area* of a triangle (x_1, y_1) (x_2, y_2) (x_3, y_3) is defined by

$$\pm 2\triangle = \begin{vmatrix} x_1 & y_1 & 1 \\ x_2 & y_2 & 1 \\ x_3 & y_3 & 1 \end{vmatrix}$$

or by means of the formula $\sqrt{\{s(s-a)(s-b)(s-c)\}}$ combined with the definition of distance.

Two lines $a_1 x + b_1 y + c_1 = 0$ and $a_2 x + b_2 y + c_2 = 0$ are said to be *parallel* if $a_1 : b_1 = a_2 : b_2$ and *perpendicular* if $a_1 a_2 + b_1 b_2 = 0$.

A conic is defined to be a locus

and so on. $$ax^2 + 2hxy + by^2 + 2gx + 2fy + c = 0,$$

We shall not develop G_3 in detail. The essential point is that, by the method that has been indicated, it is possible to create a new geometry which will be for the most part in exact correspondence with G_2. The correspondence is only incomplete in that certain results which are regarded as elementary theorems requiring proofs in G_2 are taken as definitions or axioms in G_3. The later developments correspond exactly.

The reader may suspect that G_3 is useless. And so it is. It is introduced here as a stepping-stone from G_2 to G_4 and G_5, and to emphasise the important fact that it is possible to create an abstract geometry.

8·4. Complex Cartesian Geometry, G_4

8·41. In this geometry a point is defined to be a pair of complex numbers instead of real numbers as in G_3. The definitions in G_4 of locus, line, meet, lie on, distance, perpendicularity, conic, etc. are word for word the same as in G_3. The possibility of the above definition of point depends upon the abstract nature of the geometry.

8·42. Consider in G_4 the interpretation of the examples in 8·12, 8·13, 8·14. The circle and line of 8·12 meet in two points $(-1, 3)$, $(\frac{9}{5}, -\frac{13}{5})$ as before. In 8·13, as before, the line is a tangent and meets the circle in two coincident points $(3, 1)$. In 8·14, $(x+4)^2 = -2$ now gives $x = -4 \pm i\sqrt{2}$; hence $y = 2 \pm 2i\sqrt{2}$. Thus the circle and line meet in the two points

$$(-4+i\sqrt{2},\ 2+2i\sqrt{2}), \quad (-4-i\sqrt{2},\ 2-2i\sqrt{2}).$$

In G_3 some lines meet a circle in two points and others do not meet it. The examples that have been given illustrate the way in which the distinction usually disappears when G_3 is replaced by G_4. The solution of $ax+by+c = 0$ and $x^2+y^2 = r^2$ only fails to give two points when $a^2+b^2 = 0$. See Exercise 8 B, Nos. 1, 2. Apart from this exception, which cannot arise when the coefficients in the equations are x-axal numbers, a line and a circle in G_4 have two common points.

8·43. From the definition, $d^2 = (x_1 - x_2)^2 + (y_1 - y_2)^2$, of distance it follows that if two points coincide the distance between them is zero.

But the converse is not true. For the equation $d^2 = 0$ is equivalent to

$$\{(x_1 - x_2) + i(y_1 - y_2)\} \{(x_1 - x_2) - i(y_1 - y_2)\} = 0$$

and is true if (x_2, y_2) lies on either of the lines

$$x_1 - x + i(y_1 - y) = 0, \qquad x_1 - x - i(y_1 - y) = 0.$$

Hence the distance between two points is zero if and only if the points coincide or lie on a line $x \pm iy = c$.

8·44. The condition of perpendicularity for

$$a_1 x + b_1 y + c_1 = 0 \quad \text{and} \quad a_2 x + b_2 y + c_2 = 0$$

is $a_1 a_2 + b_1 b_2 = 0$. Hence $ax + by + c = 0$ is perpendicular to itself if $a^2 + b^2 = 0$, i.e. if $a = \pm ib$. Such a line is called *isotropic*.

An isotropic line has an equation of the form $x \pm iy = c$. Through any point there pass two isotropic lines. The distance between two points on an isotropic line is zero.

8·45. In G_4 as opposed to G_2 or G_3 there is no distinction of type between such curves as $x^2 + y^2 = r^2$ (circle) and $x^2 - y^2 = r^2$ (rectangular hyperbola) or between $x^2/a^2 + y^2/b^2 = 1$ (ellipse) and $x^2/a^2 - y^2/b^2 = 1$ (hyperbola). The classification of conics is simpler. The parabola remains a distinct type of conic.

The distinction between an ordinary double point and an isolated point, such as the origin on the curves

$$(x^2 + y^2)^2 = a^2 x^2 \pm b^2 y^2$$

in G_3, also disappears in G_4.

It is not worth while pursuing these matters at this stage because further simplifications are introduced in G_5 and G_6.

EXERCISE 8B

1. Find the points of intersection in G_4 of the line $ax + by + c = 0$ and the circle $x^2 + y^2 = 1$, assuming $a^2 + b^2 \neq 0$. What is the condition for the points to coincide?

2. Find the point of intersection in G_4 of $x + iy = c$ and $x^2 + y^2 = r^2$.

In Nos. 3–9, find the points of intersection in G_3 and in G_4 of the pair of loci.

3. $x^2 + y^2 = 3$, $x = 2$.   **4.** $x^2 + y^2 = 25$, $x + y = 7$.

5. $y^2 = 8x$, $x^2 = y$.   **6.** $x^3 = y$, $x + y = 2$.

7. $2(x^3 + y^3) + 7xy = 0$, $x - y = 3$.

8. $(x^2 + y^2)^2 = x^2 - y^2$, $y = i$.   **9.** $y = x^2$, $y = 2px - q$.

10. Verify that the lines joining $(7, 4i)$ to $(-2, -5i)$ and $(5 + i, 1)$ to $(3, 2i)$ are isotropic and find their point of intersection.

11. Write down the equations of the isotropic lines through $(3 + 2i, 4 - 5i)$.

12. Verify that $(c, 0)$, $(-c, 0)$, $(0, ci)$, $(0, -ci)$ are the vertices of a rhombus. What is the length of its side?

13. Prove that in G_3 there are two distinct intersections of $y - b = t(x - a)$ and $y^2 = x$ if (i) $b^2 < a$, $t \neq 0$, (ii) $b^2 = a$, $t \neq 0$, $t \neq 1/2b$, (iii) $0 < a < b^2$, $t \neq 0$, t not between $\{b \pm \sqrt{(b^2 - a)}\}/2a$, or (iv) $a < 0$, $t \neq 0$, t between $\{b \pm \sqrt{(b^2 - a)}\}/2a$.

Also prove that there are two coincident intersections if $b^2 = a$, $t = 1/2b$, or $b > a$, $t = \{b \pm \sqrt{(b^2 - a)}\}/2a$, and one intersection only if $t = 0$, and no intersections if $0 < a < b^2$, t between $\{b \pm \sqrt{(b^2 - a)}\}/2a$, or $a < 0$, t not between $\{b \pm \sqrt{(b^2 - a)}\}/2a$.

State the corresponding results for G_4.

14. Show that in G_4 the condition for the circle

$$x^2 + y^2 + 2gx + 2fy + c = 0$$

to degenerate into a line-pair is that its radius should be zero.

8·5. Real Homogeneous Cartesian Geometry, G_5

8·51. We now return to consider another kind of exception that arises in the real geometries G_2 and G_3.

Two lines $a_1 x + b_1 y + c_1 = 0$ and $a_2 x + b_2 y + c_2 = 0$ usually have a point of intersection. There is an exception if $a_1 : b_1 = a_2 : b_2$, when the lines are parallel.

Consider, for example, the lines $y = 1$, $y = tx$. They meet at $(1/t, 1)$. If $t \to 0$, $y = tx$ tends to a limit $y = 0$, but the point $(1/t, 1)$ does not tend to a limit; it moves farther and farther away on the line $y = 1$. Thus there is an irregularity in G_2 and G_3 of quite a different kind from that illustrated in 8·14.

Another example is given by $y = ax + b$ and $y = x^2$. These will always meet in two points, possibly coincident, if we work in G_4 instead of G_3, but the line $x = c$ only meets $y = x^2$ in one point even in G_4. $x = c$ can be obtained as the limit of $y = ax + b$ when a, b increase without limit in such a way that $b = -ac$.

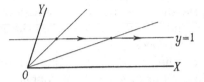

The points of intersection of $y = a(x - c)$ with $y = x^2$ are given by $2x = a \pm \sqrt{(a^2 - 4ac)}$. One of these points tends to (c, c^2) and the other tends to no limit, but moves farther and farther away on the parabola.

If we assign to every point (x, y) a third coordinate z which is always equal to 1 and call the point (x, y, z), and if we use only homogeneous equations such as

$$ax + by + cz = 0 \quad \text{instead of} \quad ax + by + c = 0,$$

$$x^2 + y^2 + 2gxz + 2fyz + cz^2 = 0$$

instead of $x^2 + y^2 + 2gx + 2fy + c = 0,$

we shall make no effective difference to the geometry. It will be unnecessary to give the actual values of the coordinates x, y, z of a point and will be sufficient to give their ratios. For $(\kappa x_1, \kappa y_1, \kappa z_1)$ satisfies every homogeneous equation that is satisfied by (x_1, y_1, z_1). Hence the point $(\frac{1}{2}, \frac{1}{3}, 1)$ may be called the point $(3, 2, 6)$ or $(3\kappa, 2\kappa, 6\kappa)$ where κ has any value except zero. A point in this geometry would be a triplet $(x, y, 1)$ or $(\kappa x, \kappa y, \kappa)$ where $\kappa \neq 0$. This geometry would not be effectively different from G_3.

In constructing G_5, a point is a triplet (x, y, z) and z is allowed to take the value zero.

Thus in G_5 an ordered triplet of real numbers (x, y, z) which are not all zero is called a *point*.

Points (x, y, z), $(\kappa x, \kappa y, \kappa z)$ where $\kappa \neq 0$ are equivalent.

There are certain points in G_5, namely those for which $z \neq 0$, which are in exact correspondence with the points of G_3. (x, y, z) corresponds to $(x/z, y/z)$. But there are also the special points for which $z = 0$, and these correspond to no points in G_3.

A point (x, y, z) for which $z = 0$ is called a *point at infinity*. The aggregate of points (x_1, y_1, z_1) which are such that $x = x_1$, $y = y_1$, $z = z_1$ satisfies a homogeneous equation $H(x, y, z) = 0$ is called a *locus* and $H(x, y, z) = 0$ is called the equation of the locus, or briefly the *locus*.

Definitions of "lie on", "meet", "pass through", etc. are the same as in G_3 and in general the language of the earlier geometries is adopted for G_5.

8·52. A locus $ax + by + cz = 0$ is called a *line*. Every line contains a point at infinity $(-b, a, 0)$.

One particular line $z = 0$ contains all points at infinity and this line is called the *line at infinity*.

The square of the *distance* from (x_1, y_1, z_1) to (x_2, y_2, z_2) is defined to be $(x_1/z_1 - x_2/z_2)^2 + (y_1/z_1 - y_2/z_2)^2$ and has no meaning when either of the points is at infinity. When neither point is at infinity the distance is the same as the distance between the corresponding points in G_3.

Lines $a_1 x + b_1 y + c_1 = 0$ and $a_2 x + b_2 y + c_2 = 0$ are called *parallel* if $a_1 : b_1 = a_2 : b_2$ and *perpendicular* if $a_1 a_2 + b_1 b_2 = 0$.

Any line parallel to $ax + by = 0$ is $ax + by + \lambda z = 0$. The set of parallel lines given by this equation when λ varies all meet in the point $(b, -a, 0)$ at infinity. Thus parallel lines meet at infinity, and a set of parallel lines is a special case of concurrent lines.

We saw in 8·51 that the point of intersection of $y = 1$ and $y = tx$ in G_3 has no limit when $t \to 0$. The corresponding lines in G_5 are $y = z$ and $y = tx$. These meet at $(1, t, t)$. When $t \to 0$, $y = tx$ tends to $y = 0$ which is parallel to $y = z$, and the point of intersection $(1, t, t)$ tends to the point at infinity $(1, 0, 0)$ on $y = 0$.

8·53. A *conic* in G_5 is a locus

$$ax^2 + 2hxy + by^2 + 2gxz + 2fyz + cz^2 = 0.$$

It meets the line at infinity in points given by

$$ax^2 + 2hxy + by^2 = 0, \qquad z = 0,$$

and conics are classified according to the nature of these points.

If $ab < h^2$, the conic has two distinct points at infinity and is called a *hyperbola*. A hyperbola can degenerate into a line-pair.

If $ab = h^2$, the points at infinity are coincident and the conic is called a *parabola*. For example, $y^2 = 4axz$ and $yz = x^2$ are parabolas. The line $x = cz$ meets the parabola $yz = x^2$ in $(c, c^2, 1)$ and $(0, 1, 0)$, of which the second is at infinity. This should be compared with the corresponding example in G_3. See 8·51. A parabola can degenerate into two parallel or two coincident lines.

If $ab > h^2$, the conic has no points at infinity and is called an *ellipse*. An ellipse can degenerate into one point or into nothing; this is illustrated by the examples $x^2 + y^2 = 0$ and $x^2 + y^2 + z^2 = 0$.

The special type of ellipse given by

$$k(x^2 + y^2) + 2gxz + 2fyz + cz^2 = 0$$

is called a *circle*. Unless $k = 0$, the equation of the circle can be written

$$(x/z + g/k)^2 + (y/z + f/k)^2 = (g^2 + f^2 - ck)/k^2$$

and so, by analogy with G_2, if $g^2 + f^2 - ck \geqslant 0$, $(-g/k, -f/k, 1)$ is called the centre and $\sqrt{\{(g^2 + f^2 - ck)/k^2\}}$ is called the radius of the circle. When $k = 0$, the circle degenerates into two lines of which one is the line at infinity. When $g^2 + f^2 < ck$, the locus has no points.

8·54. *Envelope Coordinates and Equations in* G_5. In 3·8 the envelope coordinates of a line $Xx + Yy + 1 = 0$ are defined to be X, Y, and a line through the origin has no coordinates. There is one point, the origin, which has no envelope equation.

The homogeneous envelope coordinates of a line

$$Xx + Yy + Zz = 0$$

are now defined to be X, Y, Z, and every line has coordinates.

The line $[X, Y, Z]$ passes through the point (x_1, y_1, z_1) if $Xx_1 + Yy_1 + Zz_1 = 0$. Hence $Xx_1 + Yy_1 + Zz_1 = 0$ is the envelope equation of (x_1, y_1, z_1).

In particular the line $[X, Y, Z]$ passes through the origin if $Z = 0$, and $Z = 0$ is the envelope equation of the origin.

A point at infinity has coordinates $(x_1, y_1, 0)$ and its envelope equation is $x_1 X + y_1 Y = 0$.

Corresponding to the statement in 3·93 that lines $[X, Y]$ for which $LX + MY = 0$ are parallel, we now say that, in G_5, lines $[X, Y, Z]$ for which $LX + MY = 0$ pass through the point at infinity $(L, M, 0)$ and therefore $LX + MY = 0$ is the equation of that point.

8·55. *The Points at Infinity on* $ax^2 + 2hxy + by^2 = 0$. The coordinates of these two points satisfy $ax^2 + 2hxy + by^2 = 0$ and $z = 0$. Let $[X, Y, Z]$ be any line through either of them; then its equation $Xx + Yy + Zz = 0$ is satisfied by the coordinates of that point. Thus the equations

$$ax^2 + 2hxy + by^2 = 0, \qquad z = 0, \qquad Xx + Yy + Zz = 0$$

are simultaneously true. Hence

$$aY^2 - 2hXY + bX^2 = 0.$$

This is the equation of the pair of points at infinity on $ax^2 + 2hxy + by^2 = 0$. It follows that the results of 4·32 and 4·33 to the effect that

$a + b = 0$ is the condition for $ax^2 + 2hxy + by^2 = 0$ to be at right angles

$A + B = 0$ is the condition for $AX^2 + 2HXY + BY^2 = 0$ to be in perpendicular directions

are conditions for the same geometrical property.

8·56. Consider now the dual of the important process of 4·8.

pq is the point of intersection $xX + yY + zZ = 0$ of tangents p, q to the circle $a^2(X^2 + Y^2) = Z^2$. The coordinates of p satisfy both of these equations and therefore they satisfy

$$a^2(X^2 + Y^2) z^2 = (xX + yY)^2.$$

The same is true of the coordinates of q. But this equation is of the form $uX^2 + 2vXY + wY^2 = 0$ and represents a pair of points at infinity. Hence it is the equation of the pair of points at infinity on the tangents to the circle from (x_1, y_1, z_1).

More generally the equation of the pair of points at infinity on the tangents from (x_1, y_1, z_1) to

$$AX^2 + 2HXY + BY^2 + 2GXZ + 2FYZ + CZ^2 = 0$$

is

$$(AX^2 + 2HXY + BY^2)z_1^2 - 2(GX + FY)(x_1X + y_1Y)z_1$$
$$+ C(x_1X + y_1Y)^2 = 0.$$

In 4·8 the equations are not homogeneous. But the process there used of making the equation of the conic homogeneous by means of the equation of the line corresponds to the elimination of z from the homogeneous equations.

EXERCISE 8 c

[The questions in this exercise refer to G_5]

1. Find the points of intersection of $y^2 = 4axz$ with $x = bz$, $y = cz$, $z = 0$.

2. Find the points of intersection of $xy = k^2z^2$ with $y = x$, $x = z$, $z = 0$.

3. Write down the coordinates of the points at infinity on

$$y = tx, \quad OX, \quad OY, \quad ax + by + cz = 0, \quad x/a + y/b = z.$$

4. Verify that the lines $x/a \pm y/b = kz$ meet the conic

$$x^2/a^2 - y^2/b^2 = z^2$$

in two points of which one is at infinity. What happens when $k = 0$?

5. Find for $c = 1, 2, 3, 4$ the intersections of $x = cy$ and

$$(x - y)(x - 2y)(x - 3y) = (x + 4y)z^2 - 2z^3.$$

6. Write down the envelope coordinates of $4x = 5y$, OX, $ax + cz = 0$.

7. Write down the envelope equations of $(0, 0, 1)$, $(1, 0, 0)$.

In Nos. 8–10, find the envelope equations of the points at infinity.

8. On $ax^2 + 2hxy + by^2 = 0$ where $h^2 - ab = k^2$.

9. On the lines of the envelope $aY^2 = XZ$ through $(a, 0, -1)$.

10. On the lines of the envelope $2X^2 + 3YZ - 4Z^2 = 0$ through $(1, 1, 1)$.

11. Show that

$$x^2 - 3xy + 2y^2 = 2(x + y)z \quad \text{and} \quad x^2 - 5xy + 6y^2 = 2(x + 3y)z$$

meet at $(0, 0, 1)$, $(0, 1, 1)$, $(2, 0, 1)$, and a certain point at infinity.

12. Find the four points of intersection of $3(x - y)^2 = (3x + 4y)z$ and $2(x - y)^2 = (2x + 3y)z$.

8·6. Complex Homogeneous Cartesian Geometry, G_6

8·61. This geometry is constructed so as to combine the advantages of G_4 and G_5.

In G_6 an ordered triplet of complex numbers (x, y, z) which are not all zero is called a *point*.

Points (x_1, y_1, z_1), (kx_1, ky_1, kz_1) where $k \neq 0$ are equivalent.

A point (x, y, z) for which z is zero is called a *point at infinity*.

An aggregate of points (x_1, y_1, z_1) such that $x = x_1$, $y = y_1$, $z = z_1$ satisfies a homogeneous equation $H(x, y, z) = 0$ is called a *locus*. $H(x, y, z) = 0$ is called the equation of the locus, or briefly the locus.

Definitions of "lie on", "meet", "pass through", etc. are the same as in the earlier geometries.

8·62. The particular locus $ax + by + cz = 0$ is called a *line*.

The special line $z = 0$ is called the *line at infinity* because it is composed of points at infinity.

Lines $a_1 x + b_1 y + c_1 z = 0$ and $a_2 x + b_2 y + c_2 z = 0$ are called *parallel* if $a_1 : b_1 = a_2 : b_2$ and *perpendicular* if $a_1 a_2 + b_1 b_2 = 0$.

Parallel lines meet in a point at infinity.

As in G_4 the distance between two distinct points is zero if and only if the line joining them is of the form $x \pm iy = cz$, and such lines are called *isotropic*.

Homogeneous envelope coordinates and equations are defined for G_6 as in 8·54.

8·63. *Conics.* A locus $ax^2 + 2hxy + by^2 + 2gxz + 2fyz + cz^2 = 0$ of the second degree is called a *conic*.

This conic meets the line at infinity in points given by

$$ax^2 + 2hxy + by^2 = 0 \quad \text{and} \quad z = 0.$$

When $ab = h^2$, these points coincide and the conic is called a *parabola*. For example $y^2 = 4axz$ and $yz = x^2$ are parabolas.

As in G_4 there is no distinction between ellipses and hyperbolas.

Conics for which $ab \neq h^2$ are called *central conics*.

When $a + b = 0$, the lines $ax^2 + 2hxy + by^2 = 0$ joining the origin to the points at infinity on the conic are at right angles and the conic is called a *rectangular hyperbola*.

8·64. *Circles.* A locus $k(x^2+y^2)+2gxz+2fyz+cz^2 = 0$ is called a *circle*. All circles pass through the points given by $z = 0, x^2+y^2 = 0$. These are the points $(1, i, 0), (1, -i, 0)$. They are called the *circular points* at infinity and are commonly denoted by I and J.

Any isotropic line passes through I and J. The isotropic lines through K are KI and KJ.

8·65. *Degenerate Conics.* It is proved in 4·45 that

$$\delta \equiv \begin{vmatrix} a & h & g \\ h & b & f \\ g & f & c \end{vmatrix} = 0$$

is a necessary condition for $ax^2+2hxy+by^2+2gx+2fy+c$ to factorise. It is also a necessary condition for

$$ax^2+2hxy+by^2+2gxz+2fyz+cz^2$$

to factorise. But it is not a sufficient condition for the general equation to represent a line-pair in G_3, G_4, or G_5. This is proved for G_3 and G_5 by the example $x^2+y^2 = 0$ and for G_4 by $2gx+2fy+c = 0$.

It will now be proved that in G_6, if $\delta = 0$, then

$$s \equiv ax^2+2hxy+by^2+2gxz+2fyz+cz^2 = 0$$

represents a line-pair.

This is done by expressing s as the difference between two squares, or by solving $s = 0$ as a quadratic for x, y, or z.

By the usual process of completing the square

$$as = (ax+hy+gz)^2+(ab-h^2)y^2-2(gh-af)yz+(ac-g^2)z^2$$

$$= (ax+hy+gz)^2+Cy^2-2Fyz+Bz^2, \quad (0·6)$$

$$= (ax+hy+gz)^2+(y\sqrt{C} \pm z\sqrt{B})^2, \quad \text{since } BC-F^2 = a\delta = 0,$$

$$= (ax+hy+gz)^2-\{y\sqrt{(-C)}-z\sqrt{(-B)}\}^2.$$

Unless $a = 0$ this proves that s factorises. A similar method can be applied to bs or cs unless b and c are both zero.

But if $a = b = c = 0$, s reduces to $2fyz+2gxz+2hxy$ and δ reduces to $2fgh$. Hence s also factorises when $a = b = c = 0$ if $\delta = 0$.

It is therefore proved by 4·45 and the present paragraph that $\delta = 0$ is a necessary and sufficient condition for $s = 0$ to represent a line-pair in G_6. The two lines may be coincident.

A degenerate parabola is a pair of parallel or coincident lines since it meets the line at infinity in coincident points. For instance $x^2 = a^2z^2$ is a degenerate parabola.

A degenerate central conic is a pair of intersecting lines. When these lines are perpendicular, the conic is a rectangular hyperbola.

The circle $k(x^2 + y^2) + 2gxz + 2fyz + cz^2 = 0$ degenerates when

$$0 = \delta = \begin{vmatrix} k & 0 & g \\ 0 & k & f \\ g & f & c \end{vmatrix} = k(kc - g^2 - f^2).$$

Thus it degenerates in two different ways. When $k = 0$ it consists of the line at infinity and the line $2gx + 2fy + cz = 0$. When $g^2 + f^2 = ck$, the radius is zero and the equation is

$$(kx + gz)^2 + (ky + fz)^2 = 0$$

or $\qquad \{k(x + iy) + (g + if)z\}\{k(x - iy) + (g - if)z\} = 0;$

thus the circle consists of two isotropic lines through its centre. In particular $x^2 + y^2 = 0$ consists of the isotropic lines OI, OJ. The separate equations of OI, OJ are $x + iy = 0$, $x - iy = 0$. The points at infinity I, J on these are $(1, i, 0)$, $(1, -i, 0)$ and their envelope equations are $X + iY = 0$, $X - iY = 0$. Thus $X^2 + Y^2 = 0$ is the equation of the point-pair I, J.

8·7.　Generality of G_6

The gain in generality in passing to G_6 is illustrated in 8·71–8·79.

8·71. In G_6 two distinct lines $a_1x + b_1y + c_1z = 0$ and $a_2x + b_2y + c_2z = 0$ have one and only one common point $(b_1c_2 - b_2c_1, c_1a_2 - c_2a_1, a_1b_2 - a_2b_1)$. This is also true in G_5.

8·72. In G_6 a line and a conic have two common points, because their equations lead to a quadratic in the ratios of the coordinates, and a quadratic has two roots in complex algebra. This result does not hold in the earlier geometries. In G_5, the algebra is real and the quadratic need not have roots. In G_4 the quadratic is not homogeneous, and it reduces to an equation of the first degree when the coefficient of x^2 happens to be zero. Even in G_6 the two common points may be coincident.

8·73. In G_6 a line and a curve of order n have n common points, because their equations lead to a homogeneous equation of degree n in the ratios of the coordinates. This equation has n roots. See $A.A.$ vol. II, p. 254.

The statement in 7·62 about curves of orders m and n can be simplified for G_6 in the same way.

8·74. Dually a point and an envelope conic have two common lines in G_6, and an envelope point has n lines in common with a curve of class n. Also two curves of classes m, n have mn common lines.

For the validity of these statements the conventions about multiple roots are necessary.

8·75. In 5·61 a distinction is made between different types of coaxal circles. In G_6 the distinction disappears. The radical axis of two circles is the line through the common points other than I and J of the circles. A coaxal system is a set of circles through two common points besides I and J. It is a special case of a system of conics through four points.

The intermediate type of coaxal system given by $x^2 + y^2 = \kappa xz$ can be regarded in the same way with a certain convention about coincident points. When a curve K touches a line AB at A it is said to meet AB in two coincident points at A (7·33). If another curve K' touches AB at A, we agree to say that the curves K, K' have two coincident points of intersection at A. For example in the first figure the conics are said to be conics through A, A, C, D; this implies that the tangent at A is the same for both. With this understanding, $x^2 + y^2 = \kappa xz$ is a system of conics through four points O, O, I, J; all the conics have the same tangent at O.

By the same convention concentric circles $x^2 + y^2 = \kappa z^2$ are now regarded as conics through the four points I, I, J, J because they all touch the same lines at I and J.

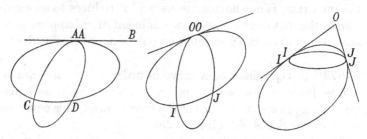

8·76. In 6·85 it is proved that the curve

$$x : y : a = t^2 - 1 : t^3 - t : t^3,$$

which is equivalent to $y^3 = a(y^2 - x^2)$, has no point of inflexion.

The corresponding curve in G_6 is $x : y : az = t^2 - 1 : t^3 - t : t^3$ or $y^3 = az(y^2 - x^2)$. The equation which gives the parameters of any inflexions is $3t + t^3 = 0$. This gives $t = 0$, $i\sqrt{3}$, $-i\sqrt{3}$.

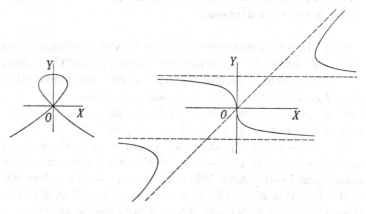

$t = 0$ gives the point $(1, 0, 0)$ at infinity on the x-axis. The other values of t give the points $(\pm 4ai\sqrt{3}, 12a, 9)$. Thus the curve has three points of inflexion.

From the condition of collinearity $t_1 + t_2 + t_3 + t_1 t_2 t_3 = 0$, it follows by the method of 6·84 that $t = \pm 1$ are the parameters of the double point, which is the origin. The G_2 diagram of this

curve is shown in the first figure. It is interesting to compare
the curve with $1:y:ax = t^2 - 1:t^3 - t:t^3$ or $y^3 = ax(y^2 - 1)$,
which is found by exchanging x and z in the corresponding
equations for G_6. This curve is easily sketched from its cartesian
equation by giving values to y. It is shown in the second
figure. In G_6, there is a double point at $(1, 0, 0)$ and there are
inflexions at $(0, 0, 1)$ and $(9, 12a, \pm 4ai\sqrt{3})$.

8·77. In 6·86, it is shown that the curve $y^2(y^2 + x) = 1$ has
no double point. The condition of collinearity for points t_1, t_2, t_3
of the curve is $t_1 t_2 t_3(t_1 + t_2 + t_3) = 1$. This cannot be an identity
in t_3 and it appears that there is
no double point even in G_6. This is
due to a defect in the parametric
representation. If the parameter is
changed to s, where $st = 1$, the condition of collinearity is

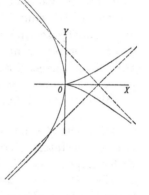

$$s_2 s_3 + s_3 s_1 + s_1 s_2 = s_1{}^2 s_2{}^2 s_3{}^3$$

and this is an identity in s_3 if
$s_1 = s_2 = 0$. Thus the point $(1, 0, 0)$
at infinity on the x-axis is a cusp.
If the parameter t is used, the cusp
is only given as a limit when $t \to \infty$.
As the point given by $t \to \infty$ is at
infinity, the defect of the parametric representation does not
appear in G_2.

The curve found from $y^2(y^2 + xz) = z^4$ by exchanging x and
z is $y^2(y^2 + xz) = x^4$. The corresponding curve in G_2 is
$y^2(y^2 + x) = x^4$. This curve has a cusp at the origin and another
branch through the origin. It should be compared with the
curve in 6·86.

8·78. In 7·36 a distinction is made between the cases $e^2 > fd$,
$e^2 < fd$ which lead to a double point and an isolated point. In
G_6 the two tangents to the curve at the point exist whether
$e^2 > fd$ or $e^2 < fd$. The process of finding parametric equations

given in 6·9 is applicable when the point is an isolated point. See 6·94; similar work applies to $y^2 = x^2(x-1)$.

The remark at the end of 6·95 is only universally valid in G_6.

8·79. EXAMPLE. Discuss the double points of $2xy^2 = 1 - 3x$ and $2xy^2 = z^3 - 3xz^2$.

The first equation may be written $x = 1/(2y^2 + 3)$ and the curve can be traced by giving values to y. It has no double point. $y = t$, $x = 1/(2t^2 + 3)$ are parametric equations.

In G_6 the curve $2xy^2 = z^3 - 3xz^2$ has a double point $(1, 0, 0)$ just as $2zy^2 = x^3 - 3zx^2$ has a double point $(0, 0, 1)$. In G_5 these points are isolated points. In G_3 there are no points at infinity; nevertheless parametric equations of $2xy^2 = 1 - 3x$ are found by solving the equation with $y = t$ which represents a line parallel to OX. This corresponds to solving with $y = tz$ in G_6, and $y = tz$ is an arbitrary line through the double point $(0, 0, 1)$.

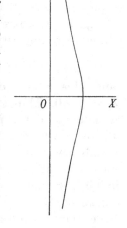

8·8. Some of the later chapters of this book can be interpreted in different ways, with reference to G_6 or to earlier geometries. If they are taken to refer to G_3, which will not always be possible, they will be independent of the present chapter. Certain parts of the book are valid only in G_6.

It is not always necessary to state explicitly which geometry is being assumed. It may be irrelevant or it may be sufficiently clear from the context. For example, in the statement

$y = k$ meets $(x^2 + y^2)^2 = x^2 - y^2$ in exactly four points if and
  only if $0 < k^2 < \frac{1}{8}$,

it is G_3 that is implied. For the geometry is not homogeneous and conditions such as $0 < k^2 < \frac{1}{8}$ are meaningless in complex algebra.

Again, in the proposition

$$x^2/(a^2 - \lambda) + y^2/(b^2 - \lambda) = 1 \qquad (a^2 > b^2.)$$

is an ellipse, hyperbola, or nothing according as

$$\lambda < b^2, \quad b^2 < \lambda < a^2, \quad \text{or} \quad a^2 < \lambda,$$

the geometry is G_3. In proving the proposition G_5 may be used, so that an ellipse and hyperbola can be distinguished by a consideration of the points at infinity on the curve. G_6 would not be used: there is no distinction between an ellipse and a hyperbola in G_6.

A justification of the transition from one geometry to another is attempted in the following section.

8·9. Applications of G_6

8·91. The manner in which G_6 can be used to prove results in G_3 will become clearer when examples arise.

The elements of G_6 are points $(a + a'i, b + b'i, c + c'i)$. Let those points for which $a' = b' = c' = 0$, $c \neq 0$ be called *special* points. A special point (a, b, c) may be said to correspond to the point $(a/c, b/c)$ in G_2 or G_3. There is an exact correspondence between all the points of G_3 and some of the points of G_6, namely the special points, so that to any theorem of G_3 there corresponds a theorem of G_6 in which only the special points are involved. An ordinary theorem of G_1 may be expressed analytically in terms of coordinates (x, y) in G_2 or G_3. There will be a corresponding theorem of G_6 concerned with special points only. These theorems will be both true or both false. But it may happen that the theorem of G_6 is most conveniently proved with the help of some construction involving points of G_6 other than the special points. Even so, once the theorem is proved, it is known that the corresponding result in G_3 is true. Thus the greater resources of G_6 are available for proving results of G_3 and hence of G_2 and G_1.

Both in G_3 and G_6 the results are purely algebraic and diagrams are theoretically out of place. The G_1 or G_2 diagram may however help the imagination in dealing with the corresponding theorem in G_3. If a theorem of G_6 only involves special points, the G_1 or G_2 diagram may also help in G_6. Theorems and proofs in G_6 will, however, in general involve points which do not correspond to any points in G_3, and it is not then possible to obtain a complete diagram. But it is often helpful to represent the facts diagrammatically as far as possible. For this purpose general points are represented as if they were ordinary points of G_1 or G_2. For example, in 8·75 diagrams are used to represent the facts about $x^2 + y^2 = \kappa xz$ and $x^2 + y^2 = \kappa z^2$.

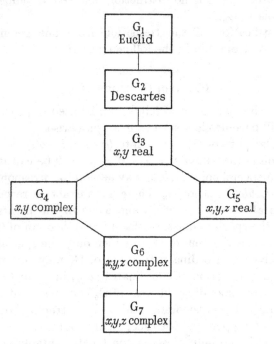

8·92. The application of G_4 or G_5 to a problem of G_3 is justified by arguments similar to those in 8·91. It is useful sometimes to work in G_5 rather than G_6. For example the conic $ax^2 + 2hxy + by^2 = 1$ in G_3 is a hyperbola if $ab < h^2$, and this is the same as the condition for the corresponding conic $ax^2 + 2hxy + by^2 = z^2$ in G_5 to possess two distinct points at infinity.

8·93. The application of complex geometry to obtain results of real geometry may be compared with the application of complex numbers to real algebra. See *A.T.* Chapter VII.

8·94. The special properties of the z-coordinate and of the points for which $z = 0$ make the geometry G_6 unsymmetrical.

In the ideal geometry G_7 a point is a triplet of complex numbers (x, y, z) and there is nothing peculiar about the points for which $z = 0$.

8·95. Abstract geometries of n dimensions can be constructed. In three dimensions, if the geometry is to be homogeneous, a point is defined to be a set of four numbers (x, y, z, t) and points such that $X_1 x + Y_1 y + Z_1 z + T_1 t = 0$ are said to form a plane $[X_1, Y_1, Z_1, T_1]$.

(x, y, z, t) may also denote a point in non-homogeneous geometry of four dimensions. The distance between two such points may be defined by

$$d^2 = (x_1 - x_2)^2 + (y_1 - y_2)^2 + (z_1 - z_2)^2 + (t_1 - t_2)^2.$$

In homogeneous geometry of four dimensions a point is a set of five numbers, and so on.

EXERCISE 8D

In Nos. 1, 2, find the points at infinity on the curve. (G_6).

1. $x^2/a^2 + y^2/b^2 = z^2$.   **2.** $x(x^2 + y^2) = y^2 z$.

3. For what values of m do $y = mx$ and $x^2/a^2 + y^2/b^2 = z^2$ meet at infinity, and what are the envelope coordinates of the tangents at infinity to the curve? (G_6).

In Nos. 4–8, find the common points.

4. $x^2 + y^2 = 3$, $x = k$. (G_4).

5. $y = x^3$, $x + y + 2 = 0$. (G_3, G_4).

6. $yz^2 = x^3$, $x = z$. (G_5, G_6).

7. $x^3 = y^3$, $x = 1$, (G_3, G_4); $x^3 = y^3$, $x = z$, (G_5, G_6).

8. $x^2 = 27y$, $y^2 = x$, (G_3, G_4); $x^2 = 27yz$, $y^2 = xz$, (G_5, G_6).

In Nos. 9, 10, discuss the intersections for various values of k.

9. $x = k$, $4x^2 + 9y^2 = 36$. (G_3, G_4).

10. $x - y = k$, $x^2 - y^2 = a^2$, (G_3); $x - y = kz$, $x^2 - y^2 = a^2z^2$, (G_5).

In Nos. 11–13, discuss the intersections of the curve with the line $x = at$, $y = bt$ in G_6. Are the results the same in G_5?

11. $x^2 - 4y^2 = (x + y)z$. Compare with Exercise 7A, No. 4.

12. $x^3 = (x - y)z^2$. Compare with Exercise 7A, No. 5.

13. $yz(y^2 - x^2) = x^4$. Compare with Exercise 7A, No. 10.

14. Find the intersections of OX with $x^4 + y^5 = x^2y$, (G_3), and with $x^4z + y^5 = x^2yz^2$, (G_5).

In Nos. 15–17, find the envelope equations of the points at infinity on the lines.

15. Of $k^2XY = Z^2$ through $(0, 0, 1)$.

16. Of $a^2X^2 + b^2Y^2 = Z^2$ through $(0, 0, 1)$.

17. Of $aY^2 = XZ$ through $(a, b, 1)$.

18. Find the equation of the points at infinity on the tangents from (x_1, y_1, z_1) to $xy = k^2z^2$.

19. Verify that the points at infinity on $r^2 = a^2 \cos 2\theta$ lie on the isotropic lines through the origin.

20. Find where the inverse of $x^2/a^2 - y^2/b^2 = z^2$ wo $x^2 + y^2 = k^2z^2$ meets the line at infinity.

21. Give the general equation of the cubic curve through I, J touching the axes at the origin.

22. Give the general equation of a parabola that is a rectangular hyperbola. Where does it meet the line at infinity?

23. Sixteen points in space of four dimensions are given by

$$x^2 = y^2 = z^2 = t^2 = 1.$$

Of the 120 distances between pairs of these points, how many are of length 2, and what are the lengths of the others?

Chapter 9

CONICAL PROJECTION

9·1. Projection in G_1

9·11. If V is a fixed point and α, α' are two planes (not through V) meeting in a line l, then the *conical projection* of a point P of α on to α' is the point P' where VP meets α'. V is called the *vertex of projection*. If P describes a figure F in α, P' describes another figure F' in α', and F' is called the conical projection of F. The projecting lines VPP' form a cone in the general sense of a set of concurrent lines in space, and this accounts for the name conical projection.

9·12. When the planes α, α' are parallel, the figures F, F' are similar. Some properties of a conical projection in which α, α' are not parallel will now be investigated.

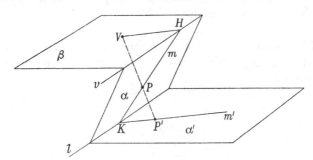

9·13. The plane β through V parallel to α' meets α in a line v which is parallel to l. If P lies in v, VP does not meet the plane α' and so P has no projection on α'. Every point P of α that is not on v has a projection. Points P in l coincide with their projections.

v is called the *vanishing line* and l is called the *axis of projection*.

9·14. The projection of a line m other than v is a line. For let m meet v and l in H and K, and let P' be the projection of an arbitrary point P in m. Then P' lies in the plane of m and V, and also lies in the plane α'; therefore it lies in the line of intersection of these two planes. Since the plane of m and V meets α and α' in parallel lines, the projection of m is parallel to VH. See also Exercise 9 A, No. 1.

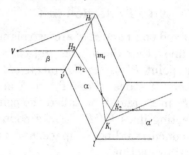

9·15. If two intersecting lines m_1, m_2 in α meet v in H_1, H_2, and meet l in K_1, K_2, their projections on α' are the lines through K_1, K_2 parallel to VH_1, VH_2 (by 9·13). Therefore the angle between the projections of $m_1 m_2$ is equal to the angle $H_1 V H_2$. See also Exercise 9 A, No. 2.

9·16. EXAMPLE. Show how to choose a conical projection so that four given points in a plane α may project into the corners of a square.

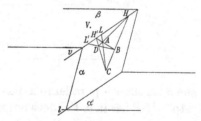

Let A, B, C, D be the points and let DA, CB meet at H and BA, CD at H'. Take HH' to be v, and any line in α parallel to HH' to be l. Let two parallel planes through l, v be taken as β, α'. With any point V in β as vertex of projection, DA, CB will project into lines in α' parallel to VH and BA, CD will project into lines parallel to VH'. Thus the projection $A'B'C'D'$ will be a parallelogram. If V is chosen on the circle in the plane β on HH' as diameter, AB, AD will project into lines parallel to VH, VH', i.e. into perpendicular lines. Thus $A'B'C'D'$ will be a rectangle. Let AC, BD meet v in L, L'. If V is chosen at a point of intersection of the circles on HH' and LL' as diameters, $A'B'C'D'$ will be a rectangle with perpendicular diagonals, i.e. a square.

9·2. Formulae for Projection

9·21. In the figure a plane γ is drawn through V perpendicular to v and l, meeting α in OX and α' in OX'. P is any point in α and XP meets l at M'. The projection P' is found as the point of intersection of VP and the line through M' parallel to VX.

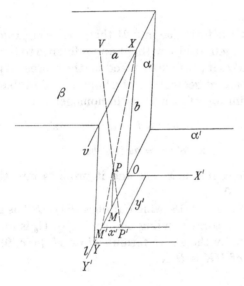

With OX and l as axes in α let P be (x, y), and with OX' and l as axes in α' let P' be (x', y'). Also let $VX = a$ and $XO = b$. Then from the figure, by similar triangles,

$$\frac{x}{b} = \frac{MP}{OX} = \frac{M'P}{M'X} = \frac{x'}{x'+a}$$

and
$$\frac{x}{b} = \frac{MP}{OX} = \frac{M'M}{M'O} = \frac{y'-y}{y'}.$$

$$\therefore \quad x = \frac{bx'}{x'+a}, \qquad y = \frac{ay'}{x'+a},$$

$$x' = \frac{ax}{b-x}, \qquad y' = \frac{by}{b-x}.$$

These equations determine P or P' when P' or P is given. By means of them a projection can be carried out analytically.

The denominators $b-x$ correspond to the fact that a point on the vanishing line $x = b$ has no projection.

If the roles of a and a' are exchanged, the vanishing line is $x'+a = 0$, and this explains the occurrence of the denominator $x'+a$.

9·22. By the formulae in 9·21 the process of projection can be defined algebraically without any reference to the figures of G_1. This makes it possible to introduce the process of projection into the abstract geometries of Chapter 8. Suitable formulae for the definition of projection in homogeneous geometry are

$$x:y:z = bx':ay':x'+az',$$

$$x':y':z' = ax:by:bz-x.$$

The vanishing line is $x = bz$ and it projects into the line at infinity $z' = 0$.

In the figure of 9·13, which applies to G_1, H has no projection. But the corresponding point in G_5 or G_6 is $(b, c, 1)$ and this projects by the above formulae into H' $(a, c, 0)$; also the projection of HK is $H'K$.

9·23. A homogeneous equation of degree n in x, y, z is transformed by the formulae of 9·22 into a homogeneous equation of the same degree in x', y', z'. Hence a curve of order n projects into a curve of order n. Partly on account of this property, conical projection is one of the most important methods of deriving one figure from another.

9·3. Transformation

9·31. The transformation in 9·22 is a special case of the general linear transformation

$$x:y:z = a_{11}x' + a_{12}y' + a_{13}z' : a_{21}x' + a_{22}y' + a_{23}z'$$
$$: a_{31}x' + a_{32}y' + a_{33}z',$$

in which the determinant
$\begin{vmatrix} a_{11} & a_{12} & a_{13} \\ a_{21} & a_{22} & a_{23} \\ a_{31} & a_{32} & a_{33} \end{vmatrix}$
formed by the

coefficients is not zero. And since it is not zero, the equations can be solved for $x':y':z'$. This gives the reverse transformation (*A.A.* vol. III, Chapter XVII)

$$x':y':z' = A_{11}x + A_{21}y + A_{31}z : A_{12}x + A_{22}y + A_{32}z$$
$$: A_{13}x + A_{23}y + A_{33}z.$$

These equations show that the degree of an equation and the order of a curve are unaltered by the general linear transformation. A special case of this result is proved in 9·23.

9·32. The method of orthogonal projection is explained in 6·3. The formulae for orthogonal projection in homogeneous geometry are

$$x:y:z = kx':y':z'.$$

This also is a special case of the general linear transformation. It is not a special case of conical projection, but is the limiting case obtained from it by putting $b = ka$ and making $a \to \infty$ in the formulae of 9·22.

9·33. The formulae obtained in 2·54 for change of axes, when adapted for homogeneous geometry, become

$$x:y:z = ax' + by' + ez' : cx' + dy' + fz' : z'.$$

The determinant of the coefficients is $ad - bc$ and is not zero. This is another case of the general transformation in 9·31.

It was pointed out in Chapter 2 that the equations could be used in two different ways: to give the coordinates of an arbitrary point P when referred to new axes; and to define the position of a corresponding point P' referred to the original axes. The general transformation can also be used in these two ways.

9·4. Projective and Metrical Geometry

In conical projection, points $P, Q, ...$ become points P', $Q', ...,$ and lines $p, q, ...$ become lines $p', q', ...$ Collinear points become collinear points and concurrent lines become concurrent lines. Points and lines are therefore said to be *projective* elements and collinearity and concurrence are called projective properties. (To avoid exceptions in which concurrent lines become parallel lines, G_5 or G_6 must be used.)

Usually the projection $P'Q'$ of a segment PQ is not equal to PQ, the mid-point of PQ does not project into the mid-point of $P'Q'$, a circle does not project into a circle and a right angle does not project into a right angle. Properties and elements like this, which are concerned with measurement, are not preserved in projection. They are called *metrical* properties or elements.

9·5. Conic Sections

9·51. The projection of a circle is a curve of order 2. In a later chapter it is found that the only non-degenerate curves of order 2 are the ellipse, parabola, and hyperbola; the circle is regarded as a special case of the ellipse. These curves were first investigated by the Greeks as sections of right circular cones.

At first the three types were obtained as sections of acute-angled, right-angled, and obtuse-angled cones by planes perpendicular to a generating line. Later it became customary to obtain all three curves from one cone by taking different sections of it. In 14·44 and 15·44 it is shown how this can be done.

Any section of a circular cone is the projection of a circle, and a circle is a curve of order 2. Hence by 9·23 all sections of any circular cone are curves of order 2. No new curves are obtained by taking sections of an oblique cone instead of a right circular cone.

9·52. So far as projective properties are concerned all curves of order 2 are equivalent. Properties of conics might be deduced from properties of the circle. But in fact it is the tendency of modern geometry to make the metrical properties appear as special cases of the projective properties rather than to derive the projective ones from the metrical. This is because the axioms of projective geometry are essentially simpler than those of metrical geometry.

EXERCISE 9A

1. If m is a line in α parallel to v, prove that its projection on α' is parallel to v.

2. If m is a line in α parallel to v, and n is a line in α meeting v at N, prove that the angle between the projections of m and n is equal to the angle between VN and v.

3. Show how to project two lines which cut at an angle θ into two lines which cut at the same angle θ.

4. Show that the projection of collinear points in the order $ABCD$ can be in any of the orders $A'B'C'D'$, $B'C'D'A'$, $C'D'A'B'$, $D'A'B'C'$.

5. Show how to project a given triangle into an equilateral triangle.

In Nos. 6–18, a and b have the meanings of 9·2. In Nos. 6–11, use the formulae of 9·21 and 9·22 to find the projections from α to α' of:

6. $(c, 0)$, $(c, 0, 1)$.   **7.** (c, d), $(c, d, 1)$.

8. $y = \lambda(x - c)$, $y = \lambda(x - cz)$.

9. $(y - d) = \lambda(x - b)$, $(y - dz) = \lambda(x - bz)$.

10. $x^2 + y^2 = b^2 - c^2$, $x^2 + y^2 = b^2 + c^2$.

11. $x^2 + y^2 = bx + k$, for $k = b^2$, 0, $-\tfrac{1}{8}b^2$.

12. In the figure of 9·2, if C is a given point in OX, find a point D in OX such that $C'D' = CD$.

13. Use the formulae of 9·21 to find the projections of $y = x^3$ and $y^2 = x^3$. Sketch the projections.

14. Use the formulae of 9·22 to find the projections of $yz^2 = x^3$ and $zy^2 = x^3$.

15. Use the formulae of 9·21 and 9·22 to find the projections of

$$a_1 x^2 + 2h_1 xy + b_1 y^2 + 2g_1 x + 2f_1 y + c_1 = 0$$

and

$$a_2 x^2 + 2h_2 xy + b_2 y^2 + 2g_2 xz + 2f_2 yz + c_2 z^2 = 0.$$

16. What points of α' are not the projections of any points of α in G_3? Of what points are the corresponding points in G_5 the projections?

17. What lines of α project into the concurrent lines

$$y' + cz' = k(x' + az')$$

where k varies?

18. Show that the cone with vertex V and base $x^2 + y^2 = r^2$ in the figure of 9·2 is a right circular cone if $b = a \cos \theta$. If this condition holds and $r = b$, prove that the section of the cone by α' is $y'^2 = (b^2/a)(2x' + a)$ and that for other values of r it is of the form $A^2(x' - k)^2 + B^2 y'^2 = 1$ or $-A^2(x' - k)^2 + B^2 y'^2 = 1$ according as $r^2 <$ or $> a^2 \cos^2 \theta$.

Chapter 10

CROSS-RATIO

10·1. Cross-ratio in G_3

10·11. Not only a length AB but also the ratio $AP:PB$ of two lengths in the same line is in general altered by conical projection. But it will be found that the ratio $(AP:PB):(AQ:QB)$ of two ratios of lengths in the same line is not altered.

10·12. The *cross-ratio* of four numbers x_1, x_2, x_3, x_4 is defined to be $\dfrac{(x_1 - x_2)(x_3 - x_4)}{(x_1 - x_4)(x_3 - x_2)}$ and is denoted by $(x_1 x_2 x_3 x_4)$.

10·13. Points that lie on a fixed line l are said to form a *range*. A point P of a range has one degree of freedom. It is determined by one coordinate or parameter.

The cross-ratio of four points A, B, C, D of a range is defined to be the cross-ratio of their one-dimensional coordinates x_1, x_2, x_3, x_4 which are their distances from a fixed point O of the line l. Thus the cross-ratio

$$= \frac{(x_1 - x_2)(x_3 - x_4)}{(x_1 - x_4)(x_3 - x_2)} = \frac{BA.DC}{DA.BC} = \frac{AB.CD}{AD.CB}.$$

This is denoted by $(ABCD)$. It is independent of the particular position of the fixed point O on l.

10·14. Since $(x_1 - x_2)(x_3 - x_4) = \kappa(x_1 - x_4)(x_3 - x_2)$ is linear in each of x_1, x_2, x_3, x_4, it follows that if three of these numbers are given, the fourth is in general uniquely determined by a given value κ of the cross-ratio.

Hence if three of the points A, B, C, D and the value of $(ABCD)$ are given, the position of the fourth point is determinate.

Exceptions arise when certain of the numbers x_1, x_2, x_3, x_4 are equal or certain of the points A, B, C, D are coincident.

10·2. Various Cross-ratios of Four Elements

10·21. Four numbers or four points of a line can be arranged in 24 different orders. But these do not give rise to 24 different values of the cross-ratio.

10·22. The value of $\dfrac{AB \cdot CD}{AD \cdot CB}$ is unaltered by the simultaneous exchange of any two of the letters and of the other two, i.e.

$$(ABCD) = (BADC) = (CDAB) = (DCBA).$$

This is verified by making the actual exchanges in

$$(AB \cdot CD)/(AD \cdot CB).$$

It follows that there are not more than six different cross-ratios of four given points.

10·23. From 1·12, $BC \cdot AD + CA \cdot BD + AB \cdot CD = 0.$

$$\therefore \quad 1 + \frac{CA \cdot BD}{BC \cdot AD} + \frac{AB \cdot CD}{BC \cdot AD} = 0,$$

$$\therefore \quad (ABCD) + (ACBD) = 1,$$

$$\therefore \quad \text{if } (ABCD) = \kappa, \quad (ACBD) = 1 - \kappa.$$

Three other cross-ratios equal to $1 - \kappa$ can be obtained from $(ACBD)$ by 10·22.

10·24. Since $\dfrac{AB \cdot CD}{AD \cdot CB} = \dfrac{AB/BC}{AD/DC}$, the exchange of B and D changes the cross-ratio into its reciprocal.

$$\therefore \quad \text{if } (ABCD) = \kappa, \quad (ADCB) = 1/\kappa.$$

Three other cross-ratios equal to $1/\kappa$ can be obtained from $(ADCB)$ by 10·22.

10·25. Combining 10·23 and 10·24 it follows that

$$1 - 1/\kappa = (ACDB)$$

and

$$1/(1 - \kappa) = (ADBC),$$

and therefore

$$1 - 1/(1 - \kappa) = (ABDC).$$

By applying 10·22 to these last three results, the values of the remaining cross-ratios can be found. The 24 cross-ratios are expressed in terms of one of them, κ, as follows:

$$\kappa = (ABCD) = (BADC) = (CDAB) = (DCBA).$$

$$1 - \kappa = (ACBD) = (BDAC) = (CADB) = (DBCA).$$

$$1/\kappa = (ADCB) = (BCDA) = (CBAD) = (DABC).$$

$$1 - 1/\kappa = (ACDB) = (BDCA) = (CABD) = (DBAC).$$

$$1/(1 - \kappa) = (ADBC) = (BCAD) = (CBDA) = (DACB).$$

$$\kappa/(\kappa - 1) = (ABDC) = (BACD) = (CDBA) = (DCAB).$$

10·3. Special Values of (ABCD)

It is of interest to discuss the exceptional cases in which $\kappa, 1 - \kappa, 1/\kappa, 1 - 1/\kappa, 1/(1 - \kappa), \kappa/(\kappa - 1)$ are not all different.

10·31. $\kappa = 1 - \kappa$ gives $\kappa = \frac{1}{2}$, and the six values are $\frac{1}{2}, \frac{1}{2}, 2, -1, 2, -1$.

10·32. $\kappa = 1/\kappa$ gives $\kappa = +1$ or -1 of which $\kappa = -1$ gives the values obtained in 10·31. If $\kappa = +1$, $(ACBD) = 0$; therefore A coincides with C or B with D. $1/(1 - \kappa)$ and $\kappa/(\kappa - 1)$ do not exist, but they may be denoted by ∞ if this is understood to mean that the denominators of the expressions for $(ADBC)$ and $(ABDC)$ are zero. Thus the six values are $1, 0, 1, 0, \infty, \infty$.

10·33. $\kappa = 1 - 1/\kappa$ gives $\kappa^2 - \kappa + 1 = 0$, $\kappa^3 = -1$, $\kappa \neq -1$. Thus $\kappa = -\omega = \operatorname{cis}(-\frac{1}{3}\pi)$ or $\kappa = -\omega^2 = \operatorname{cis}(\frac{1}{3}\pi)$. The six values are $-\omega, -\omega^2, -\omega^2, -\omega, -\omega, -\omega^2$.

10·34. Other equalities amongst the six values only lead to repetitions of these results. In applications of cross-ratios the four points are usually distinct, and therefore 10·32 is unimportant. There is no interpretation of 10·33 in G_3. 10·31 is investigated in Chapter 11.

10·41. $(APBQ)(AQBR) = (APBR)$.

For $(APBQ) = \dfrac{AP/PB}{AQ/QB}$ and $(AQBR) = \dfrac{AQ/QB}{AR/RB}$.

Hence the product $= \dfrac{AP/PB}{AR/RB} = (APBR)$.

10·42. *If* $(ABCD) = (ADCB)$ *and* A, B, C, D *are distinct, then the value of the equal cross-ratios is* -1.

For $(ABCD) = \dfrac{AB/BC}{AD/DC} = 1/(ADCB) = 1/(ABCD)$.

$$\therefore \quad (ABCD)^2 = 1.$$

But if $(ABCD) = +1$, the points are not distinct. Therefore $(ABCD) = -1$.

EXERCISE 10 A

1. Give the cross-ratio of the numbers 1, 3, 6, 10.

2. Given A, B, C such that $AB = BC$, find P such that $(ABCP) = 3$.

3. If A is the mid-point of BC and $(APBC) = \frac{4}{3}$, find the position of P.

4. If $(ABXD) = (ABYD)$, prove that X coincides with Y unless A coincides with B.

5. For what positions of X, Y relative to A, B is $(AXBY)$ positive?

6. Use the equation $(x_1 - x_2)(x_3 - x_4) = k(x_1 - x_4)(x_3 - x_2)$ to find x_4 in terms of x_1, x_2, x_3 when $k = 0$ and when $k = 1$. Interpret the results in terms of points A, B, C, D.

7. If $AB : BC : CD = p : q : r$, what is the value of $(DABC)$?

8. Show that the cross-ratio of four collinear points is equal to the cross-ratio of their x-coordinates or y-coordinates.

9. Show that the cross-ratio of four numbers x is equal to that of the corresponding values of y when (i) $y = x + a$, (ii) $y = ax + b$, (iii) $y = a/x$, where a and b are constants.

10. By writing $(ax + b)/(cx + d)$ in the form $e + f/(x + g)$ and using the results of No. 9, prove that the cross-ratio of four numbers x is equal to that of the corresponding numbers $(ax + b)/(cx + d)$.

11. Calculate the cross-ratios of the numbers 1, 2, 3, 4 arranged in different orders.

12. Calculate the cross-ratios when one of them is

    (i) 5,   (ii) $\frac{5}{4}$,   (iii) $\frac{1}{5}$,   (iv) -4,

    (v) $-\frac{1}{4}$,   (vi) $\frac{4}{5}$,   (vii) 100,   (viii) $-\frac{9}{10}$.

13. If $(ABXY) = (A'B'X'Y')$ and $(ABYZ) = (A'B'Y'Z')$, prove that $(ABXZ) = (A'B'X'Z')$.

14. If $(XPYQ) = (XP'YQ')$, prove that $(XPYP') = (XQYQ')$.

15. If $(ABXY) = (ABX'Y')$, prove that $(ABXX') = (ABYY')$.

16. If $\kappa/(\kappa-1) = \lambda$, express the other five cross-ratios in terms of λ.

17. If x_1 satisfies $(x^2-x+1)^3 = c(x^2-x)$, show that $1-x_1$, $1/x_1$, $1-1/x_1$, $1/(1-x_1)$, $x_1/(x_1-1)$ also satisfy it.

18. If $ax_r y_r + bx_r + cy_r + d = 0$ is true for $r = 1, 2, 3, 4$, prove that $(x_1 x_2 x_3 x_4) = (y_1 y_2 y_3 y_4)$.

10·5. Homogeneous Parameters

10·51. In a parametric representation of a curve, the co-ordinates of a point on the curve are given by

$$x:y:1 = f(t):g(t):h(t) \quad \text{or} \quad x:y:z = f(t):g(t):h(t).$$

If f, g, h are polynomials, and if each value of t determines just one point of the curve, and each point of the curve determines just one value of the parameter, the representation is called *proper*.

$x:y:k = t^2:t:1$ is a proper representation of $y^2 = kx$, but for a curve in G_3 it is not usually possible to find a proper representation. Consider for example the representation of $x^2 - y^2 = 1$ by the parametric equations

$$x:y:1 = 1+t^2:2t:1-t^2.$$

The values ± 1 of t give no points of the curve and the point $(-1, 0)$ of the curve is given by no value of t although it can be obtained as a limit by making $t \to \infty$. The first exception $(t = \pm 1)$ does not arise in the corresponding work in G_5 or G_6: the values $t = \pm 1$ give the points at infinity on

$$x:y:z = 1+t^2:2t:1-t^2.$$

To avoid the second exception a new kind of parameter is needed.

Even to represent the points of a line in G_5 or G_6 a single coordinate is insufficient. For example, the points of OX are given by (x, z) as in 8·5, and not by x alone. In this way the point at infinity of the line is included. (x, z) is an example of the new kind of parameter.

The curve $x^2 - y^2 = z^2$ can be represented by

$$x : y : z = s^2 + t^2 : 2st : s^2 - t^2.$$

Any value of $t : s$ gives just one point of the curve and any point of the curve gives just one value of $t : s$. The parameter is the number-pair (t, s) with the conventions that (kt, ks) is the same as (t, s) unless $k = 0$ and that $(0, 0)$ has no meaning. The point $(-1, 0, 1)$, which is the point of the curve $x^2 - y^2 = z^2$ corresponding to the exceptional point on $x^2 - y^2 = 1$, is now given by the value $(1, 0)$ of the parameter.

10·52. If the coordinates of a point on any curve are given by

$$x : y : z = f(t, s) : g(t, s) : h(t, s),$$

where f, g, h are homogeneous polynomials of degree n with no common factor, and if each point of the curve determines just one value of (t, s), then (t, s) is called a *homogeneous parameter*. Since f, g, h are polynomials with no common factor, each value of (t, s) necessarily determines just one point of the curve. The representation is proper. The convention of 0·5 is assumed in connexion with $t : s$.

By putting $s = 1$, the parametric representation in terms of t, s is reduced to a representation in terms of the single parameter t. In practice the single parameter is generally used until it is necessary to deal with exceptional points and then the parameter is made homogeneous.

10·53. If a point P of a curve is properly represented in different ways by two different parameters, then to a given value of either parameter there corresponds just one value of the other. For the first parameter determines just one point P of the curve and this point determines just one value of the second parameter.

If there is an algebraic relation between two such parameters, it must be linear in t and linear in t'. It appears therefore that it can be only of the form

$$att' + bt + ct' + d = 0.$$

A proof is given by J. A. Todd, *Mathematical Gazette*, vol. XXIII, p. 58.

$att' + bt + ct' + d = 0$ is known as the *bilinear* relation. It is necessary that $ad \neq bc$. For if $ad = bc$, the relation is of one of the forms

$$(at + c)(at' + b) = 0, \quad bt + d = 0, \quad ct' + d = 0, \quad d = 0$$

and it ceases to determine t uniquely in terms of t' or t' in terms of t.

10·54. Even when $ad \neq bc$, there are exceptions to the unique determination. The bilinear relation gives

$$t = -(ct' + d)/(at' + b) \quad \text{unless} \quad at' + b = 0,$$
$$t' = -(bt + d)/(at + c) \quad \text{unless} \quad at + c = 0.$$

It therefore determines t, t' uniquely in terms of t', t save for the exceptional values of t, t' given by $at + c = 0$ and $at' + b = 0$.

By using homogeneous parameters (t, s) these exceptions can be avoided. The modified bilinear relation which holds between $t : s$ and $t' : s'$ is

$$att' + bts' + cst' + dss' = 0 \qquad (ad \neq bc).$$

10·6. Cross-ratio of Homogeneous Parameters

10·61. The cross-ratio of four values (t_1, s_1), (t_2, s_2), (t_3, s_3), (t_4, s_4) of a parameter is defined to be

$$\frac{(t_1 s_2 - t_2 s_1)(t_3 s_4 - t_4 s_3)}{(t_1 s_4 - t_4 s_1)(t_3 s_2 - t_2 s_3)}.$$

10·62. The *homogeneous cross-ratio* of the four values of a parameter (t, s) is the number-pair

$$\{(t_1 s_2 - t_2 s_1)(t_3 s_4 - t_4 s_3), \quad (t_1 s_4 - t_4 s_1)(t_3 s_2 - t_2 s_3)\}.$$

Homogeneous cross-ratios can be applied in 10·32. The values $\kappa, 1-\kappa$, $1/\kappa$, $1-1/\kappa$, $1/(1-\kappa)$, $\kappa/(\kappa-1)$ are then replaced by $\{\kappa, \lambda\}$, $\{\lambda-\kappa, \lambda\}$, $\{\lambda, \kappa\}$, $\{\kappa-\lambda, \kappa\}$, $\{\lambda, \lambda-\kappa\}$, $\{\kappa, \kappa-\lambda\}$ and the values corresponding to those in 10·32 are $\{1, 1\}$, $\{0, 1\}$, $\{1, 1\}$, $\{0, 1\}$, $\{1, 0\}$, $\{1, 0\}$.

10·63. If (t, s) and (t', s') are two homogeneous parameters of the same point P which satisfy a relation

$$att' + bts' + cst' + dss' = 0 \qquad (ad \neq bc),$$

then the cross-ratio of four values of (t, s) is equal to that of the corresponding values of (t', s').

For
$$(at_1' + bs_1')(at_2' + bs_2') \times (t_1 s_2 - t_2 s_1)$$
$$= -(at_2' + bs_2')(ct_1' + ds_1')s_1 s_2 + (at_1' + bs_1')(ct_2' + ds_2')s_1 s_2$$
$$= (ad - bc)s_1 s_2(t_1' s_2' - t_2' s_1'),$$

and similar results hold for $t_3 s_4 - t_4 s_3$, $t_1 s_4 - t_4 s_1$, $t_3 s_2 - t_2 s_3$.

Hence
$$\frac{(t_1 s_2 - t_2 s_1)(t_3 s_4 - t_4 s_3)}{(t_1 s_4 - t_4 s_1)(t_3 s_2 - t_2 s_3)} = \frac{(t_1' s_2' - t_2' s_1')(t_3' s_4' - t_4' s_3')}{(t_1' s_4' - t_4' s_1')(t_3' s_2' - t_2' s_3')}.$$

10·64. In the same way it may be proved from

$$att' + bt + ct' + d = 0 \qquad (ad \neq bc)$$

that
$$\frac{(t_1 - t_2)(t_3 - t_4)}{(t_1 - t_4)(t_3 - t_2)} = \frac{(t_1' - t_2')(t_3' - t_4')}{(t_1' - t_4')(t_3' - t_2')},$$

i.e.
$$(t_1 t_2 t_3 t_4) = (t_1' t_2' t_3' t_4').$$

A simpler proof of this result is given in Exercise 10 A, Nos. 9 and 10.

10·7. Ranges

10·71. A point on a line may be represented by a parameter in various ways. For instance, a point on the line $ax + by + c = 0$ may be represented by its coordinate x (unless $b = 0$) or by its coordinate y (unless $a = 0$). It can also be represented by its distance t from the fixed point where the line meets OY (unless $b = 0$) or by its distance from some other fixed point on the line. x and y are connected by the relation $ax + by + c = 0$;

x and t are connected by a relation $x = kt$; y and t are connected by $akt + by + c = 0$. All these relations are special cases of the bilinear relation in 10·53.

10·72. Cross-ratios in G_3 and G_6

The cross-ratio of four points A, B, C, D on a line in G_3 has been defined in 10·13 as the cross-ratio of OA, OB, OC, OD. By 10·64 it is equal to the cross-ratios of the x-coordinate, or of the y-coordinate, or of any other parameter related to OP by a bilinear relation.

In G_6 the cross-ratio of four collinear points is defined to be the cross-ratio of the homogeneous parameters of the points.

The coordinates (x, z) may be used as the homogeneous parameter. If the points are not at infinity, the cross-ratio is equal to

$$\frac{(x_1/z_1 - x_2/z_2)(x_3/z_3 - x_4/z_4)}{(x_1/z_1 - x_4/z_4)(x_3/z_3 - x_2/z_2)},$$

and this is the same as the cross-ratio of the ordinary co-ordinates of the corresponding points A, B, C, D in G_3. Hence it is equal to $(ABCD)$ as defined in 10·13.

The cross-ratio of points $A(x_1, 1)$, $B(x_2, 1)$, $C(x_3, 1)$, and $D(x_4, 0)$, of which D is at infinity, by 10·61,

$$= \frac{(x_1 - x_2)(-x_4)}{(-x_4)(x_3 - x_2)} = \frac{x_1 - x_2}{x_3 - x_2}.$$

In G_3 there is no point corresponding to D. Also $\dfrac{x_1 - x_2}{x_3 - x_2} = \dfrac{AB}{CB}$.

The same result is found by taking the limit of the cross-ratio of x_1, x_2, x_3, x_4 when $x_4 \to \infty$, or of $(ABCK)$ when the co-ordinate of K tends to infinity.

10·73. The points of a line may be represented by other homogeneous parameters instead of the coordinates. By 10·63, the cross-ratio may be calculated from any homogeneous parameter which is connected with the coordinate by a bilinear relation.

For example, $P(\kappa_1 x_1 + \kappa_2 x_2,\ \kappa_1 y_1 + \kappa_2 y_2,\ \kappa_1 z_1 + \kappa_2 z_2)$

and $\qquad P'(\kappa_1' x_1 + \kappa_2' x_2,\ \kappa_1' y_1 + \kappa_2' y_2,\ \kappa_1' z_1 + \kappa_2' z_2)$

are two points on the join of (x_1, y_1, z_1) to (x_2, y_2, z_2). (κ_1, κ_2) is a homogeneous parameter of points on this line. The points P_1, P, P_2, P' are given by the values $(1, 0)$, (κ_1, κ_2), $(0, 1)$, (κ_1', κ_2') of the parameter.

$$\therefore \quad (P_1 P P_2 P') = \frac{\kappa_2 \kappa_1'}{\kappa_1 \kappa_2'} \quad \text{by 10·61.}$$

The bilinear relation (10·54) between (x, z) and (κ_1, κ_2) is

$$x(\kappa_1 z_1 + \kappa_2 z_2) = z(\kappa_1 x_1 + \kappa_2 x_2).$$

10·74. The cross-ratio of four points of a range is called, shortly, the cross-ratio of the range, provided that there is no doubt about which points are intended.

10·8. Pencils

10·81. Coplanar lines through a point V are said to form a *pencil* with *vertex* V. The lines are sometimes called *rays* of the pencil.

A line of a pencil may be represented by a parameter in various ways.

10·82. If V is the origin, an arbitrary line of the pencil is $y = tx$. There is one line which is not of this form, namely OY. Thus if t is used as the parameter, there is one line of the pencil that is not represented; it is given by $t \to \infty$. A line through the origin has an equation $Xx + Yy = 0$ and there is no exception to this. Thus (X, Y) is a homogeneous parameter which determines a line of the pencil.

10·83. If V is not the origin, suppose that it is the point of intersection (a, b) of the two lines

$$\alpha \equiv X_1 x + Y_1 y + 1 = 0,$$
$$\beta \equiv X_2 x + Y_2 y + 1 = 0.$$

An arbitrary line of the pencil is

$$\alpha = \kappa\beta \quad \text{or} \quad s\alpha = t\beta,$$

where κ and (t, s) are parameters. It is usual to replace κ by the homogeneous parameter only when the ray $\beta = 0$ is involved.

10·84. An alternative parameter is the X-coordinate of the line (unless $b = 0$) or the Y-coordinate (unless $a = 0$). The X-coordinate of

$$X_1 x + Y_1 y + 1 = \kappa(X_2 x + Y_2 y + 1)$$

is given by $\qquad (1 - \kappa) X = X_1 - \kappa X_2.$

The parameter may also be the x-coordinate of the point in which the line meets a fixed line. If the fixed line is OX, x is given by $x(X_1 - \kappa X_2) = \kappa - 1.$

More generally if the fixed line is $X_3 x + Y_3 y + 1 = 0$, the elimination of y leads to

$$\begin{vmatrix} (X_1 - \kappa X_2) x + 1 - \kappa & Y_1 - \kappa Y_2 \\ X_3 x + 1 & Y_3 \end{vmatrix} = 0,$$

and this is a bilinear relation between x and κ.

10·85. Cross-ratio of a Pencil

The cross-ratio of four lines of a pencil is defined to be the cross-ratio of their parameters. For example, the cross-ratio of

$$\alpha = \kappa_1 \beta, \quad \alpha = \kappa_2 \beta, \quad \alpha = \kappa_3 \beta, \quad \alpha = \kappa_4 \beta$$

is $\qquad (\kappa_1 - \kappa_2)(\kappa_3 - \kappa_4)/(\kappa_1 - \kappa_4)(\kappa_3 - \kappa_2).$

By 10·64 this is the same as the cross-ratio of any other parameter which determines the line, such as X, Y, or x in 10·84. The bilinear relations between κ and the parameters X, x are given in 10·84.

The cross-ratio of four lines of a pencil is called, shortly, the cross-ratio of the pencil, provided that there is no doubt about which four lines are intended.

10·86. When a homogeneous parameter, such as (t, s) in 10·83, is used, the cross-ratio is

$$(t_1 s_2 - t_2 s_1)(t_3 s_4 - t_4 s_3)/(t_1 s_4 - t_4 s_1)(t_3 s_2 - t_2 s_3).$$

An important special case is the pencil

$$\alpha - \beta = 0, \quad \alpha = 0, \quad \alpha + \beta = 0, \quad \beta = 0$$

whose cross-ratio is -1. This is proved by putting $s_1 = 1$, $t_1 = 1$; $s_2 = 1$, $t_2 = 0$; $s_3 = 1$, $t_3 = -1$; $s_4 = 0$, $t_4 = 1$. Alternatively it may be verified from 10·85 by treating $\beta = 0$ as the limit of $\alpha = \kappa_4\beta$ when $\kappa_4 \to \infty$. When $\kappa_1 = 1$, $\kappa_2 = 0$, $\kappa_3 = -1$, $(\kappa_1 - \kappa_2)(\kappa_3 - \kappa_4)/(\kappa_1 - \kappa_4)(\kappa_3 - \kappa_2)$ becomes $(1 + \kappa_4)/(1 - \kappa_4)$, which tends to -1 when $\kappa_4 \to \infty$.

If $x_1 : y_1$ and $x_2 : y_2$ are the values of $x : y$ given by

$$ax^2 + 2hxy + by^2 = 0,$$

the cross-ratio of $(1, 0)$, (x_1, y_1), $(0, 1)$, (x_2, y_2) is $x_2 y_1 / x_1 y_2$. This is equal to -1 when $x_1/y_1 + x_2/y_2 = 0$, i.e. when $h = 0$.

10·9. Projective Property of Cross-ratio

10·91. Each value of a parameter κ determines a line $\alpha = \kappa\beta$ of a pencil vertex V. Hence each value of κ also determines a unique point P where the line $\alpha = \kappa\beta$ meets a fixed line not passing through V. Also if P is given, this determines VP, and hence κ, uniquely. Hence κ can also be used as a parameter to determine the position of P. The bilinear relation between κ and the coordinate of P is given in 10·84.

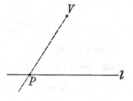

If four positions a, b, c, d of the line $\alpha = \kappa\beta$, given by $\kappa = \kappa_1, \kappa_2, \kappa_3, \kappa_4$, meet the line l in points A, B, C, D, then

$$(ABCD) = (\kappa_1\kappa_2\kappa_3\kappa_4) = (abcd).$$

Similarly, if l' is another fixed line,

$$(A'B'C'D') = (abcd),$$

$$\therefore \quad (ABCD) = (A'B'C'D').$$

A', B', C', D' may be the projections of A, B, C, D from V on to any plane. Hence the result may be stated in the form:

Cross-ratios are unaltered by projection

or

The cross-ratio of a pencil is equal to that of the section of the pencil by any line.

10·92. When a conclusion $(ABCD) = (A'B'C'D')$ is drawn by taking two sections of a pencil vertex V by lines l, l' as in 10·91, this is indicated by writing $(ABCD) \underset{V}{=} (A'B'C'D')$.

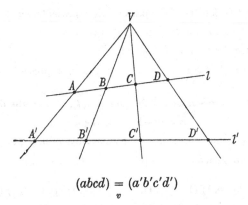

Dually $(abcd) \underset{v}{=} (a'b'c'd')$

means that the cross-ratios are equal because the line v meets a, b, c, d and a', b', c', d' in the same four points.

10·93. A pencil of lines projects from a vertex V outside its plane α into another pencil of lines in a plane α', and if the projections of the lines a, b, c, d of the first pencil are the lines

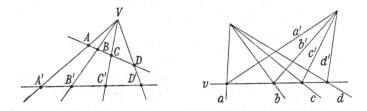

a', b', c', d' of the second pencil, then $(abcd) \underset{l}{=} (a'b'c'd')$, where l is the axis of projection. Hence, in this sense,

Cross-ratios of pencils are projective.

Examples of the use of these results will be found in books on pure geometry. ($P.G.$ Chapter VIII.)

EXERCISE 10B

1. Show how the graph of $xy + 3x - 2y - 12 = 0$ illustrates the cases in which unique values of x, y do not correspond to given values of y, x.

2. State the defects of the parametric representations

$$x : y : 1 = t^2 : 1 : t \quad \text{and} \quad x : y : z = t^2 : 1 : t$$

and show how they can be remedied.

3. Is $x/a : y/b : 1 = 1 - t^2 : 2t : 1 + t^2$ a proper representation of the ellipse $x^2/a^2 + y^2/b^2 = 1$?

4. Find the cross-ratios of the points t_1, t_2, t_3, t_4 on the line

$$x = at + b, \quad y = ct + d.$$

5. If $p \to \infty$, find the limit of the cross-ratio of $0, a, p, b$.

6. In homogeneous geometry of one dimension, give the cross-ratio of

(i) $(0, 1) \, (p, 1) \, (1, 1) \, (q, 1)$,   (ii) $(0, 1) \, (p, 1) \, (1, 1) \, (1, 0)$,

(iii) $(0, 1) \, (a, 1) \, (p, 1) \, (b, 1)$,   (iv) $(0, 1) \, (a, 1) \, (1, 0) \, (b, 1)$.

7. If P_1, P_2, P, P' are (x_1, y_1, z_1), (x_2, y_2, z_2),

$$(k_1 x_1 + k_2 x_2, \ k_1 y_1 + k_2 y_2, \ k_1 z_1 + k_2 z_2),$$

$$(k_1' x_1 + k_2' x_2, \ k_1' y_1 + k_2' y_2, \ k_1' z_1 + k_2' z_2),$$

what is the value of $(P_1 P P_2 P')$?

8. Give the cross-ratios of

(i) $y = 0, \ y = tx, \ x = 0, \ y = t'x$,

(ii) $\alpha = 0, \ \beta = 0, \ \alpha - \beta = 0, \ \alpha + \beta = 0$,

(iii) $y = 0, \ y = kx, \ y = x, \ y = -x$,

and find the limit when $k \to \infty$ of the cross-ratio in (iii).

9. Calculate directly the coordinates and cross-ratio of the points in which $y = t_1 x, \ y = t_2 x, \ y = t_3 x, \ y = t_4 x$ meet $ax + by + c = 0$.

10. Four lines through (h, k) are $[X_p, Y_p]$, $p = 1$ to 4. Prove that their cross-ratio is equal to that of the numbers X_1, X_2, X_3, X_4 or Y_1, Y_2, Y_3, Y_4.

11. Concurrent lines $[X_p, Y_p], p = 1$ to 4, meet $[X, Y]$ in A, B, C, D. Find the x-coordinates of A, B, C, D and the value of $(ABCD)$ in terms of X_p.

12. Find the point of concurrence of lines $X = aT + b, \ Y = cT + d$, where $ad \neq bc$ and T varies. Also find the cross-ratio of the lines

$$[b, d], \quad [a + b, c + d], \quad [2a + b, 2c + d], \quad [b - a, d - c].$$

Chapter 11

HARMONIC SECTION

11·1. Harmonic Section

11·11. When $(x_1x_2x_3x_4) = -1, (ABCD) = -1$, or $(abcd) = -1$, the numbers, the range, or the pencil is said to be *harmonic*.

If $-1 = (ABCD) = (BADC) = (CDAB) = (DCBA)$,

it follows from 10·2 that

$$-1 = (ADCB) = (BCDA) = (CBAD) = (DABC).$$

This shows that the harmonic property is a relation between two pairs A, C and B, D. Also A and C are interchangeable, and so are B and D. And the order of the pairs is immaterial. Each pair is said to be harmonically separated by the other.

Cross-ratios $(x_1x_2x_3x_4)$ and $(abcd)$ are also called harmonic when the value is -1.

11·12. Since $(ABCD) = \dfrac{AB \cdot CD}{AD \cdot CB} = \dfrac{AB}{BC} \Big/ \dfrac{AD}{DC}$,

$(ABCD) = -1$ is equivalent to $\dfrac{AB}{BC} = -\dfrac{AD}{DC}$,

i.e. AC is divided at B, D internally and externally in the same ratio.

When $(ABCD) = -1$, $\dfrac{AB \cdot CD}{AD \cdot CB} = -1$,

$$\therefore \ \frac{AB(AD - AC)}{AD(AB - AC)} = -1, \qquad \therefore \ \frac{1}{AC} - \frac{1}{AD} = \frac{1}{AB} - \frac{1}{AC},$$

i.e. AB, AC, AD are in harmonic progression. This is the origin of the name "harmonic" as applied to a range.

11·13. By 10·91 the pencil formed by joining any point V to the points of a harmonic range is a harmonic pencil. And any other section of this pencil is another harmonic range.

By 10·86 the pencil formed by $\alpha - \beta = 0$, $\alpha = 0$, $\alpha + \beta = 0$, $\beta = 0$ is harmonic.

In virtue of 10·61 a system (t_1, s_1) (t_2, s_2) (t_3, s_3) (t_4, s_4) is called harmonic if

$$\frac{(t_1 s_2 - t_2 s_1)(t_3 s_4 - t_4 s_3)}{(t_1 s_4 - t_4 s_1)(t_3 s_2 - t_2 s_3)} = -1.$$

11·2. In taking a section of the pencil $V(ABCD)$ by a line l parallel to VD no point D' is found in G_3. As in 10·72 the cross-ratio $(ABCD)$ is equal to $(x_1 - x_2)/(x_3 - x_2)$, i.e. to $A'B'/C'B'$. Thus if $(ABCD) = -1$, $A'B' = B'C'$.

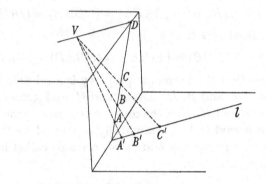

Hence the projection of a harmonic range $(ABCD)$ with D on the vanishing line is a segment $A'C'$ bisected at B'.

In this way a metrical concept, the bisected segment, appears as a special case of a projective concept, a harmonic range. (The definition of a harmonic range in 11·11 is not projective, but in 11·55 it is shown how a projective treatment can be given.) In 9·16 it is shown that a square can be derived by projection as a special case of an arbitrary quadrangle; and in volume II examples of generalisation by projection are given.

11·3. The Analytical Harmonic Condition

11·31. The necessary and sufficient condition for the roots of $ax^2 + 2hx + b = 0$ to separate those of $a'x^2 + 2h'x + b' = 0$ harmonically is $ab' + a'b = 2hh'$.

Let $ax^2 + 2hx + b \equiv a(x - x_1)(x - x_3)$

and $a'x^2 + 2h'x + b' \equiv a'(x - x_2)(x - x_4).$

$$\therefore \quad a(x_1 + x_3) = -2h, \quad ax_1x_3 = b, \tag{1}$$

and
$$a'(x_2 + x_4) = -2h', \quad a'x_2x_4 = b'.$$

The harmonic condition is

$$\frac{(x_1 - x_2)(x_3 - x_4)}{(x_1 - x_4)(x_3 - x_2)} = -1,$$

i.e.
$$2(x_1x_3 + x_2x_4) = (x_1 + x_3)(x_2 + x_4).$$

$\therefore$ by (1)
$$ab' + a'b = 2hh'.$$

11·32. It follows from 11·31 that the condition for the lines $ax^2 + 2hxy + by^2 = 0$ to separate the lines $a'x^2 + 2h'xy + b'y^2 = 0$ harmonically is $ab' + a'b = 2hh'$; for x/y is a parameter which determines the lines and it has the same values as the parameter x in 11·31.

11·33. Also the harmonic condition for the points at infinity $AX^2 + 2HXY + BY^2 = 0$ and $A'X^2 + 2H'XY + B'Y^2 = 0$ is $AB' + A'B = 2HH'$. This follows from the fact that the lines $Bx^2 - 2Hxy + Ay^2 = 0$ and $B'x^2 - 2H'xy + A'y^2 = 0$ are harmonic under that condition (8·55).

11·34. The same condition $ab' + a'b = 2hh'$ applies to the line-pairs $a\alpha^2 + 2h\alpha\beta + b\beta^2 = 0$, $a'\alpha^2 + 2h'\alpha\beta + b'\beta^2 = 0$, where $\alpha = 0$, $\beta = 0$ are any two lines, because α/β is a parameter which determines the lines and it has the same values as x in 11·31.

11·35. It also applies to the point-pairs

$$a\mathrm{A}^2 + 2h\mathrm{AB} + b\mathrm{B}^2 = 0, \quad a'\mathrm{A}^2 + 2h'\mathrm{AB} + b'\mathrm{B}^2 = 0,$$

where $\mathrm{A} = 0$ and $\mathrm{B} = 0$ are envelope equations of any two points.

EXERCISE 11A

1. If $AB = 5$, $BC = 3$, $(ABCD) = -1$, find CD.

2. If $(ABCD) = -1$, what are the values of $(ABDC)$, $(ACBD)$, $(ADBC)$?

3. If $(ABCD) = -1$, prove that DA, DB, DC are in harmonic progression.

4. What is the condition for $y = 0$, $y = kx$, $x = 0$, $y = k'x$ to form a harmonic pencil?

5. Verify that every pair of lines $ax^2 + 2hxy + by^2 = \lambda(x^2 + y^2)$ is harmonically separated by the lines $hx^2 - (a - b)xy - hy^2 = 0$, and interpret the result.

6. If b, d are the bisectors of the angles between a, c, prove that $(abcd) = -1$.

7. If $(ABCD) = -1$ and BVD is a right angle, prove that VB, VD are the bisectors of the angles between VA, VC.

8. Find a pair of lines harmonically separated both by

$$a_1x^2 + 2h_1xy + b_1y^2 = 0 \quad \text{and by} \quad a_2x^2 + 2h_2xy + b_2y^2 = 0.$$

11·4. Quadrangle and Quadrilateral

11·41. A set of four points A, B, C, D is said to form a *quadrangle*. The three pairs of joins AB and CD, AC and BD, AD and BC are called opposite *sides* of the quadrangle. The opposite sides meet in points X, Y, Z called the *diagonal points* of the quadrangle.

A quadrangle therefore consists of four points and six sides. Through each point pass three sides. On each side lie two of the points.

11·42. Let the join XY of two diagonal points meet the opposite sides BC, AD at P, Q.
Then

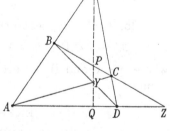

$$(AQDZ) \underset{X}{=} (BPCZ) \underset{Y}{=} (DQAZ).$$

$\therefore$ by 10·42 $\quad (AQDZ) = -1$.

This is one of the harmonic properties of the quadrangle:

Each pair of points of the quadrangle is separated harmonically by one diagonal point and a point collinear with the other two.

and

Each pair of diagonal points is separated harmonically by points on the sides through the third.

11·43. The *quadrilateral* is the dual of the quadrangle. It is formed by a set of four lines a, b, c, d. The three pairs of meets of ab and cd, ac and bd, ad and bc are called opposite *vertices*. The lines x, y, z joining opposite vertices are called *diagonal lines*.

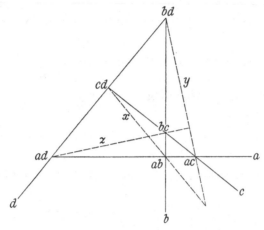

A quadrilateral therefore consists of four lines and six vertices. On each line lie three vertices. Through each vertex pass two of the lines.

11·44. The harmonic properties of the quadrilateral are:

Each pair of lines of the quadrilateral is separated harmonically by one diagonal line and a line concurrent with the other two.

and

Each pair of diagonal lines is separated harmonically by lines through the vertices on the third.

In the figure of 11·42, which represents a quadrangle, it has been proved that $(AQDZ) = -1$. If the figure is regarded as a quadrilateral formed by the lines BX, XC, CY, YB, it follows from $(AQDZ) = -1$ that the pair of diagonals PY, PZ is separated harmonically by the lines PA, PD through the vertices on the third diagonal. Also the pair of sides BX, BY is separated harmonically by the diagonal BC and the line

BQ concurrent with the other diagonals. Thus it is unnecessary to prove the harmonic properties of the quadrilateral independently. They can however be proved by the dual of the method used in 11·42 for the quadrangle.

11·45. EXAMPLE. Show that the sides of any quadrilateral may be taken to be $\alpha = 0$, $\beta = 0$, $\gamma = 0$, $\delta = 0$, where

$$\alpha + \beta + \gamma + \delta \equiv 0,$$

and find the equations of its diagonals.

In the figure the opposite vertices are A, A' and B, B' and C, C', and XYZ is the diagonal triangle.

Let BA', $B'A'$, $B'A$ be $\alpha = 0$, $\beta' = 0$, $\gamma' = 0$

$$\therefore \quad AA' \text{ is } \alpha = k\beta'. \qquad \therefore \quad AB \text{ is } \alpha - k\beta' = l\gamma'.$$

Hence AB is $\delta = 0$, where $\delta \equiv -\alpha + k\beta' + l\gamma'$.

Replace $k\beta'$, $l\gamma'$ by $-\beta$, $-\gamma$. Then the equations of BA', $A'B'$, $B'A$, AB are

$$\alpha = 0, \quad \beta = 0, \quad \gamma = 0, \quad \delta = 0.$$

Also $\qquad\qquad\qquad \alpha + \beta + \gamma + \delta \equiv 0.$

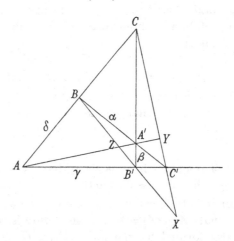

Since $\alpha + \delta = 0$ is a line through B and $\beta + \gamma = 0$ is a line through B', and these equations are the same in virtue of $\alpha + \beta + \gamma + \delta \equiv 0$,

$$BB' \text{ is } \alpha + \delta = 0 \quad \text{or} \quad \beta + \gamma = 0,$$

Similarly CC' is $\beta + \delta = 0$ or $\gamma + \alpha = 0$,

and AA' is $\gamma + \delta = 0$ or $\alpha + \beta = 0$.

This notation also leads to the harmonic property. For $\alpha - \gamma = 0$ is a line through C', and since it is the same line as $\alpha + \delta = \gamma + \delta$, it is a line through Z. Hence $C'C, C'B, C'Z, C'A$ are $\alpha + \gamma = 0$, $\alpha = 0$, $\alpha - \gamma = 0$, $\gamma = 0$, and these form a harmonic pencil.

11·5. The Harmonic Construction

11·51. The harmonic property of the quadrangle suggests a ruler construction to find the fourth point P' which with a given point P separates two given points A, B harmonically.

Draw any two lines a, b through A, B meeting in X. Take any point Z in XP. Let AZ, BZ meet b, a in C, D. Let CD meet AB at P'.

Then P' is one of the diagonal points of the quadrangle $ABCD$, and $(APBP') = -1$. This shows that the construction leads to the same point P' no matter how a, b are drawn through the given points and no matter what point Z is taken in XP.

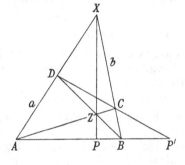

11·52. It is of interest to see that the uniqueness of the point P' can be proved, without the use of cross-ratio or of analytical methods, from a few elementary assumptions. These assumptions are:

1. There is just one line which passes through two given points.
2. Two lines in the same plane meet in a point.
3. Just one plane passes through a line and a point not on the line.

1'. There is just one line which lies in two given planes.
2'. Two lines through the same point lie in a plane.
3'. Just one point lies on a line and a plane not through the line.

Certain deductions can be made without further assumptions. There is a three-dimensional duality between the assumptions and consequently also between the deductions that can be made from them. Corresponding elements in three-dimensional duality are given in the following table:

Point	Plane
Line joining two points	Line of intersection of two planes
Lines through a point	Lines in a plane
Coplanar lines through a point	Concurrent lines in a plane
Collinear points	Collinear planes
Two lines with no common point on them (i.e. skew lines).	Two lines with no common plane through them (i.e. skew lines).

Thus in three dimensions the point and plane are dual elements and the line is self-dual.

11·53. EXAMPLE.

If a, b are two skew lines not through the point P, show how to determine the unique line through P meeting a and b.

If a, b are two skew lines not in the plane π, show how to determine the unique line in π meeting a and b.

Let ξ, η be the planes which (by 3) contain a, P and b, P.
Then (by 1') there is a unique line in ξ, η. This is the required line.

Let X, Y be the two points in which (by 3') a, b meet π.
Then (by 1) there is a unique line through X, Y. This is the required line.

11·54. If three lines meet in pairs, they must either be concurrent or else coplanar.

For suppose that a, b, c are the lines; and let b, c meet at X; let c, a meet at Y; and a, b at Z.

Then if two of these points coincide (at P) the three lines concur at P. And, if not, the unique plane, which (by 3) contains X and a, also contains Y and Z. Since it contains X and Y, it also contains c (by 1). Similarly it contains b. Hence the lines are coplanar.

The dual is: If three lines are coplanar in pairs, they must either be coplanar or else concurrent. This is equivalent to the original result.

11·55. The Uniqueness Theorem

In the figure the construction of 11·51 is supposed to have been carried out in two different planes α, α' through APB. It is to be proved that the points Y, Y' coincide.

Since DD', XX' lie in the plane of a, a', they intersect. Similarly it may be proved that DD', ZZ' intersect and that XX', ZZ' intersect. Hence, by 11·54, DD', XX', ZZ' meet in a point O.

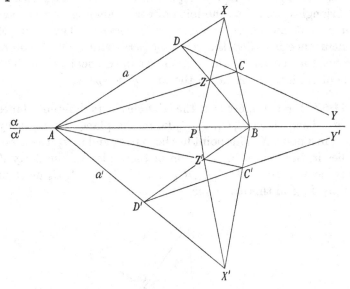

Similarly it may be proved that CC', XX', ZZ' meet in a point; and this must be the same point O.

Now DX, $D'X'$ meet at A; CX, $C'X'$ meet at B; and DC, $D'C'$ meet at a point K in the plane $OCC'DD'$. Since K lies in the planes XDC, $X'D'C'$, it lies in their line of intersection APB. Hence the points Y, Y' where DC, $D'C'$ meet APB coincide with K.

For every construction in α no matter how a, b, Z are chosen, it is proved above that Y coincides with Y'. Hence the construction in α always leads to the same point Y.

This theorem makes it possible to develop the theory of

harmonic section on a purely projective basis. A harmonic range $APBY$ can be defined by the property that if X is any point, Z is any point on PX, AZ meets BX at C, and BZ meets AX at D, then Y is in CD.

11·6. Desargues' Perspective Theorem

11·61. If $A_1B_1C_1$ and $A_2B_2C_2$ are triangles such that A_1A_2, B_1B_2, C_1C_2 meet in a point O, then B_1C_1 and B_2C_2, C_1A_1 and C_2A_2, A_1B_1 and A_2B_2, meet in collinear points.

Triangles such that the joins of corresponding vertices are concurrent are called *triangles in perspective*. The point of concurrence is called the *centre of perspective*. The theorem proves that corresponding sides meet in collinear points. The line of collinearity is called the *axis of perspective*.

11·62. First suppose that the triangles lie in different planes σ_1, σ_2. Then since B_1C_1, B_2C_2 lie in the plane OB_1C_1, they meet in a point X. This point lies both in σ_1 and in σ_2; therefore it lies in the line of intersection of these planes. Similarly it may be proved that C_1A_1, C_2A_2 and A_1B_1, A_2B_2 meet in points Y, Z in this same line.

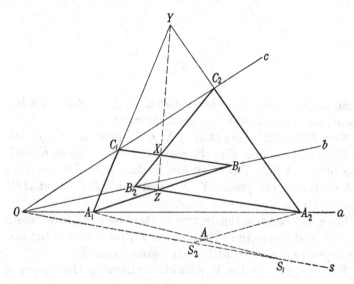

11·63. Now suppose that the triangles lie in the same plane σ.

Draw a line s through O outside σ, and take any two points S_1, S_2 on it. Since $S_1 A_1$, $S_2 A_2$ lie in the plane of s and $O A_1 A_2$, they meet in a point A. Let B, C be similarly defined.

Then the plane τ of ABC is distinct from σ. Also $B_1 B_2$, $C_1 C_2$, $S_1 S_2$ meet at O; hence, by 11·62, $B_1 C_1$ and $B_2 C_2$, $C_1 S_1$ and $C_2 S_2$, $S_1 B_1$ and $S_2 B_2$ meet in collinear points, i.e. $B_1 C_1$ and $B_2 C_2$ meet on BC which is in τ. Similarly $C_1 A_1$, $C_2 A_2$ and also $A_1 B_1$, $A_2 B_2$ meet at points in τ. The three points of intersection are therefore in the line of intersection of σ and τ.

11·64. Conversely: if $A_1 B_1 C_1$ and $A_2 B_2 C_2$ are two triangles such that $B_1 C_1$ and $B_2 C_2$, $C_1 A_1$ and $C_2 A_2$, $A_1 B_1$ and $A_2 B_2$ meet in collinear points, then $A_1 A_2$, $B_1 B_2$, $C_1 C_2$ meet in a point.

First suppose that the triangles are in different planes. Then since $B_1 C_1$, $B_2 C_2$ meet at X, $B_1 B_2 C_1 C_2$ is a plane. Also $C_1 C_2 A_1 A_2$ and $A_1 A_2 B_1 B_2$ are planes.

$A_1 A_2$, $B_1 B_2$, $C_1 C_2$ are the lines of intersection of these three planes taken in pairs. Hence they meet at the common point of the three planes.

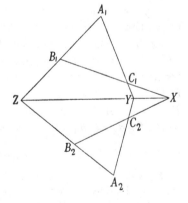

11·65. Now suppose that the triangles lie in one plane. Then the triangles $B_1 B_2 Z$, $C_1 C_2 Y$ have $B_1 C_1$, $B_2 C_2$, ZY concurrent (at X); therefore, by 11·63, $B_1 B_2$ and $C_1 C_2$, $B_1 Z$ and $C_1 Y$, $B_2 Z$ and $C_2 Y$ meet in collinear points; i.e. $B_1 B$ and $C_1 C_2$ meet in a point collinear with A_1 and A_2. This proof holds also when the triangles are not coplanar, but 11·62 is used instead of 11·63.

11·66. The figure of Desargues' Theorem consists of 10 points and 10 lines. Through each point pass three of the lines and on each line lie three of the points. Any of the points can play the part of O; the three lines through it join the vertices of the triangles, which each have three sides; the tenth line is the line of collinearity. For example, if X is the point of concurrence, the triangles are $B_1 B_2 Z$, $C_1 C_2 Y$, and the line of collinearity is $O A_2 A_1$.

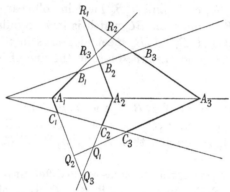

11·67. EXAMPLE. Three coplanar triangles are in perspective two by two, with the same centre of perspective for each pair. Prove that the three axes of perspective are concurrent.

Let $B_2 C_2$ and $B_3 C_3$ meet in P_1, $B_3 C_3$ and $B_1 C_1$ in P_2, $B_1 C_1$ and $B_2 C_2$ in P_3; and let Q_1, Q_2, Q_3, R_1, R_2, R_3 be similarly defined. The axes of perspective are $P_1 Q_1 R_1$, $P_2 Q_2 R_2$, $P_3 Q_3 R_3$. Since $Q_2 Q_3$ and $R_2 R_3$, $Q_3 Q_1$ and $R_3 R_1$, $Q_1 Q_2$ and $R_1 R_2$ meet in the collinear points A_1, A_2, A_3, it follows that $Q_1 R_1$, $Q_2 R_2$, $Q_3 R_3$ meet in a point.

EXERCISE 11 B

1. A figure of a points and b lines has x of the points on each line and y of the lines through each point. What is the relation between a, b, x, y? Draw the figure for

   (i) $a = b = 3$, $x = y = 2$;   (ii) $a = b = 9$, $x = y = 3$.

2. The equations of four lines are

$$\lambda \equiv 2x + 3y + 4z = 0, \qquad \mu \equiv x + 2y + z = 0,$$
$$\nu \equiv 3x + 5y + 6z = 0, \qquad \rho \equiv 3x + y + 16z = 0;$$

find u, v, w such that $\rho \equiv u\lambda + v\mu + w\nu$. What must α, β, γ, δ be if $\alpha = 0$, $\beta = 0$, $\gamma = 0$, $\delta = 0$ are the equations of the lines and

$$\alpha + \beta + \gamma + \delta \equiv 0\,?$$

3. In the figure of 11·51 give a construction for the point A' such that A, A' separate P, B harmonically.

4. Given a triangle and a point P not on any of its sides, give a ruler construction for the line through P meeting the sides in Q, R, S, so that $(PQRS) = -1$.

5. In the figure of Desargues' Theorem, if A_1 is the point of concurrence, which is the line of collinearity?

6. Three coplanar triangles are in perspective two by two with the same axis of perspective; prove that the three centres of perspective are collinear.

7. $ABCD$ is a quadrangle; AB and CD, AC and BD, AD and BC meet in F, G, H; also CD and HG, BD and HF, BC and GF meet in P, Q, R. Prove that P, Q, R are collinear.

8. AB and CD, AC and BD meet in U, V; also UV meets AD at F and BC at G, and BF meets AC at K. Prove that GK, FC, UA are concurrent.

9. $A_1 B_1 C_1$, $A_2 B_2 C_2$ are in perspective. $B_1 C_2$ and $B_2 C_1$, $C_1 A_2$ and $C_2 A_1$, $A_1 B_2$ and $A_2 B_1$ meet in X, Y, Z. Prove that XYZ is in perspective with $A_1 B_1 C_1$ and with $A_2 B_2 C_2$. (Consider the triangles $B_1 C_1 A_2$, $B_2 C_2 A_1$.)

10. State the three-dimensional dual of the following: If $VABC$, $V'ABC$ are pyramids on the same base ABC, then the planes VAV', VBV', VCV' are collinear.

11. $l_r \equiv a_r x + b_r y + c_r z = 0$ ($r = 1$ to 4) are the equations of four planes in space. Prove that numbers p_1, p_2, p_3, p_4 exist such that $p_1 l_1 + p_2 l_2 + p_3 l_3 + p_4 l_4 \equiv 0$, and state the condition for $p_1 : p_2 : p_3 : p_4$ to be unique.

12. State some correspondences in four-dimensional duality.

11·7. Polar of a Point wo a Line-pair

11·71. If four lines v, a, p, b through a point O form a harmonic pencil, any line not through O meets these lines in a harmonic range. Two lines $VAPB$, $VA'P'B'$ through a point V on v give two harmonic ranges with a common point V.

Now suppose that a and b are two fixed lines and V is a fixed point not on a or b. A variable line through V meets a and b

in points A and B and a point P can be taken in that line such that $(VAPB) = -1$. The locus of P is the line through O which with OV separates a and b harmonically. This line is called the *polar* of V wo the line-pair.

Any point V' in OV has the same polar as V. The point O has no polar. The idea of a polar is generalised in the next chapter.

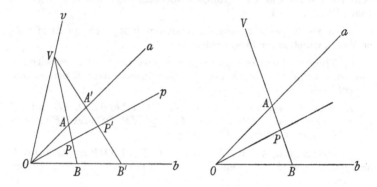

11·72. *Polar of* (x_1, y_1) *wo the line-pair composed of the axes.* The equation of OP_1 is $xy_1 - yx_1 = 0$. This line and its polar separate $y = 0$, $x = 0$ harmonically. Hence the polar is

$$xy_1 + yx_1 = 0.$$

11·73. *Polar of* (x_1, y_1) *wo the line-pair* $ax^2 + 2hxy + by^2 = 0$.

1st method. OP is $xy_1 = yx_1$. Let the polar be $xy_2 = yx_2$. Then the harmonic condition must be satisfied by

$$ax^2 + 2hxy + by^2 = 0 \quad \text{and} \quad (xy_1 - yx_1)(xy_2 - yx_2) = 0.$$

$$\therefore \quad ax_1x_2 + by_1y_2 + h(x_1y_2 + x_2y_1) = 0.$$

Hence the polar is

$$ax_1x + by_1y + h(yx_1 + xy_1) = 0.$$

2nd method. Let P_2 be an arbitrary point on the locus. Any point on P_1P_2 is

$$\left(\frac{\kappa_1 x_1 + \kappa_2 x_2}{\kappa_1 + \kappa_2}, \ \frac{\kappa_1 y_1 + \kappa_2 y_2}{\kappa_1 + \kappa_2} \right).$$

This lies on $ax^2 + 2hxy + by^2 = 0$ if

$$a(\kappa_1 x_1 + \kappa_2 x_2)^2 + 2h(\kappa_1 x_1 + \kappa_2 x_2)(\kappa_1 y_1 + \kappa_2 y_2) + b(\kappa_1 y_1 + \kappa_2 y_2)^2 = 0,$$

i.e.

$$(ax_1^2 + 2hx_1 y_1 + by_1^2)\kappa_1^2 + 2\{ax_1 x_2 + h(x_1 y_2 + x_2 y_1) + by_1 y_2\}\kappa_1 \kappa_2$$
$$+ (ax_2^2 + 2hx_2 y_2 + by_2^2)\kappa_2^2 = 0,$$

which determines the points of intersection of $P_1 P_2$ with $ax^2 + 2hxy + by^2 = 0$. Since these points separate P_1, P_2 harmonically, the sum of the values of κ_2/κ_1 given by the quadratic equation must be zero. Thus

$$ax_1 x_2 + h(x_1 y_2 + x_2 y_1) + by_1 y_2 = 0.$$

Thus P lies on $ax_1 x + h(yx_1 + xy_1) + by_1 y = 0.$

11·74. *Polar of* (x_1, y_1) *wo the line-pair* $\alpha\beta = 0$. Here α, β may be any expressions linear in the coordinates. The equation of OP_1 is of the form $\alpha = t\beta$ and that of the polar is $\alpha + t\beta = 0$. Since $\alpha = t\beta$ passes through P, t is given by $\alpha_1 = t\beta_1$ where α_1 and β_1 are the values found by substituting the coordinates of P_1 in α and β. Hence the equation of the polar is

$$\alpha\beta_1 + \beta\alpha_1 = 0.$$

11·8. Pole of a Line wo a Point-pair

11·81. If A, B are points which lie on a line o, and v is a fixed line not through A or B, then the point on o which with ov separates A, B harmonically is called the *pole* of v wo the point-pair.

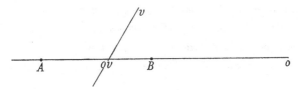

The idea of pole is generalised in the next chapter.

11·82. EXAMPLE. The medians of a triangle meet at (p, q) and two sides are $ax^2 + 2hxy + by^2 = 0$. Find the equation of the third side.

The mid-point M of the base is $(\frac{3}{2}p, \frac{3}{2}q)$. The polar of M wo the line-pair is, by 11·73,

$$(ap + hq)\,x + (hp + bq)\,y = 0.$$

The base is parallel to this line and passes through $(\frac{3}{2}p, \frac{3}{2}q)$. Hence the base is

$$(ap + hq)\,(2x - 3p) + (hp + bq)\,(2y - 3q) = 0.$$

EXERCISE 11c

1. Use the equation of the polar of (x_1, y_1) wo $ax^2 + 2hxy + by^2 = 0$ to verify that if the polar of P passes through Q, then the polar of Q passes through P.

2. Give the polar of any point on OX wo $ax^2 + 2hxy + by^2 = 0$.

3. Give the pole of $[X_1, Y_1, Z_1]$ wo the point-pair formed by the points at infinity on the axes.

4. Find the pole of $[X_1, Y_1, Z_1]$ wo the point-pair

$$AX^2 + 2HXY + BY^2 = 0.$$

5. Find the equation of the pole of $[X_1, Y_1]$ wo the point-pair composed of (x_1, y_1) and (x_2, y_2).

6. Prove that the three lines each of which forms a harmonic pencil with $ax^2y + 2hxy^2 + by^3 = 0$ are

$$(ax + hy)\,\{a(ax^2 + 2hxy + by^2) + 8(ab - h^2)\,y^2\} = 0.$$

7. A variable line through (f, g) meets the fixed lines

$$ax^2 + 2hxy + by^2 = 0$$

in P and Q. Prove that the vertex R of the parallelogram $POQR$ lies on the conic $(ax + hy)\,(x - 2f) + (hx + by)\,(y - 2g) = 0.$

8. Find the polar of (x_1, y_1, z_1) wo the line-pair

$$(l_1 x + m_1 y + n_1 z)\,(l_2 x + m_2 y + n_2 z) = 0$$

and express the result in terms of a, b, c, f, g, h where

$$a = l_1 l_2, \quad 2f = m_1 n_2 + m_2 n_1, \text{ etc.}$$

Chapter 12

THE GENERAL CONIC

12·11. In 7·41 and 8·63 the general equation of a conic is taken to be

$$ax^2 + 2hxy + by^2 + 2gx + 2fy + c = 0$$

or

$$ax^2 + 2hxy + by^2 + 2gxz + 2fyz + cz^2 = 0.$$

It is proved in 4·45 and 8·65 that $\delta = 0$ is a necessary and sufficient condition for the conic to be degenerate in G_6, and in 7·43 that a non-degenerate conic has a single tangent at every point.

The general conic meets the line at infinity $z = 0$ where

$$ax^2 + 2hxy + by^2 = 0 \text{ and } z = 0.$$

Therefore in G_3, the conic has 2, 2 coincident, or 0 points at infinity, according as $ab - h^2 <, =, \text{ or } > 0$.

The corresponding types of curve are investigated in Chapters 15, 13, 14. In the present chapter the general properties of the conic are discussed. Some of the work will refer only to G_3.

12·12. EXAMPLE. Find the ratio in which the join of $(1, -3)$ and $(6, 7)$ is divided by the line $2x + 5y = 6$.

The point that divides the join in the ratio $\kappa_2 : \kappa_1$ is

$$\left(\frac{\kappa_1 + 6\kappa_2}{\kappa_1 + \kappa_2}, \frac{-3\kappa_1 + 7\kappa_2}{\kappa_1 + \kappa_2} \right).$$

It lies on $2x + 5y = 6$ if $\dfrac{2(\kappa_1 + 6\kappa_2)}{\kappa_1 + \kappa_2} + \dfrac{5(-3\kappa_1 + 7\kappa_2)}{\kappa_1 + \kappa_2} = 6$,

i.e. $\qquad 2(\kappa_1 + 6\kappa_2) + 5(-3\kappa_1 + 7\kappa_2) = 6(\kappa_1 + \kappa_2).$

or $\qquad\qquad\qquad 41\kappa_2 = 19\kappa_1.$

$$\therefore \quad \kappa_2 : \kappa_1 = 19 : 41.$$

This is the required ratio.

12·13. EXAMPLE. If the join of $P_1(x_1, y_1)$ and $P_2(x_2, y_2)$ is cut by the curve $xy = c^2$, determine the ratios in which it is divided at the points of intersection.

The point
$$\left(\frac{\kappa_1 x_1 + \kappa_2 x_2}{\kappa_1 + \kappa_2}, \frac{\kappa_1 y_1 + \kappa_2 y_2}{\kappa_1 + \kappa_2}\right)$$

lies on the curve if

$$(\kappa_1 x_1 + \kappa_2 x_2)(\kappa_1 y_1 + \kappa_2 y_2) = c^2(\kappa_1 + \kappa_2)^2,$$

i.e. $(x_1 y_1 - c^2)\kappa_1^2 + (x_1 y_2 + x_2 y_1 - 2c^2)\kappa_1 \kappa_2 + (x_2 y_2 - c^2)\kappa_2^2 = 0.$

This is a quadratic equation for $\kappa_2 : \kappa_1$ whose roots are the ratios in which $P_1 P_2$ is divided at the two points where $xy = c^2$ meets it.

12·2. Joachimsthal's Method (G₃)

12·21. The method used in 12·12 and 12·13 will now be applied to the conic in G_3 given by the general equation $ax^2 + 2hxy + by^2 + 2gx + 2fy + c = 0$. It depends only on the formulae $x = \dfrac{\kappa_1 x_1 + \kappa_2 x_2}{\kappa_1 + \kappa_2}$, $y = \dfrac{\kappa_1 y_1 + \kappa_2 y_2}{\kappa_1 + \kappa_2}$, and is applicable whether the axes are rectangular or oblique.

Suppose that the general conic meets $P_1 P_2$ in the point A such that $P_1 A : A P_2 = \kappa_2 : \kappa_1$. Then since A is

$$\left(\frac{\kappa_1 x_1 + \kappa_2 x_2}{\kappa_1 + \kappa_2}, \frac{\kappa_1 y_1 + \kappa_2 y_2}{\kappa_1 + \kappa_2}\right)$$

and lies on the conic,

$$a\left(\frac{\kappa_1 x_1 + \kappa_2 x_2}{\kappa_1 + \kappa_2}\right)^2 + 2h\frac{\kappa_1 x_1 + \kappa_2 x_2}{\kappa_1 + \kappa_2}\frac{\kappa_1 y_1 + \kappa_2 y_2}{\kappa_1 + \kappa_2} + b\left(\frac{\kappa_1 y_1 + \kappa_2 y_2}{\kappa_1 + \kappa_2}\right)^2$$

$$+ 2g\frac{\kappa_1 x_1 + \kappa_2 x_2}{\kappa_1 + \kappa_2} + 2f\frac{\kappa_1 y_1 + \kappa_2 y_2}{\kappa_1 + \kappa_2} + c = 0.$$

$$\therefore \quad a(\kappa_1 x_1 + \kappa_2 x_2)^2 + 2h(\kappa_1 x_1 + \kappa_2 x_2)(\kappa_1 y_1 + \kappa_2 y_2)$$

$$+ b(\kappa_1 y_1 + \kappa_2 y_2)^2 + 2g(\kappa_1 + \kappa_2)(\kappa_1 x_1 + \kappa_2 x_2)$$

$$+ 2f(\kappa_1 + \kappa_2)(\kappa_1 y_1 + \kappa_2 y_2) + c(\kappa_1 + \kappa_2)^2 = 0,$$

$$\therefore \quad (ax_1{}^2 + 2hx_1y_1 + by_1{}^2 + 2gx_1 + 2fy_1 + c)\,\kappa_1{}^2$$

$$+ 2\{ax_1x_2 + h(x_1y_2 + x_2y_1) + by_1y_2 + g(x_1 + x_2) + f(y_1 + y_2) + c\}\,\kappa_1\kappa_2$$

$$+ (ax_2{}^2 + 2hx_2y_2 + by_2{}^2 + 2gx_2 + 2fy_2 + c)\,\kappa_2{}^2 = 0.$$

This is a quadratic equation for $\kappa_2 : \kappa_1$. Its roots are the values of $P_1A : AP_2$ for the points of intersection A of the line and conic. Since the equation is a quadratic, it has 2, 2 coincident, or 0 roots, and consequently a line meets a conic in 2, 2 coincident, or 0 points A. The equation is called *Joachimsthal's ratio equation*.

12·22. We denote $ax^2 + 2hxy + by^2 + 2gx + 2fy + c$ by s, and

$$ax_mx_n + h(x_my_n + x_ny_m) + by_my_n + g(x_m + x_n) + f(y_m + y_n) + c$$

by s_{mn}.

For example,

$$s_{11} \equiv ax_1{}^2 + 2hx_1y_1 + by_1{}^2 + 2gx_1 + 2fy_1 + c,$$

$$s_{12} \equiv ax_1x_2 + h(x_1y_2 + x_2y_1) + by_1y_2 + g(x_1 + x_2) + f(y_1 + y_2) + c,$$

$$s_{22} \equiv ax_2{}^2 + 2hx_2y_2 + by_2{}^2 + 2gx_2 + 2fy_2 + c.$$

Dropping the suffix m from s_{mn} gives s_n, and so

$$axx_n + h(xy_n + x_ny) + byy_n + g(x + x_n) + f(y + y_n) + c$$

is denoted by s_n.

For example,

$$s_1 \equiv axx_1 + h(xy_1 + x_1y) + byy_1 + g(x + x_1) + f(y + y_1) + c,$$

$$s_2 \equiv axx_2 + h(xy_2 + x_2y) + byy_2 + g(x + x_2) + f(y + y_2) + c,$$

and, since these are of the first degree in x, y, $s_1 = 0$ and $s_2 = 0$ are lines.

$s = 0$ is the equation of the conic,

$s_{11} = 0$ is the condition for P to lie on the conic,

$s_{12} = 0$ is the condition for P_1 to lie on $s_2 = 0$ or for P_2 to lie on $s_1 = 0$.

With this notation the results for the general conic are as

compact for special cases such as those in 12·12 and 12·13. Joachimsthal's ratio equation becomes

$$s_{11}\kappa_1^2 + 2s_{12}\kappa_1\kappa_2 + s_{22}\kappa_2^2 = 0.$$

12·23. Internal and External Points. When s_{11} and s_{22} have opposite signs, the product of the roots of Joachimsthal's equation is negative. Hence one and only one root is positive and the conic meets P_1P_2 internally just once. P_1 and P_2 are then said to be on opposite sides of the conic. It can be proved that tangents can be drawn to the conic from points on one side; and that side is called the *outside*. See Exercise 12B, No. 39. Thus internal and external points can be defined. The distinction disappears in G_4.

EXERCISE 12A

In Nos. 1–4, find the ratio in which the join of the points is divided at its points of intersection with the locus.

1. $(1,3)$, $(7, -2)$; $x + 2y + 1 = 0$.

2. (x_1, y_1), (x_2, y_2); $ax + by + c = 0$.

3. $(-1, 1)$, $(5, -2)$; $x^2 + 3xy + y^2 = 1$.

4. $(-3, 2)$, $(1, 5)$; $2x^2 - 2xy + y^2 + 10x - 3y = 0$.

5. Write in full the work of obtaining $s_{11}\kappa_1^2 + 2s_{12}\kappa_1\kappa_2 + s_{22}\kappa_2^2 = 0$ for the conic $y^2 - 4ax = 0$.

In Nos. 6–10, give the value of $s_{11}\kappa_1^2 + 2s_{12}\kappa_1\kappa_2 + s_{22}\kappa_2^2$ in full.

6. $ax^2 + by^2 = c$.　　　7. $xy = k^2$.　　　8. $ax^2 + 2hxy + by^2 = k$.

9. $3x^2 - 5xy + y^2 - 4x + 7y - 11 = 0$.　　10. $(x-a)^2 + (y-b)^2 = c^2$.

11. State the condition for the general conic to pass through the mid-point of P_1P_2.

12. Find the ratio in which the line joining $(6, 30)$ to $(11, -50)$ is divided at its points of intersection with $2(x^3 + y^3) = 63xy$.

13. Find whether the points $(4, 2\frac{1}{2})$, $(-10\frac{1}{2}, 10)$ are on the same or opposite sides of the curve $x^2 - y^2 = 10$.

14. State the condition for $(0, a)$ and (b, c) to be on the same side of $x^2 = 4ay$.

15. Express the condition for the point whose coordinates are

$$(\kappa_1 x_1 + \kappa_2 x_2)/(\kappa_1 + \kappa_2), \quad (\kappa_1 y_1 + \kappa_2 y_2)/(\kappa_1 + \kappa_2)$$

to lie on the curve

$$s \equiv ax^3 + 3bx^2y + 3cy^2 + 3dx + e = 0$$

in the form $s_{111}\kappa_1{}^3 + 3s_{112}\kappa_1{}^2\kappa_2 + 3s_{122}\kappa_1\kappa_2{}^2 + s_{222}\kappa_2{}^3 = 0$,

giving the values of s_{112} and s_{122}. What kind of loci are represented by $s_{11} = 0$ and $s_1 = 0$? Interpret the condition $s_{111} = 0$.

12·3. Deductions from Joachimsthal's Equation

12·31. Suppose that P_1 is a fixed point on the conic. Then $s_{11} = 0$, and one root of the equation, regarded as a quadratic for κ_2/κ_1, is zero. The second root is also zero if $s_{12} = 0$. Hence $s_{12} = 0$ is the condition for P_2 to lie on the tangent at P_1.

Hence the equation of the tangent at P_1 is $s_1 = 0$.

12·32. Suppose that P_1 is a fixed point outside the conic. The quadratic equation has equal roots when $s_{11}s_{22} = s_{12}{}^2$. This happens when P_1P_2 meets the conic in two coincident points. Hence $s_{11}s_{22} = s_{12}{}^2$ is the condition for P_2 to lie on any tangent from P_1.

Hence the equation of the tangents from P_1 is $s_{11}s = s_1{}^2$. This equation is of the second degree: there may be two tangents from P_1 to the conic, but in G_3 there may also be none.

12·33. The points where the tangents from P_1 touch the conic belong both to $s = 0$ and to $s_{11}s = s_1{}^2$. Therefore they also belong to $s_1 = 0$. But this equation is of the first degree. Therefore it represents the join of the points of contact of the tangents from P_1. This line is called the *chord of contact* of P_1.

12·34. *The Polar.* Suppose that P_1 is any fixed point. Let P_1AA' be an arbitrary chord of the conic through P_1 and let P_2 be taken on this line so that $(P_1A\,P_2A') = -1$. Then the locus of P_2 is called the *polar* of P_1 wo the conic.

Since $P_1A/AP_2 = -P_1A'/A'P_2$, the quadratic for κ_2/κ_1 has equal roots with opposite signs.

$$\therefore \quad s_{12} = 0.$$

This shows that P_2 lies on $s_1 = 0$ which is a fixed line.

Hence the polar of a point is a line, and the polar of P_1 is $s_1 = 0$. This proof only shows that every point of the polar lies on the line $s_1 = 0$, not that every point of $s_1 = 0$ belongs to the polar. With the definition given, and in G_3, the polar may actually be only part of the line. If, for example, the conic is

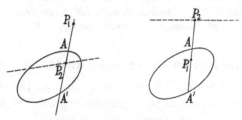

closed, and P_1 is outside it, P_2 must be inside it, and so the locus consists of the part of $s_1 = 0$ that is inside the conic. This however is not of great importance as the distinction between internal and external points disappears in the later geometries. In the meantime we shall use the term "polar" to denote the whole line.

It follows from 12·33 and 12·34 that the polar of an external point P_1 coincides with the chord of contact of P_1. Also from 12·31 the polar of a point on the conic coincides with the tangent at the point.

The argument of 12·21 is applicable even when the conic is a line-pair. Hence the polar of P_1 wo a line-pair $s = 0$ is $s_1 = 0$. See 11·73.

The equation of the polar, in full, is

$$s_1 \equiv (ax_1 + hy_1 + g)\,x + (hx_1 + by_1 + f)\,y + (gx_1 + fy_1 + c) = 0.$$

No such line exists if $ax_1 + hy_1 + g = hx_1 + by_1 + f = 0$, and there is in general in G_3 one point (x_1, y_1) for which these equations are true. The existence of a point which has no polar is explained in 12·5.

12·35. EXAMPLE. Find the locus of the meet of tangents at the ends of chords through a fixed point P_1.

Let P_2 be any point on the locus. Then the chord of contact of P_2 passes through P_1.

$$\therefore \quad s_2 = 0 \text{ is satisfied by } x = x_1, y = y_1.$$

Thus $s_{12} = 0$ and P_2 lies on $s_1 = 0$. Hence the locus is the polar of P or, strictly, is the part of the polar which lies outside the conic.

12·36. EXAMPLE. Find the point whose polar wo $2x^2 + 7y^2 = 6$ is $3x + 4y = 5$.

Let (x_1, y_1) be the point. Then

$$2xx_1 + 7yy_1 = 6 \quad \text{and} \quad 3x + 4y = 5$$

are the same line.

$$\therefore \quad 2x_1 : 7y_1 : 6 = 3 : 4 : 5.$$

Thus the point is $(\frac{9}{5}, \frac{24}{35})$.

12·37. EXAMPLE. If P_1 and P_2 lie on $s = 0$, verify that $s_1 + s_2 = s_{12}$ is the chord $P_1 P_2$.

Since P_1 lies on $s = 0$, $s_{11} = 0$, and therefore when x_1, y_1 are substituted for x, y, $s_1 + s_2$ becomes $s_{11} + s_{12}$ which is equal to s_{12}. Thus the coordinates of P_1 satisfy $s_1 + s_2 = s_{12}$.

Similarly the coordinates of P_2 satisfy it. But the equation is of the first degree in x, y. Hence it represents a line; and this line passes through P_1 and P_2.

12·38. EXAMPLE. Show that the tangents from the fixed point (f, g) to the conic $x^2/a + y^2/b = 1$ have fixed angle-bisectors when a, b vary in such a way that $a - b$ is constant.

The equation of the pair of tangents is

$$\left(\frac{f^2}{a} + \frac{g^2}{b} - 1\right)\left(\frac{x^2}{a} + \frac{y^2}{b} - 1\right) = \left(\frac{fx}{a} + \frac{gy}{b} - 1\right)^2.$$

The equation of the parallel lines through the origin is

$$\left(\frac{f^2}{a} + \frac{g^2}{b} - 1\right)\left(\frac{x^2}{a} + \frac{y^2}{b}\right) = \left(\frac{fx}{a} + \frac{gy}{b}\right)^2,$$

or

$$x^2(g^2 - b) - 2fgxy + y^2(f^2 - a) = 0.$$

By 4·34 the equation of the angle-bisectors of these lines is $\dfrac{x^2 - y^2}{g^2 - f^2 + a - b} + \dfrac{xy}{fg} = 0$, and, since $a - b$ is constant, these are fixed. So also are the parallel lines through (f, g).

EXERCISE 12B

In Nos. 1–3, write down the condition for (x_1, y_1) to lie on the conic, and assuming that this condition is satisfied give the equation of the tangent at (x_1, y_1).

    1. $ax^2 + by^2 = c$.   **2.** $xy = 1$.   **3.** $3x^2 + 4xy = 5$.

In Nos. 4, 5, write down the equation of the tangent at the point to the conic.

    4. $(1, -1)$, $x^2 + 3y^2 = 4$.   **5.** $(2, 0)$, $4xy + y^2 - 6x + 12 = 0$.

In Nos. 6–9, find the equation of the pair of tangents from the point to the conic.

    6. (x_1, y_1), $y^2 = ax + by$.   **7.** $(1, 10)$, $y^2 = 4x$.

    8. $(0, -b)$, $x^2 = 4a(y - c)$.   **9.** $(-1, -2)$, $x^2 = 4x + 5y$.

In Nos. 10–16, write down the equation of the polar of the point wo the conic.

    10. (x_1, y_1), $x^2 + y^2 = a^2$.   **11.** (x_1, y_1), $x^2/a^2 - y^2/b^2 = 1$

    12. (x_1, y_1), $y = x^2$.

    13. (x_1, y_1), $2x^2 + xy - 3y^2 - x + 2y + 1 = 0$.

    14. $(4, 7)$, $y^2 = 5x$.   **15.** $(3, -2)$, $y^2 - xy = 1$.

    16. $(-3, 1)$, $3x^2 + 4xy - 2y^2 + x - 8y + 7 = 0$.

In Nos. 17–23, find the point whose polar wo the conic is the given line.

    17. $x^2 + y^2 = 1$, $2x + 3y = 4$.   **18.** $y^2 = 4x$, $2x - 3y = 10$.

    19. $2xy = 1$, $2x + y = 5$.   **20.** $3x^2 + 4xy + y^2 = 1$, $5x + 3y = 1$.

    21. $x^2 = y$, $ax + by + c = 0$.

    22. $2x^2 - xy + 3x - 5y + 4 = 0$, $2x - 7y = 1$.

    23. $a + by + cy^2 = x^2$, $y = 0$.

    24. Show that the tangent at (x_1, y_1) to $xy = k^2$ is $x/x_1 + y/y_1 = 2$ and state the equation of the polar of (x_1, y_1).

    25. Write down the equation of the pair of tangents from (x_1, y_1) to $y^2 = 4ax$. To what does it reduce when

$$\text{(i) } x_1 = b, \ y_1 = 0, \quad \text{(ii) } x_1 = -a, \ y_1 = b?$$

Verify in (ii) that the tangents are at right angles.

    26. Show that O has no polar wo $ax^2 + by^2 = c$, but that every point has a polar wo $y^2 = 4ax$.

    27. Show that no line through O is the polar of any point wo $ax^2 + by^2 = c$ and that no line parallel to OX is a polar wo $y^2 = 4ax$.

28. Are there any points which have no polar wo

$$\text{(i)} \ x^2 = y^2, \quad \text{(ii)} \ x^2 = 1?$$

29. Use the result of 12·37 to write down the equation of the chord $P_1 P_2$ of $xy = k^2$. Deduce the tangent at $(kt, k/t)$.

30. Show how the tangent at P to $s = 0$ can be deduced from the pair of tangents from an external point.

31. If OP, OQ are the tangents from O to the circle

$$x^2 + y^2 + 2gx + 2fy + c = 0,$$

find the equations of PQ and the circle OPQ.

32. If $s \equiv x^2 + y^2 - 4a^2$ and $x_1 = a$ and $y_1 = 0$, verify that $ss_{11} = s_1^2$ represents the point $(a, 0)$ only, and explain the result.

33. Verify that the polar of P wo $x^2 + y^2 = a^2$ is the perpendicular to OP at the inverse of P.

34. Find the inverse of $(6, 2)$ wo $x^2 + y^2 - 4x + 2y = 5$.

35. Find the equation of the lines through the origin parallel to the tangents from P_1 to $ax^2 + by^2 = 1$.

36. Find the condition of perpendicularity for the tangents from (x_1, y_1) to $x^2/a^2 + y^2/b^2 = 1$. Deduce that the locus of the point of intersection of perpendicular tangents to this ellipse is $x^2 + y^2 = a^2 + b^2$.

37. Find the locus of the point of intersection of perpendicular tangents to (i) $x^2/a^2 - y^2/b^2 = 1$, (ii) $y^2 = 4ax$, (iii) $y^2 = 4axz$.

38. If the tangents from P to $ax^2 + by^2 = 1$ harmonically separate those to $cx^2 + dy^2 = 1$, prove that P lies on

$$acx^2(b+d) + bdy^2(a+c) = bc + ad.$$

39. Show that the condition for $ss_{11} = s_1^2$ to represent lines in real geometry is $s_{11} \delta < 0$. [Hence for a non-degenerate conic it represents lines if P_1 is on one side of the conic.]

40. To what form does $ss_{11} = s_1^2$ reduce when $\delta = 0$?

12·4. Joachimsthal's Method (G_6)

12·41. When the method of 12·2 is applied in G_5 or G_6 the conic is

$$ax^2 + 2hxy + by^2 + 2gxz + 2fyz + cz^2 = 0$$

and A is the point $(\kappa_1 x_1 + \kappa_2 x_2, \kappa_1 y_1 + \kappa_2 y_2, \kappa_1 z_1 + \kappa_2 z_2)$. This is a point on $P_1 P_2$ determined by the homogeneous parameter (κ_1, κ_2). It coincides with P_1 when (κ_1, κ_2) is $(1, 0)$ and coincides with P_2 when (κ_1, κ_2) is $(0, 1)$. But it is not necessarily the point which divides $P_1 P_2$ in the ratio $\kappa_2 : \kappa_1$.

The quadratic equation which expresses that A lies on the conic is

$$s_{11}\kappa_1^2 + 2s_{12}\kappa_1\kappa_2 + s_{22}\kappa_2^2 = 0,$$

where

$$s_{mn} \equiv ax_m x_n + h(x_m y_n + x_n y_m) + by_m y_n$$
$$+ g(x_m z_n + x_n z_m) + f(y_m z_n + y_n z_m) + cz_m z_n.$$

In G_6 this equation has two roots, and therefore $P_1 P_2$ meets the conic in two points. The roots may be equal and the points are then coincident.

If $s_{11} = 0$, P_1 is on the conic, and $s_1 = 0$ is the equation of the tangent at P.

12·42. There is no distinction between external and internal points. If $s_{11} \neq 0$, there is a pair of tangents $s_{11}s = s_1^2$ from P_1 to the conic.

12·43. $s_1 = 0$ is the equation of the chord of contact of P_1.

12·44. The *polar* of P_1 is defined as the locus of P_2 such that $(P_1 A P_2 A') = 1$. The homogeneous parameters of A, A' have the values (κ_1, κ_2), (κ_1', κ_2') found by solving the equation

$$s_{11}\kappa_1^2 + 2s_{12}\kappa_1\kappa_2 + s_{22}\kappa_2^2 = 0$$

and those of P_1, P_2 are $(1, 0)$, $(0, 1)$. Hence, by 10·86, the harmonic condition is $s_{12} = 0$. Therefore the polar of P_1 is $s_1 = 0$. It is the whole of this line and coincides with the chord of contact of P_1 and with the locus defined in 12·35.

The equation of the polar of P_1 is

$$(ax_1 + hy_1 + gz_1)x + (hx_1 + by_1 + fz_1)y + (gx_1 + fy_1 + cz_1) = 0. \quad (1)$$

The only case of failure arises when

$$ax_1 + hy_1 + gz_1 = hx_1 + by_1 + fz_1 = gx_1 + fy_1 + cz_1 = 0 \quad (2)$$

and this requires $\delta = 0$. Therefore every point has a polar wo a non-degenerate conic in G_6. This should be compared with 12·34. In general (1) represents the polar of P_1 even when the conic is a line-pair: the point given by (2) is the point of intersection of the lines of the pair and this point has no polar. It is called the *centre* of the line-pair.

12·45. *Reciprocal Property of the Polar.* If the polar of P_1 passes through P_2, the symmetry of the property (11·11)

$$(P_1 A P_2 A') = -1$$

shows that the polar of P_2 passes through P_1. Two points such that the polar of one passes through the other are called *conjugate* points. The condition for P_1 and P_2 to be conjugate points is $s_{12} = 0$.

12·46. *Polars of collinear Points.* If the polars of P_1, P_2 meet at Q, then by the reciprocal property the polar of Q passes through P_1 and P_2. Hence $P_1 P_2$ is the polar of Q. If P is any other point on $P_1 P_2$, since the polar of Q passes through P, the polar of P passes through Q. Hence the polars of collinear points are concurrent, and the line of collinearity is the polar of the point of concurrence.

12·47. EXAMPLE. Find the equation of the chord of $s = 0$ whose mid-point is P_1.

1st method (G_3). Let AA' be the chord and if possible let it meet the polar of P_1 in P_2. Then $AP_1 = P_1 A'$,

$$AP_2 = -P_2 A' = A'P_2,$$

which is absurd. Therefore the chord cannot meet the polar of P_1. Hence it is parallel to the polar and has an equation $s_1 = k$. But it passes through P_1.

$\therefore s_{11} = k$. Hence the chord is

$$s_1 = s_{11}.$$

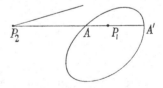

2nd method (G_6). Let the chord be AA' and suppose that it meets the line at infinity in P_2. Then $(AP_1 A' P_2) = -1$. $\therefore P_2$ lies on the polar $s_1 = 0$ of P_1. Hence the chord passes through the point of intersection of $s_1 = 0$ and $z = 0$. Hence its equation is $s_1 = kz$. It also passes through P_1. $\therefore s_{11} = kz_1$. Hence the equation of the chord is $s_1 z_1 = s_{11} z$.

The equation obtained in the first method can be obtained from $s_1 z_1 = s_{11} z$ by putting $z = z_1 = 1$.

12·48. Joachimsthal's method can also be applied in G_4 and G_5. The results of 12·45 and 12·46 are valid in any geometry G_3, G_4, G_5, or G_6, subject to the existence of the polars.

12·5. Diameters

12·51. The polar of P_1 wo $s = 0$ is

$$s_1 \equiv x_1(ax + hy + gz) + y_1(hx + by + fz) + z_1(gx + fy + cz) = 0. \quad (1)$$

If P_1 is at infinity, $z_1 = 0$ and the equation of the polar reduces to

$$x_1(ax + hy + gz) + y_1(hx + by + fz) = 0. \quad (2)$$

But when P_1 is at infinity, the chords through it are parallel chords, and the point P_2 such that $(P_1 A\, P_2 A') = -1$ is the mid-point of AA'. Hence

The locus of the mid-points of parallel chords through $(x_1, y_1, 0)$ is a line whose equation is (2).

Such a line is called a *diameter*.

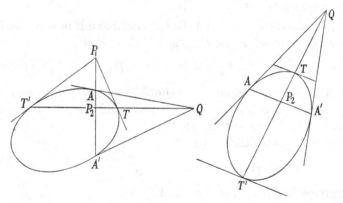

12·52. For different values of $x_1 : y_1$ the corresponding diameters are the polars of collinear points (at infinity) and they are therefore concurrent by 12·46. The point of concurrence is called the *centre* of the conic because it is the mid-point of every chord through it.

From 12·51 (2) the centre is the point of intersection of

$$ax + hy + gz = 0 \quad \text{and} \quad hx + by + fz = 0$$

and this is the point given by $x:y:z = hf-bg:gh-af:ab-h^2$. With the notation of 0·6 the point is (G, F, C).

The equations $ax+hy+gz = 0$, $hx+by+fz = 0$ may be written

$$\frac{\partial s}{\partial x} = 0, \qquad \frac{\partial s}{\partial y} = 0.$$

12·53. The corresponding point in G_3 is $(G/C, F/C)$. It does not exist when $C = 0$. The terms centre and diameter are also used in G_3. Thus the centre of

$$ax^2 + 2hxy + by^2 + 2gx + 2fy + c = 0$$

is given by $ax+hy+g = 0$, $hx+by+f = 0$ provided that $ab-h^2 \neq 0$. It is the mid-point of the chords that pass through it, which are called diameters. The centre is the point which has no polar (12·34). Also there are no points which have the diameters for polars.

12·54. It is not possible to illustrate all the polar properties in G_6 in a single diagram. The figure shows the polar of P_1 as a chord of contact TT', as the harmonic locus, and as the locus of the meet Q of tangents at the ends of chords through P_1 (12·35). It does not show the points at which the conic meets P_1Q or the chord through P_1 which has the tangents at its ends meeting at P_2.

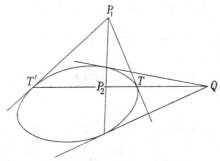

If P_1 is at infinity, no point of G_3 corresponds to it, but the tangents at T, T' are parallel, and TT' is the diameter bisecting chords parallel to AA'; also the tangents at A, A' meet on the diameter. See the figure in 12·51.

12·55. EXAMPLE. Find the locus of the centre of the conic

$$2x^2 - 2t^2xy + y^2 = 2t^3(x+y) \qquad \text{when } t \text{ varies.}$$

The centre is given by $\dfrac{\partial s}{\partial x} = 0$, $\dfrac{\partial s}{\partial y} = 0$,

i.e. $\qquad\qquad\qquad 2x - t^2y = t^3$,

and $\qquad\qquad\qquad -t^2x + y = t^3$,

$$t^3 : t^2 : 1 = y^2 - 2x^2 : 2x - y : y - x,$$

$$(2x - y)^3 = (y^2 - 2x^2)^2 (y - x).$$

Hence the locus is a quintic curve.

EXERCISE 12 c

In Nos. 1, 2, obtain ab initio the tangent, polar, and pair of tangents for the conic.

1. $x^2/a^2 + y^2/b^2 = 1$. **2.** $y^2 = 4axz$.

In Nos. 3–7, write down the polar of the point wo the conic.

3. (x_1, y_1, z_1), $x^2 - y^2 = a^2z^2$. **4.** (x_1, y_1, z_1), $xy = k^2z^2$.

5. (x_1, y_1, z_1), $k(x^2 + y^2) + 2gxz + 2fyz + cz^2 = 0$.

6. $(-1, 1, 0)$, $y^2 - x^2 = 2axz$.

7. $(3, 0, -2)$, $x^2 + 3y^2 + 2z^2 - 4yz - 3zx + 5xy = 0$.

In Nos. 8, 9, write down the conditions for P_1 and P_2 to be conjugate points wo the conics.

8. $x^2/a^2 - y^2/b^2 = 1$; $x^2/a^2 - y^2/b^2 = z^2$..

9. $xy = k^2$; $xy = k^2z^2$.

10. Write in full the work of obtaining $s_{11}k_1^2 + 2s_{12}k_1k_2 + s_{22}k_2^2 = 0$ for the conic $xy = k^2z^2$.

In Nos. 11, 12, give in full the value of s_{12}.

11. $ax^2 + 2hxy + by^2 = k$. **12.** $k(x^2 + y^2) + 2gxz + 2fyz + cz^2 = 0$.

13. Find the points of intersection of $y = bz$ and $y^2 = 4axz$, and the point of concurrence of the polars wo the conic of points on the line.

14. Find the point of concurrence of polars of points on $x = kz$ wo $ax^2 + by^2 = cz^2$.

15. Find the point whose polar wo $4x^2 - 6xy + 9y^2 + 12yz = 0$ is $2x - 3y + z = 0$.

16. Find the point whose polar wo $ax^2 + 2hxy + by^2 = 1$ is

$$lx + my + n = 0$$

and discuss the case of $(ax + hy)^2 = 1$.

17. Answer the same questions as in No. 16 for $ax^2 + 2hxy + by^2 = z^2$ and $lx + my + nz = 0$ and discuss the case of $(ax + hy)^2 = z^2$.

18. Give the polar of P_1 wo the parallel line-pair

$$(lx + my + 2n_1)(lx + my + 2n_2) = 0,$$

stating when it does not exist. Answer the corresponding questions in G_6.

19. Write down the equation of the chord of $y^2 = 4ax$ whose mid-point is (x_1, y_1). Find the locus of the mid-points of chords through the fixed point (f, g).

20. Find the locus of the mid-points of chords of $xy = k^2$ through the fixed point (f, g).

21. Find the locus of the mid-points of chords of $ax^2 + by^2 = c$ that are parallel to $lx + my + n = 0$.

In Nos. 22–29, find the centres of the conics.

22. $ax^2 + by^2 = cz^2$; $ax^2 + by^2 = c$.

23. $xy = k^2 z^2$; $xy = k^2$.   **24.** $y^2 = 4axz$; $y^2 = 4ax$.

25. $x^2 + 4xy + 3y^2 + 6x + 8y + 5 = 0$.

26. $2x^2 + 5xy + 7y^2 - 10x + 20y - 1 = 0$.

27. $2x^2 + xy + y^2 - 6xz + 11z^2 = 0$.

28. $9x^2 - 24xy + 16y^2 - 12x + 16y + 3 = 0$.

29. $2gxz + 2fyz + cz^2 = 0$.

30. Write down the equation of the polar of (k, mk) wo $s = 0$ and find its limit when $k \to \infty$. Interpret the result.

31. Find the points whose polars wo $s = 0$ are $ax + hy + gz = 0$, $hx + by + fz = 0$, and $gx + fy + cz = 0$. Explain the meaning of the condition of concurrence of these lines.

12·6. Envelope Equations

12·61. The method of Joachimsthal can be applied to the general envelope equation

$$AX^2 + 2HXY + BY^2 + 2GX + 2FY + C = 0 \quad \text{in } G_3$$

or

$$AX^2 + 2HXY + BY^2 + 2GXZ + 2FYZ + CZ^2 = 0 \quad \text{in } G_6.$$

When $\Delta = 0$, the left side of the equation is the product of linear factors and the equation represents a point-pair. We assume now that $\Delta \neq 0$.

The envelope equation is denoted by $S = 0$ and the meanings of S_1, S_{11}, S_{12}, etc. are defined, as in 12·22, by

$$S_{mn} \equiv AX_mX_n + H(X_mY_n + X_nY_m) + BY_mY_n$$
$$+ G(X_m + X_n) + F(Y_m + Y_n) + C \quad \text{in } G_3$$

and in G_6 by

$$S_{mn} \equiv AX_mX_n + H(X_mY_n + X_nY_m) + BY_mY_n$$
$$+ G(X_mZ_n + X_nZ_m) + F(Y_mZ_n + Y_nZ_m) + CZ_mZ_n.$$

It is convenient to work in G_6.

Let p_1, p_2 be two lines $[X_1, Y_1, Z_1]$, $[X_2, Y_2, Z_2]$. An arbitrary line through their point of intersection $p_1 p_2$ is

$$[\kappa_1 X_1 + \kappa_2 X_2, \ \kappa_1 Y_1 + \kappa_2 Y_2, \ \kappa_1 Z_1 + \kappa_2 Z_2],$$

where (κ_1, κ_2) is a homogeneous parameter. This line a belongs to the envelope $S = 0$ if

$$A(\kappa_1 X_1 + \kappa_2 X_2)^2 + 2H(\kappa_1 X_1 + \kappa_2 X_2)(\kappa_1 Y_1 + \kappa_2 Y_2)$$
$$+ B(\kappa_1 Y_1 + \kappa_2 Y_2)^2 + 2G(\kappa_1 X_1 + \kappa_2 X_2)(\kappa_1 + \kappa_2)$$
$$+ 2F(\kappa_1 Y_1 + \kappa_2 Y_2)(\kappa_1 + \kappa_2) + C(\kappa_1 + \kappa_2)^2 = 0,$$

and, as in 12·22, this reduces to

$$S_{11}\kappa_1{}^2 + 2S_{12}\kappa_1\kappa_2 + S_{22}\kappa_2{}^2 = 0.$$

This quadratic equation has two roots, and these give the parameters (κ_1, κ_2), (κ_1', κ_2') of the two lines a, a' of the envelope which pass through $p_1 p_2$. The roots may be equal and the lines a, a' are then coincident.

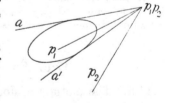

Suppose that one of the lines p_1, p_2 belongs to the envelope. If this is the line p_1, then $S_{11} = 0$, and one value of (κ_1, κ_2) found by solving the quadratic equation is $(1, 0)$. The second value is also $(1, 0)$ if $S_{12} = 0$. But $S_{12} = 0$ is the condition for p_2 to pass through the point $S_1 = 0$ (which lies on p_1 in virtue of $S_{11} = 0$). Hence on any line p_1 of the envelope there is a point $S_1 = 0$ having the property that the two lines of the envelope which pass through it are coincident with p_1. This point is the contact of the line p_1 with its envelope. The contact of a line with its envelope is the dual of the tangent at a point to a locus. The contact of a line is introduced in 4·1 as a limit. The equivalence of the two aspects is explained as in 7·33.

12·62. If $S_{11} \neq 0$, there is a pair of contacts of the envelope which lie on p_1. These are the points $p_1 p_2$ found by expressing that the lines of the envelope which pass through $p_1 p_2$ are coincident. The condition is $S_{11} S_{22} = S_{12}{}^2$ and it holds if p_2 belongs to the envelope $S_{11} S = S_1{}^2$. This is the equation of the point-pair of contacts on p_1.

The contacts on p_1 are the duals of the tangents from P_1.

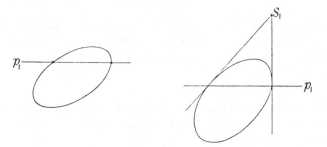

12·63. The lines of the envelope $S = 0$ whose contacts are on p_1 belong both to the envelope $S = 0$ and to the envelope $S_{11} S = S_1{}^2$. They therefore also belong to $S_1 = 0$. But this equation is of the first degree. Therefore it represents the meet of the lines of the envelope whose contacts are on p_1. This is the dual of the chord of contact of P_1.

12·64. *The Pole.* Suppose that p_1 is any fixed line. Let a, a' be the lines of the envelope $S = 0$ through an arbitrary point on p_1, and let p_2 be the line through this point such that $(p_1 a p_2 a') = -1$. Then the envelope of p_2 is called the *pole* of p_1 wo the envelope conic $S = 0$.

Proper parameters of a, a' are the values of $(\kappa_1, \kappa_2), (\kappa_1', \kappa_2')$ given by
$$S_{11} \kappa_1{}^2 + 2 S_{12} \kappa_1 \kappa_2 + S_{22} \kappa_2{}^2 = 0$$

and those of p_1, p_2 are $(1,0), (0,1)$. Hence by 10·86 the harmonic condition is

$$\kappa_2'/\kappa_1' = -\kappa_2/\kappa_1.$$

$$\therefore \quad S_{12} = 0.$$

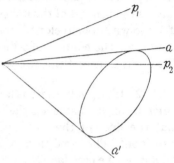

This shows that p_2 passes through $S_1 = 0$, which is a fixed point. Hence the pole is a point, and the pole of p_1 is the point $S_1 = 0$.

The equation of the pole of p_1 wo $S = 0$ in G_6 is

$$(AX_1 + HY_1 + GZ_1) X + (HX_1 + BY_1 + FZ_1) Y$$
$$+ (GX_1 + FY_1 + CZ_1) Z = 0. \quad (1)$$

The only case of failure arises when the coefficients of X, Y, Z are all zero, and this cannot happen unless $\Delta = 0$. Thus every line has a pole wo a non-degenerate envelope conic. See Exercise 12 D, No. 17.

12·65. By the symmetry of the property $(p_1 a p_2 a') = -1$ with respect to p_1 and p_2, if the pole of p_1 lies on p_2 then the pole of p_2 lies on p_1.

Two lines such that the pole of one lies on the other are called *conjugate* lines.

The poles of concurrent lines are collinear and the point of concurrence is the pole of the line of collinearity. This is proved by the dual of the method in 12·45.

12·66. EXAMPLE. Find the envelope of the join of the contacts of lines of $S = 0$ which meet on a given line p_1.

Let p_2 be any line of the envelope. Then the pole of p_2 lies on p_1. $\therefore S_{12} = 0$. Hence p_2 belongs to the envelope $S_1 = 0$. Therefore the required envelope is the pole of p_1.

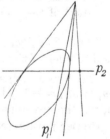

12·67. It is not possible to illustrate all the properties of the pole in G_6 in a single diagram. The figure shows the pole of p_1 as the meet of the lines t, t' whose contacts lie on p_1, as the harmonic envelope, and as the envelope of the join q of the contacts of lines of $S = 0$ which meet on p_1. It does not show the lines of the envelope through qp_1 or the point on p_1 the lines through which have their contacts on p_2.

The figures in 12·54 and 12·67 are dual, but they can also be regarded as the same figure with a different interpretation.

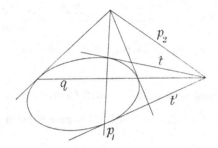

12·68. EXAMPLE. Find the condition that $Xx + Yy + 1 = 0$ should touch $x^2/a^2 + y^2/b^2 = 1$.

Let (x_1, y_1) be the point of contact.

This lies on $Xx + Yy + 1 = 0$. $\therefore Xx_1 + Yy_1 + 1 = 0$.

Since

$$s \equiv x^2/a^2 + y^2/b^2 - 1, \ s_1 \equiv xx_1/a^2 + yy_1/b^2 - 1,$$

and the tangent is

$$xx_1/a^2 + yy_1/b^2 = 1.$$

This must be the same as

$$Xx + Yy + 1 = 0.$$

$$\therefore \ \frac{x_1}{a^2 X} = \frac{y_1}{b^2 Y} = -1,$$

$\therefore$ from $Xx_1 + Yy_1 + 1 = 0$, $a^2 X^2 + b^2 Y^2 = 1$, which is the required condition.

12·69. EXAMPLE. Find the condition for $Xx + Yy + Zz = 0$
to be a tangent to $y^2 = 4axz$.

Let (x_1, y_1, z_1) be the point of contact.

This lies on $Xx + Yy + Zz = 0$. $\therefore Xx_1 + Yy_1 + Zz_1 = 0$.
Since $s \equiv y^2 - 4axz$, $s_1 \equiv yy_1 - 2a(xz_1 + zx_1)$, and the tangent is

$$yy_1 = 2a(xz_1 + zx_1).$$

This must be the same as $Xx + Yy + Zz = 0$,

$$\therefore \frac{2az_1}{X} = \frac{-y_1}{Y} = \frac{2ax_1}{Z},$$

$\therefore$ from $Xx_1 + Yy_1 + Zz_1 = 0$, $\quad \dfrac{XZ}{2a} - Y^2 + \dfrac{ZX}{2a} = 0,$

i.e. $aY^2 = XZ$, which is the required condition.

12·7. Tangents to s = 0 and Contacts of S = 0

12·71. If $[X, Y, Z]$ is a tangent to the general non-degenerate
locus conic

$$ax^2 + 2hxy + by^2 + 2gxz + 2fyz + cz^2 = 0,$$

let (x_1, y_1, z_1) be the point of contact. Then

$$(ax_1 + hy_1 + gz_1)x + (hx_1 + by_1 + fz_1)y + (gx_1 + fy_1 + cz_1)z = 0$$

must be identical with $Xx + Yy + Zz = 0$.

$$\therefore \quad \frac{ax_1 + hy_1 + gz_1}{X} = \frac{hx_1 + by_1 + fz_1}{Y} = \frac{gx_1 + fy_1 + cz_1}{Z} = k,$$

$$\therefore \quad ax_1 + hy_1 + gz_1 - kX = 0,$$

$$hx_1 + by_1 + fz_1 - kY = 0,$$

$$gx_1 + fy_1 + cz_1 - kZ = 0.$$

But also $Xx_1 + Yy_1 + Zz_1 = 0$ because the point of contact lies
on the tangent. Hence

$$\begin{vmatrix} a & h & g & X \\ h & b & f & Y \\ g & f & c & Z \\ X & Y & Z & 0 \end{vmatrix} = 0,$$

i.e., with the notation of 0·6,

$$AX^2 + 2HXY + BY^2 + 2GXZ + 2FYZ + CZ^2 = 0. \quad (1)$$

This is the condition for $[X, Y, Z]$ to be a tangent to $s = 0$.

The same proof with z and Z replaced by unity shows that the condition for $[X, Y]$ to be a tangent to

$$ax^2 + 2hxy + by^2 + 2gx + 2fy + c = 0$$

is
$$AX^2 + 2HXY + BY^2 + 2GX + 2FY + C = 0. \quad (2)$$

12·72. When $s = 0$ is a line-pair, with the notation of 4·45

$$A \equiv bc - f^2 = m_1 m_2 n_1 n_2 - \tfrac{1}{4}(m_1 n_2 + m_2 n_1)^2,$$
$$= -\tfrac{1}{4}(m_1 n_2 - m_2 n_1)^2,$$
$$F \equiv gh - af = \tfrac{1}{4}(n_1 l_2 + n_2 l_1)(l_1 m_2 + l_2 m_1) - \tfrac{1}{2}l_1 l_2 (m_1 n_2 + m_2 n_1)$$
$$= -\tfrac{1}{4}(n_1 l_2 - n_2 l_1)(l_1 m_2 - l_2 m_1).$$

Hence 12·71 (1) becomes

$$\{(m_1 n_2 - m_2 n_1) X + (n_1 l_2 - n_2 l_1) Y + (l_1 m_2 - l_2 m_1) Z\}^2 = 0,$$

and so it represents the point of intersection of the lines $s = 0$, twice.

12·73. Let (x, y, z) be a contact of a general non-degenerate envelope conic given by the equation

$$AX^2 + 2HXY + BY^2 + 2GXZ + 2FYZ + CZ^2 = 0,$$

where A, B, C, F, G, H have any values such that $\Delta \neq 0$.

Let $[X_1, Y_1, Z_1]$ be the line of the envelope of which it is the contact.

Then $S_1 = 0$ is the same as $xX + yY + zZ = 0$. Therefore, as in 12·71,
$$AX_1 + HY_1 + GZ_1 - kx = 0,$$
$$HX_1 + BY_1 + FZ_1 - ky = 0,$$
$$GX_1 + FY_1 + CZ_1 - kz = 0,$$

and also $xX_1 + yY_1 + zZ_1 = 0$ because $[X_1, Y_1, Z_1]$ passes through the contact. Hence

$$\begin{vmatrix} A & H & G & x \\ H & B & F & y \\ G & F & C & z \\ x & y & z & 0 \end{vmatrix} = 0,$$

i.e. $\quad A'x^2 + 2H'xy + B'y^2 + 2G'xz + 2F'yz + C'z^2 = 0,$ $\quad$ (1)

where $A' = BC - F^2$, $F' = GH - AF$, etc.

This is the condition for (x, y, z) to be a contact of $S = 0$.

The same proof with z and Z replaced by unity shows that the condition for (x, y) to be a contact of

$$AX^2 + 2HXY + BY^2 + 2GX + 2FY + C = 0$$

is $\quad A'x^2 + 2H'xy + B'y^2 + 2G'x + 2F'y + C' = 0.$ $\quad$ (2)

12·74. When $S = 0$ is a point-pair, it can be proved as in 12·72 that 12·73 (1) becomes the equation of the join of the points of the pair, twice.

12·75. Double Generation of the Conic

Consider a non-degenerate conic $s = 0$ which is the locus of a point P_1. For each position of P_1 there is a tangent at P_1. The tangents form a curve envelope whose equation is proved in 12·71 to be

$$AX^2 + 2HXY + BY^2 + 2GXZ + 2FYZ + CZ^2 = 0, \quad (1)$$

where $A = bc - f^2$, $F = gh - af$, etc.

Each line p_1 of this envelope has a contact and the locus of these contacts is proved in 12·73 to be

$$A'x^2 + 2H'xy + B'y^2 + 2G'xz + 2F'yz + C'z^2 = 0, \quad (2)$$

where $A' = BC - F^2$, $F' = GH - AF$, etc.

In 12·73 the coefficients A, F, ... are arbitrary, but in 12·75 (1) they are given by $A = bc - f^2$, $F = gh - af$, etc.

Also, by 0·6, $A' = BC - F^2 = a\delta$, $F' = GH - AF = f\delta$, etc. Hence the locus 12·75 (2) of the contacts of p_1 is the original locus conic $s = 0$.

A conic may therefore be regarded as a locus of points P having a tangent at each point or it can be regarded as an envelope of lines whose contacts are the points P.

$$ax^2 + 2hxy + by^2 + 2gxz + 2fyz + cz^2 = 0$$

is called the locus equation of the conic and

$$AX^2 + 2HXY + BY^2 + 2GXZ + 2FYZ + CZ^2 = 0$$

is called the envelope equation of the conic.

Starting from the locus equation the envelope equation is derived from it as in 12·71 and the values of A, F, etc. are $bc - f^2$, $gh - af$, etc.

But if the envelope equation is given, the locus equation is derived in the form 12·73 (1).

12·76. The polar of P_1 wo a locus conic is defined in 12·34 and 12·44; and the pole of p_1 wo an envelope conic is defined in 12·64. When a conic is regarded from both aspects, the polar of P_1 is the join q of the contacts of the tangents through P_1, and the pole of q is the meet of the tangents whose contacts are on q. Thus the pole of l is the point whose polar is l, and the polar of L is the line whose pole is L.

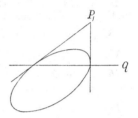

12·77. The Centre

The centre is defined in 12·52 as the point of concurrence of diameters, and this is, for G_6, the point whose polar is the line at infinity. When the envelope equation is available, the centre is found most conveniently by 12·64 (1) as the pole of the line at infinity.

Putting $X_1 = 0$, $Y_1 = 0$, $Z_1 = 1$, the centre is

$$GX + FY + CZ = 0.$$

Its coordinates are (G, F, C), in G_6. In G_3 the centre is $(G/C, F/C)$ if $C \neq 0$.

The centre of a point-pair, which is mid-way between the two points, can be found in this way. See Exercise 12 D, No. 18.

In Nos. 1–5, obtain the envelope or locus equation by an ab initio method.

1. $ax^2 + by^2 = c$.
2. $aX^2 + bY^2 = cZ^2$.
3. $xy = 1$.
4. $X^2 + 2YZ = 0$.
5. $axy + bx + cy + d = 0$.

In Nos. 6–10, calculate the values of A, B, C, F, G, H and write down the envelope equation.

6. $x^2/a^2 - y^2/b^2 = 1$.   **7.** $xy = k^2$.

8. $k(x^2 + y^2) + 2gxz + 2fyz + cz^2 = 0$.   **9.** $x^2 - xy - 2y^2 + 3x + 1 = 0$.

10. $55x^2 - 30xy + 39y^2 - 40xz - 24yz - 464z^2 = 0$.

In Nos. 11, 12, find the locus equation.

11. $X^2 + Y^2 = k^{-2}$.   **12.** $X^2 - Y^2 + 3XZ + YZ + 2Z^2 = 0$.

In Nos. 13–15, show that the conic has no envelope equation and find what form is taken by $S = 0$.

13. $(x+y)^2 = z^2$.   **14.** $(lx + my + 1)(l'x + m'y + 1) = 0$.

15. $(ax + by + cz)^2 = 0$.

16. Prove that $3X^2 - 4XY - 4Y^2 + 2X + 4Y - 1 = 0$ has no locus equation. What is the relation of $(y+2)^2 = 0$ to the given equation?

17. Give the equation of the pole of p_1 wo $S = 0$ in G_3 and examine the case of failure. Compare with 12·64.

18. Verify by 12·64(1) that the centre of

$$(x_1 X + y_1 Y + 1)(x_2 X + y_2 Y + 1) = 0 \quad \text{is} \quad (\tfrac{1}{2}(x_1 + x_2), \tfrac{1}{2}(y_1 + y_2)).$$

12·8. Self-Polar Triangles

12·81. If each side of a triangle is the polar of the opposite vertex wo a conic, the triangle is said to be *self-polar* wo the conic.

∞^1 triangles self-polar wo a given conic can be drawn with a given vertex A. For if B is any point on the polar of A, a third point C is determined as the point of intersection of the polars of A and B and AB is the polar of C. Hence ABC is self-polar.

12·82. In the figure, by 11·42, $(AQDZ) = -1$. Also

$$(BPCZ) \underset{X}{=} (AQDZ), \quad \therefore \quad (BPCZ) = -1.$$

Therefore the polar of Z wo any conic through A, B, C, D passes through Q and P and is therefore XY. Similarly the polar of X is ZY. Hence:

The diagonal points of a quadrangle are the vertices of a

triangle which is self-polar wo any conic through the points of the quadrangle.

It follows that the polar of any point X wo a conic is the locus of the point of intersection of AC and BD when AB and CD are two arbitrary lines through X.

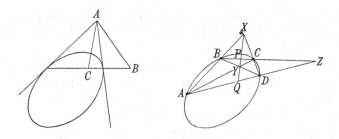

12·83. Dually the diagonal lines of a quadrilateral are the sides of a triangle self-polar wo any conic inscribed in the quadrilateral.

For in the figure, by 11·44, $(hzky) = -1$.

∴   the pole of y lies on z.

Also $(h'xk'y) = -1$. Therefore the pole of y lies on x.

Hence xz is the pole of y.

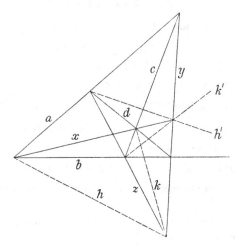

EXERCISE 12e

In Nos. 1, 2, use Joachimsthal's method to find the lines of the envelope which pass through the meet of the two given lines.

1. $[1, 3]$, $[5, -3]$; $2X^2 - 6XY + Y^2 = 18$.

2. $[3, 0, 2]$, $[4, -4, 0]$; $3Z^2 - 2YZ + 2ZX + XY = 0$.

In Nos. 3, 4, write down the condition for the line to belong to the envelope and, assuming that the condition is satisfied, give the equation of the contact.

3. $[X_1, Y_1]$, $PX^2 + QY^2 = 1$.   **4.** $[X_1, Y_1, Z_1]$, $XY + kZ^2 = 0$.

In Nos. 5, 6, verify that the line belongs to the envelope and find its contact.

5. $[1, -2]$, $X^2 + XY + Y^2 - 2X + 3Y + 5 = 0$.

6. $[1, 2, 3]$, $4X^2 - XY - 2Y^2 + 3XZ + 4YZ - 3Z^2 = 0$.

In Nos. 7, 8, use the envelope equation to verify that the line touches the conic and find the coordinates of the contact.

7. $3x + y = 1$, $14x^2 - 2y^2 = 7$.   **8.** $x - 3y + 5 = 0$, $y^2 = x + y + 1$.

9. Find the equation of the pair of points in which $[-1, 2]$ meets $X^2 + 3Y^2 = 1$.

10. Find the equation of the points of intersection of $y^2 = 4ax$ and $[-1/a, 0]$.

11. Find the contacts of the axes and the line at infinity with
$$aYZ + bZX + cXY = 0.$$

In Nos. 12–15, write down the pole of the line wo the conic.

12. $[X_1, Y_1]$, $aY^2 = X$; $[X_1, Y_1, Z_1]$, $aY^2 = XZ$.

13. $[1, -1]$, $2XY + Y^2 - 3X + 4 = 0$.

14. $[1, -3, 2]$, $X^2 + Y^2 + Z^2 = 4YZ$.

15. $[0, 0, 1]$, $X^2 + Y^2 + 2GXZ + 2FYZ + CZ^2 = 0$.

In Nos. 16, 17, find the polar of the point wo the conic.

16. $X - Y = 1$, $X^2 + 2Y^2 = 2$.

17. $LX + MY + NZ = 0$, $FYZ + GZX + HXY = 0$.

18. Write down the condition for p_1 and p_2 to be conjugate lines wo $X^2 + Y^2 = YZ$.

19. Find the condition for $p_1 x + q_1 y + r_1 = 0$ and $p_2 x + q_2 y + r_2 = 0$ to be conjugate lines wo $x^2 + 2y^2 = 1$.

In Nos. 20–24, find the centre of the conic.

20. $X^2 + 3XY + 5Z^2 = 0$.

21. $X^2 + Y^2 + Z^2 = 2YZ + 2ZX + 2XY$.

22. $X^2 + XY - Y^2 + 2X - 3Y + 1 = 0$.

23. $a/X + b/Y + c/Z = 0$.   **24.** $axy + bx + cy + d = 0$.

In Nos. 25–28, find the locus equation of the conic.

25. $aX^2 + bY^2 = c$.   **26.** $Y^2 = 2kXZ$.

27. $r^2(X^2 + Y^2) = (aX + bY + Z)^2$.

28. $5X^2 - 2XY + 11Y^2 - 14XZ - 26YZ + 17Z^2 = 0$.

29. If p_1 and p_2 are tangents to $S = 0$, what is represented by $S_1 + S_2 = S_{12}$?

30. Find the envelope of the chords of the general conic that subtend a right angle at the origin.

31. Write down the equation of the points of intersection of $S = 0$ and $[X_1, Y_1, Z_1]$. Show that if $X_1 = 0$, $Y_1 = 0$, $Z_1 = 1$, it reduces to $bX^2 - 2hXY + aY^2 = 0$. Deduce the condition for the conic to be a rectangular hyperbola.

32. Give the equation of the points of intersection of $[X_1, Y_1, Z_1]$ and $ax^2 + by^2 = z^2$, and deduce the equation of the points at infinity on the lines joining the origin to those points.

33. The points in which a line cuts $ax^2 + by^2 = 1$ harmonically separate the points in which it cuts $bx^2 + ay^2 = 1$. Prove that the line touches $(a^2 + b^2)(x^2 + y^2) = a + b$.

34. Find the eight points of contact of common tangents of $ax^2 + by^2 = 1$ and $cx^2 + dy^2 = 1$ and prove that they lie on a conic.

12·9. Parametric Equations

12·91. If any point on a conic is taken as origin, the equation of the conic in G_3 is

$$ax^2 + 2hxy + by^2 + 2gx + 2fy = 0.$$

An arbitrary line through the origin meets the curve where

$$x^2(a + 2ht + bt^2) + 2x(g + ft) = 0 \quad \text{and} \quad y = tx,$$

i.e. at the origin and at the point

$$x : y : 1 = -2(g + ft) : -2t(g + ft) : a + 2ht + bt^2. \quad (1)$$

The line $x = 0$ meets the curve at the origin and at $(0, -2f/b)$ and this is the limit when $t \to \infty$ of the point (1). This exception can be avoided by the use of a homogeneous parameter instead of t. Every other point of the curve is given by (1), the parameter of P being the value of t for which $y = tx$ is the equation of OP. The origin itself is given by $g + ft = 0$ unless $g + ft$ is a factor of $a + 2ht + bt^2$; in that case $2gx + 2fy$ is a factor of $ax^2 + 2hxy + by^2$ and so the conic is degenerate. It has therefore been proved that any non-degenerate conic through the origin has parametric equations of the form (1).

12·92. For a conic not passing through the origin, if (p, q) is any point on the curve, the equations are
$$x - p : y - q : 1 = -2(g + ft) : -2t(g + ft) : a + 2ht + bt^2$$
and these are of the form
$$x : y : 1 = a_1 t^2 + 2b_1 t + c_1 : a_2 t^2 + 2b_2 t + c_2 : a_3 t^2 + 2b_3 t + c_3.$$
In G_6 the equations are
$$x : y : z = a_1 t^2 + 2b_1 t + c_1 : a_2 t^2 + 2b_2 t + c_2 : a_3 t^2 + 2b_3 t + c_3$$
or, to avoid exceptions,
$$x : y : z = a_1 t^2 + 2b_1 ts + c_1 s^2 : a_2 t^2 + 2b_2 ts + c_2 s^2 : a_3 t^2 + 2b_3 ts + c_3 s^2.$$

12·93. *Envelope Equation.* The point t given by the equations in 12·92 lies on the line $[X, Y, Z]$ if
$$X(a_1 t^2 + 2b_1 t + c_1) + Y(a_2 t^2 + 2b_2 t + c_2) + Z(a_3 t^2 + 2b_3 t + c_3) = 0,$$
i.e. $t^2(a_1 X + a_2 Y + a_3 Z) + 2t(b_1 X + b_2 Y + b_3 Z)$
$$+ (c_1 X + c_2 Y + c_3 Z) = 0.$$
This quadratic equation gives the parameters of the points in which the line meets the conic. $[X, Y, Z]$ touches the conic if the quadratic has equal roots, i.e. when
$$(a_1 X + a_2 Y + a_3 Z)(c_1 X + c_2 Y + c_3 Z) = (b_1 X + b_2 Y + b_3 Z)^2.$$
This is therefore the envelope equation of the conic.

When $\begin{vmatrix} a_1 & a_2 & a_3 \\ b_1 & b_2 & b_3 \\ c_1 & c_2 & c_3 \end{vmatrix} = 0,$

there is a set of values X, Y, Z, not all zero, for which
$$a_1 X + a_2 Y + a_3 Z = b_1 X + b_2 Y + b_3 Z = c_1 X + c_2 Y + c_3 Z = 0.$$

For this particular line $[X, Y, Z]$ the quadratic in t becomes an identity. Hence every point of the line lies on the conic. Therefore when the determinant is zero the equations in 12·92 do not represent a non-degenerate conic.

12·94. The tangent to the conic

$$x:y:z = a_1 t^2 + 2b_1 t + c_1 : a_2 t^2 + 2b_2 t + c_2 : a_3 t^2 + 2b_3 t + c_3$$

may be obtained by the method of 6·71, which gives

$$\begin{vmatrix} x & y & z \\ a_1 t^2 + 2b_1 t + c_1 & a_2 t^2 + 2b_2 t + c_2 & a_3 t^2 + 2b_3 t + c_3 \\ a_1 t + b_1 & a_2 t + b_2 & a_3 t + b_3 \end{vmatrix} = 0.$$

$$\therefore \quad \begin{vmatrix} x & y & z \\ b_1 t + c_1 & b_2 t + c_2 & b_3 t + c_3 \\ a_1 t + b_1 & a_2 t + b_2 & a_3 t + b_3 \end{vmatrix} = 0.$$

Alternatively, from

$$a_1 t^2 + 2b_1 t + c_1 = \lambda x,$$

$$a_2 t^2 + 2b_2 t + c_2 = \lambda y,$$

$$a_3 t^2 + 2b_3 t + c_3 = \lambda z,$$

$$t^2 : t : 1 = \begin{vmatrix} 2b_1 & c_1 & x \\ 2b_2 & c_2 & y \\ 2b_3 & c_3 & z \end{vmatrix} : \begin{vmatrix} c_1 & a_1 & x \\ c_2 & a_2 & y \\ c_3 & a_3 & z \end{vmatrix} : \begin{vmatrix} a_1 & 2b_1 & x \\ a_2 & 2b_2 & y \\ a_3 & 2b_3 & z \end{vmatrix}$$

$$= A_1 x + 2B_1 y + C_1 : A_2 x + 2B_2 y + C_2 : A_3 x + 2B_3 y + C_3,$$

so that by the method of 6·24 and 6·25 the tangent at t_1 is

$$(A_1 x + 2B_1 y + C_1) - 2t_1 (A_2 x + 2B_2 y + C_2)$$
$$+ t_1^2 (A_3 x + 2B_3 y + C_3) = 0,$$

and the chord $t_1 t_2$ can be found in the same way.

The contact of an envelope conic given by

$$X : Y : Z = a_1 t^2 + 2b_1 t + c_1 : a_2 t^2 + 2b_2 t + c_2 : a_3 t^2 + 2b_3 t + c_3$$

may be found by the dual method.

Other processes which may be applied to a conic given by parametric equations are illustrated in the following example.

12·95. EXAMPLE. Discuss the conic

$$x : y : 1 = t : t^2 - 2 : t^2 + t - 1.$$

From the given equations

$$x : y : 1 - x - y = t : t^2 - 2 : 1,$$

$$x : 2 - 2x - y : 1 - x - y = t : t^2 : 1.$$

Hence the x, y equation is $x^2 = (2 - 2x - y)(1 - x - y)$.

By 12·94 the tangent at the point t is

$$(2 - 2x - y) - 2tx + t^2(1 - x - y) = 0$$

and the chord $t_1 t_2$ is

$$(2 - 2x - y) - (t_1 + t_2)x + t_1 t_2(1 - x - y) = 0.$$

$[X, Y, Z]$ is a tangent if the quadratic

$$Xt + Y(t^2 - 2) + Z(t^2 + t - 1) = 0$$

for t has equal roots, i.e. if $(Y + Z)(2Y + Z) + (X + Z)^2 = 0$.
This is the envelope equation.

Hence the pole of $[X_1, Y_1, Z_1]$ is

$$XX_1 + 2YY_1 + 2ZZ_1 + \tfrac{3}{2}(YZ_1 + ZY_1) + (ZX_1 + XZ_1) = 0,$$

i.e. $\qquad (X_1 + Z_1,\ 2Y_1 + \tfrac{3}{2}Z_1,\ X_1 + \tfrac{3}{2}Y_1 + 2Z_1).$

Therefore the centre is $(1, \tfrac{3}{2}, 2)$; and, in G_3, $(\tfrac{1}{2}, \tfrac{3}{4})$.

The asymptotes are the tangents at the points t given by $t^2 + t - 1 = 0$ and so they are

$$(2 - 2x - y) - 2tx + t^2(1 - x - y) = 0,$$

where $t = \tfrac{1}{2}(-1 \pm \sqrt{5})$. Their joint equation is found by elimination of t.

Since $t^2 = 1 - t$, $(2 - 2x - y) - 2tx + t^2(1 - x - y) = 0$ gives

$$(3 - 3x - 2y) - t(1 + x - y) = 0.$$

$$\therefore \quad (3 - 3x - 2y)(4 - 2x - 3y) = (1 + x - y)^2.$$

The axes bisect the angles between the asymptotes. Hence they are parallel to the bisectors of $x^2 + 3xy + y^2 = 0$. Since the coefficients of x and y in this equation are equal, the bisectors are the axes of coordinates. Hence the axes of the conic are $x = \tfrac{1}{2}$ and $y = \tfrac{3}{4}$.

12·96. The method of 12·73 consists in finding the envelope of a line, in the form of a curve locus, by determining the contacts of the envelope; these contacts are points through which there pass coincident lines of the envelope.

The envelope of a variable curve $f(x, y, t) = 0$ may also be found by determining these points.

If $f(x_1, y_1, t) = 0$ has a repeated root t, this also satisfies $\frac{\partial}{\partial t} f(x_1, y_1, t) = 0$. Hence the locus is found by eliminating t between these equations.

When the variable curve is $f(x, y) + tg(x, y) = 0$, the envelope is the points of intersection of the curves $f(x, y) = 0$ and $g(x, y) = 0$.

When the variable curve is $f(x, y) + 2tg(x, y) + t^2h(x, y) = 0$, the envelope is $\{g(x, y)\}^2 = f(x, y) h(x, y)$, because this is the condition for the quadratic in t to have equal roots.

EXERCISE 12<small>F</small>

In Nos. 1–4, obtain parametric equations.

1. $y(x + y) = x$.　　　　　　**2.** $(x + y)^2 = x - y$.

3. $x^2 + 2xy + 3y^2 - 4x + 5y = 0$.

4. $2x^2 + 4xy + 5y^2 + 6x + 7y - 24 = 0$.

In Nos. 5–8, find independently the locus and envelope equations.

5. $x = t^2 + 3t,\ y = t^2 - 2t + 5$.　　　**6.** $x = t + 2t^{-1},\ y = 2t + t^{-1}$.

7. $x : y : 1 = t^2 + t + 1 : 2t^2 + 3 : 4t^2 + t - 1$.

8. $x : y : z = t^2 - t : 3t^2 + 5 : 2t + 4$.

9. Find the tangent at $t = 1$ to $x : y : z = t^2 + 3t + 1 : t^2 - 1 : 4t - 3$.

10. Show that the equations
$$x : y : 1 = t^2 + 2t + 3 : 2t^2 + t - 1 : 4t^2 + 5t + 5$$
represent a line which could be represented by
$$x : y : 1 = 1 : s - 2 : s.$$

11. Find the chord $t_1 t_2$ of the conic $x : y : 1 = t^2 + 2 : t - 1 : t^2 + 3t$.

12. Find the polar of (u, v, w) wo the conic
$$x : y : z = t + 2 : t^2 + 1 : t^2 - t.$$

13. Find the pole of $[1, 1, 1]$ wo the conic
$$x : y : z = 2 + t - t^2 : 1 + t - t^2 : 2t^2 - t - 3.$$

14. Find the pole of $[1, 1, 1]$ wo the conic
$$X : Y : Z = -t^2 + 2t - 2 : (t+1)^2 : 2t - 1.$$

15. Find the asymptotes of $x : y : 1 = t^2 : t : t^2 - 3t + 2$.

16. Show that $x : y : 1 = 3t^2 - 2t + 1 : 5t^2 + 2t + 3 : t^2 + 1$ is a circle and find its radius.

17. Find the centre, asymptotes, and axes of
$$x : y : 1 = 2t^2 + t - 1 : 1 : 3t^2 + 5t - 2.$$

18. Find the centre and axes of
$$x : y : 1 = t^2 - 2t - 1 : t - 1 : t^2 - t + 1.$$

19. Find the vertex and axis of the parabola
$$x : y : 1 = t^2 + t - 1 : 2t + 3 : t^2.$$

20. If the first minors of $\begin{vmatrix} a_1 & b_1 & c_1 \\ a_2 & b_2 & c_2 \\ a_3 & b_3 & c_3 \end{vmatrix}$ are all zero, what point has coordinates x, y such that
$$x : y : 1 = a_1 t^2 + 2b_1 t + c_1 : a_2 t^2 + 2b_2 t + c_2 : a_3 t^2 + 2b_3 t + c_3?$$

21. Prove that the pole (x_1, y_1) of the chord $t_1 t_2$ of the conic
$$x : y : 1 = a_1 t^2 + 2b_1 t + c_1 : a_2 t^2 + 2b_2 t + c_2 : a_3 t^2 + 2b_3 t + c_3$$
is given by
$$x_1 : y_1 : 1 = a_1 t_1 t_2 + b_1 (t_1 + t_2) + c_1 : a_2 t_1 t_2 + b_2 (t_1 + t_2) + c_2$$
$$: a_3 t_1 t_2 + b_3 (t_1 + t_2) + c_3.$$

22. What curve is touched by the line
$$(a + bt^2) x + (c + dt) y + (e + ft)^2 = 0$$
for all values of t?

In Nos. 23–29, find the envelope when t varies.

23. $y = 3xt - t^3$.   **24.** $(x^2 - y^2) = t(x^2 - 1)$.   **25.** $y^2 = tx^2 + at^{-1}$.

26. $2xt/a - y(1 - t^2)/b = 4t(1 - t^2)/(1 + t^2)$.

27. $x/s + y/t = 1$ where $s + t = 1$.

28. $x^3/s + y^3/t = 1$ where $s^2 + t^2 = 1$.

29. $x^2 + 2t^3 xy + t^2 y^2 = 1$.

30. In what way is the line $[X, Y]$ limited by the relation
$$(X^2 - 1) = t(Y^2 - 1)?$$

Chapter 13

THE PARABOLA

13·1. In 6·2 the equation of the parabola is taken to be $y^2 = kx$.

By 12·34 the polar of P_1 wo this parabola is $yy_1 = \frac{1}{2}k(x+x_1)$ and this is the tangent at P_1 if P_1 lies on the curve.

By 12·45 the condition for P_1, P_2 to be conjugate points is

$$y_1y_2 = \frac{1}{2}k(x_1+x_2).$$

By 12·32 the equation of the pair of tangents from P is

$$(y^2 - kx)(y_1{}^2 - kx_1) = \{yy_1 - \frac{1}{2}k(x+x_1)\}^2.$$

13·2. The Parabola y² = 4ax

13·21. In 6·28 the parabola is represented by $y^2 = 4ax$ or by

$$x : y : a = t^2 : 2t : 1.$$

Every value of t determines just one point on the parabola and every point on the parabola determines just one value of t. The representation is proper (10·51).

The following results are found by the method of 6·2.

13·22. $lx + my + na = 0$ meets the parabola $x : y : a = t^2 : 2t : 1$ in points whose parameters are the roots of the equation $lt^2 + 2mt + n = 0$.

13·23. The chord t_1t_2 is $x - \frac{1}{2}(t_1+t_2)y + at_1t_2 = 0$.

13·24. The chord whose ends are given by $ut^2 + 2vt + w = 0$ is $ux + vy + wa = 0$.

13·25. The tangent at t_1 is $x - t_1y + at_1{}^2 = 0$.

13·26. The normal at t_1 is $t_1x + y = at_1{}^3 + 2at_1$.

13·27. The envelope equations are

$$X : Y : 1/a = t^2 : t : 1$$

or
$$aY^2 = X.$$

The results of 12·6 are applicable to the equation $aY^2 = X$. For example, the condition for $[X_1, Y_1]$ and $[X_2, Y_2]$ to be conjugate lines is $2aY_1Y_2 = X_1 + X_2$.

13·3. The pole of a given chord of $y^2 = 4ax$ may be found in various ways:

13·31. If the chord is $[X_1, Y_1]$, its pole wo $aY^2 = X$ is
$$aYY_1 = \tfrac{1}{2}(X + X_1), \text{ i.e. } (1/X_1, -2aY_1/X_1).$$

13·32. If the chord is $t_1 t_2$, the pole is the point of intersection of the tangents

$$x - t_1 y + at_1{}^2 = 0, \quad x - t_2 y + at_2{}^2 = 0,$$

i.e. $(at_1 t_2, at_1 + at_2)$.

13·33. If the parameters of the ends of the chord are the roots of
$$ut^2 + 2vt + w = 0$$

and (x_1, y_1) is the pole, then the equation

$$2aty_1 = 2a(at^2 + x_1),$$

which expresses that the polar of $(x_1 y_1)$ passes through $(at^2, 2at)$, must be the same as $ut^2 + 2vt + w = 0$.

$$\therefore \quad a : -y_1 : x_1 = u : 2v : w.$$

Hence the pole is $(aw/u, -2av/u)$.

13·34. If the chord is $lx + my + na = 0$ and (x_1, y_1) is the pole, then this equation is the same as $yy_1 = 2a(x + x_1)$.

$$\therefore \quad l : m : n = 2a : -y_1 : 2x_1,$$
$$\therefore \quad x_1 = an/l, \quad y_1 = -2am/l.$$

EXERCISE 13A

1. Find the tangent other than $x = 0$ from $(0, b)$ to $y^2 = 4ax$.

2. Find the tangents from $(2a, 3a)$ to $y^2 = 4ax$.

3. Find the condition for the tangents from (f, g) to $y^2 = 4ax$ to intersect at $45°$.

4. Find the locus of the mid-points of chords of $y^2 = kx$ through the vertex.

5. The tangents at the ends of the chord $t_1 t_2$ of $x : y : a = t^2 : 2t : 1$ meet at T, and V is the mid-point of the chord. Prove that TV is parallel to the axis.

6. State the condition for the chord $t_1 t_2$ of $x : y : a = t^2 : 2t : 1$ to be in a fixed direction, and show that the mid-points of such chords are collinear.

7. Find the equation of the point of intersection of the tangents t_1, t_2 of the parabola $X : Y : 1/a = t^2 : t : 1$, and deduce the equation of the contact of t_1.

8. P, Q, R are points on $y^2 = 4ax$ whose ordinates are in G.P. Prove that the tangents at P and R meet on the line through Q perpendicular to the axis.

9. A chord of $y^2 + ax = 0$ touches $y^2 - 4ax = 0$ and subtends a right angle at the origin. Find its inclination to the axis of the parabolas.

10. Two parabolas have parallel axes. A variable line parallel to these axes meets the parabolas at P and Q. Show that the mid-point of PQ lies on another parabola.

11. A polygon of $2n$ sides is inscribed in a parabola. Prove that if all the sides but one have fixed directions, so also has the remaining side.

12. A polygon of $2n$ sides is inscribed in a parabola. Prove that if all the sides but one pass through fixed points on the axis, so also does the remaining side.

13·4. Geometrical Properties

13·41. The axis, vertex, focus and directrix of a parabola are defined in 6·28 and some properties of the curve are given in Exercise 6A, Nos. 17–24.

In the figure, P is an arbitrary point t on the parabola $x : y : a = t^2 : 2t : 1$, PN is perpendicular to the axis, and the tangent and normal at P meet the axis in T and G.

Since PT is $x - ty + at^2 = 0$, T is the point $(-at^2, 0)$. Hence

$$TA = AN.$$

The subnormal $NG = y\dfrac{dy}{dx} = 2a.$

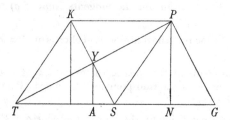

13·42. The mid-point of TG is the same for all positions of P. It is the focus $(a, 0)$.

The chord through the focus perpendicular to the axis is called the *latus rectum*.

Since S is the mid-point of the hypotenuse of the right-angled triangle PTG, $ST = SP$.

Therefore if the parallelogram $TSPK$ is completed, it is a rhombus, and SK is perpendicular to PT.

But PT is $x - ty + at^2 = 0$ and S is $(a, 0)$,

$$\therefore \quad SK \text{ is } tx + y = at.$$

But K lies on $y = 2at$. Hence K is $(-a, 2at)$.

Since K lies on the directrix $x + a = 0$, and $SP = PK$, the parabola is the locus of a point P which is equidistant from the focus and the directrix.

13·43. The diagonals of $TSPK$ bisect one another at right angles at Y, and Y is the foot of the perpendicular from S to the tangent. Y is the point $(0, at)$; it lies on the tangent at A. Hence (5·86)

The pedal of a parabola wo the focus is the tangent at the vertex.

13·44. TP bisects $\angle SPK$.

The triangles ASY, YSP are similar, each being similar to YST. Hence $SY^2 = AS . SP$. Also if PT meets the directrix at F, the triangles FKP, FSP are congruent and so

$$\angle\, FSP \text{ is a right angle.}$$

The lengths of the lines in the figure can be expressed in terms of a and θ $(= \angle\, STP)$. For example,

$$NP = NG \cot \theta = 2a \cot \theta$$

and $\quad SP = SY \cosec \theta = AS \cosec^2 \theta = a \cosec^2 \theta.$

13·45. *Pedal Equation.* The relation between the distance $r\,(=OP)$ and the perpendicular distance p from a fixed point to the tangent at an arbitrary point P of a curve is called the *pedal equation* of the curve wo the fixed point. The sign conventions that are adopted make p always positive. These are explained in *E.C.* vol. II, Chapter XVI.

The result $SY^2 = AS\,.\,SP$, proved in 13·44, can be written

$$p^2 = ar.$$

This is the pedal equation of the parabola wo the focus.

A pedal equation does not completely determine a curve. $p^2 = ar$ is satisfied not only by the parabola but also by any congruent parabola obtained by rotation about the focus. It is also satisfied by $p = 0$, $r = 0$.

13·5. The Converse Pedal Property

The parabola may be obtained as the envelope of the line p drawn at a variable point Y of a fixed line l, perpendicular to the line joining Y to a fixed point S.

Taking l as y-axis, and the perpendicular to it through S as x-axis, let p be $[X, Y]$. Then

$$AY = -1/Y, \quad TA = 1/X.$$

But $AY^2 = TA\,.\,AS.$

$$\therefore \quad aY^2 = X,$$

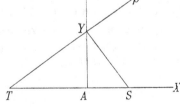

which is the envelope equation of the parabola (13·27). See also Exercise 13F, No. 2.

13·6. Focal Chords

13·61. A chord which passes through the focus is called a *focal chord*.

$x - \frac{1}{2}(t_1 + t_2)y + at_1 t_2 = 0$ passes through $(a, 0)$ if $t_1 t_2 = -1$. Hence $t_1 t_2 = -1$ is the relation between the parameters of the ends of a focal chord. Therefore the tangents at the ends of a focal chord intersect at right angles. They also meet on the directrix because the pole of the chord $t_1 t_2$ is $(at_1 t_2, at_1 + at_2)$.

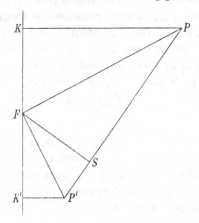

$PF, P'F$ are tangents at the ends of a focal chord. They meet at right angles at F. PK, PK' are perpendiculars to the directrix. By 13·44, $\angle FSP$ is a right angle. Also the triangles FSP, FKP are congruent and the triangles $FSP', FK'P'$ are congruent.

13·62. EXAMPLE. A circle passes through a given point and touches a given line. Find the locus of its centre.

1st method. Take the given line as x-axis and let the given point be $(0, c)$. Then the centre (x, y) of the circle is equidistant from $y = 0$ and $(0, c)$.

$$\therefore \quad y^2 = x^2 + (y - c)^2,$$
$$x^2 = 2cy - c^2 = 2c(y - \tfrac{1}{2}c).$$

This is a parabola with $x = 0$ as axis and $y = \frac{1}{2}c$ as tangent at the vertex.

2nd method. Let P be the centre of the circle through the given point A and touching the given line at Q. Then $AP = PQ$. Therefore the locus is a parabola, focus A, with the given line as directrix.

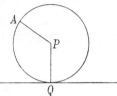

13·63. EXAMPLE. Prove that the circumcircle of a triangle formed by three tangents to a parabola passes through the focus.

By 13·43 the feet of the perpendiculars from the focus to the sides of the triangle are collinear. Hence, by the converse Simson's Line property, the focus lies on the circumcircle. See also Exercise 13 F, No. 20.

13·64. EXAMPLE. Prove that the orthocentre of a triangle formed by three tangents to a parabola lies on the directrix.

The tangents to $x : y : a = t^2 : 2t : 1$ at t_1, t_2 are

$$x - t_1 y + a t_1^2 = 0 \quad \text{and} \quad x - t_2 y + a t_2^2 = 0.$$

These meet where $(t_1 - t_2) y = a(t_1^2 - t_2^2)$;

$$\therefore \quad y = a(t_1 + t_2), \quad x = a t_1 t_2.$$

The perpendicular from this point to a third tangent

$$x - t_3 y + a t_3^2 = 0$$

is $\qquad t_3 x + y = a(t_1 + t_2 + t_1 t_2 t_3).$

And this meets the directrix $x = -a$ where

$$y = a(t_1 + t_2 + t_3 + t_1 t_2 t_3).$$

The symmetry of the result shows that the other perpendiculars meet the directrix at this same point.

13·65. EXAMPLE. Find the locus of the mid-points of focal chords of the parabola $y^2 = 4ax$.

1st method. Let (x_1, y_1) be the mid-point of the focal chord $t_1 t_2$. Then

$$2x_1 = a(t_1^2 + t_2^2), \quad 2y_1 = 2a(t_1 + t_2).$$

Also, by 13·61, $t_1 t_2 = -1$.

Thus $\qquad 2ax_1 = a^2\{(t_1+t_2)^2 - 2t_1t_2\}$
$$= y_1{}^2 + 2a^2.$$

The locus is the parabola $y^2 = 2a(x-a)$.

2nd method. Let (x_1, y_1) be the mid-point of a focal chord. As proved in 12·46 the chord is parallel to the polar

$$yy_1 = 2a(x+x_1),$$

and it passes through (x_1, y_1). Thus the chord is

$$yy_1 - 2ax = y_1{}^2 - 2ax_1.$$

Since it passes through the focus $(a, 0)$, $-2a^2 = y_1{}^2 - 2ax_1$. $\therefore$ (x_1, y_1) lies on $y^2 = 2a(x-a)$.

EXERCISE 13 B

1. Find the tangents to $y^2 = 4ax$ at the ends of the latus rectum.

2. Prove that the directrix is the polar of the focus.

3. Find the length of the focal chord of $y^2 = 4ax$ which makes an angle α with the axis and show that the latus rectum is the shortest focal chord.

In Nos. 4–6, find the equation of the parabola.

4. Focus $(3, 2)$, directrix $x + 7y = 12$.

5. Vertex $(4, 2)$, directrix $2x - y = 1$.

6. Focus $(1, 2)$, tangent at the vertex $3x + 4y = 22$.

7. Verify the result of 13·43 by using the equations of PT and the perpendicular from S to PT.

8. With the notation of 13·4 express the lengths of AY, SY, PG, SP in terms of a and the parameter of P.

9. Given the ends of a focal chord of a parabola and the direction of the axis, show how to find the focus and directrix.

10. A and B are given points. A parabola touches AB at A and its axis passes through B. Prove that the locus of the vertex is a circle.

11. From the mid-point M of a focal chord perpendiculars are drawn to the axis and the chord, meeting the axis at N and V. Prove that NV is equal to the semi latus rectum.

12. Prove that the locus of the point of intersection of tangents to $y^2 = 4ax$ which cut off a fixed length b from the directrix is

$$(x+a)^2(y^2 - 4ax) = b^2x^2.$$

13. If the nth pedal is defined to be the pedal of the $(n-1)$th pedal, find the 4th pedal of $y^2 = 4ax$ wo the focus.

14. Find the equation of the pedal of $y^2 = 4ax$ wo $(-a, 0)$.

15. A variable chord PSQ of a fixed circle passes through a fixed point S. The polar of S wo the circle on PQ as diameter cuts PQ at Y and the diameter through S at T. Prove that SY^2/ST is constant and that the envelope of the polar is a parabola.

16. A variable circle passes through two fixed points A and B and a fixed line through A meets it at P. Prove that the tangent at P envelops a parabola focus B.

13·7. Diameters

13·71. The chord $t_1 t_2$ has a fixed direction if $t_1 + t_2$ is constant. It makes an angle ω with the axis such that $\cot \omega = \frac{1}{2}(t_1 + t_2)$. The y-coordinate of the mid-point of the chord is $a(t_1 + t_2)$. Hence:

The locus of the mid-points of parallel chords is the line $y = 2a \cot \omega$, which is parallel to the axis of the parabola.

This locus is called a diameter in Chapter 12. In G_5 and G_6 it is the polar of a point at infinity. See 12·5.

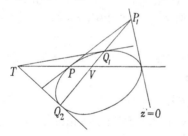 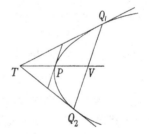

If the diameter bisecting $Q_1 Q_2$ meets the curve at P, the tangent at P goes through P_1, i.e. it is parallel to the chords. Also the tangents at Q_1, Q_2 meet on the diameter at T.

T and V are separated harmonically by the conic. Hence, for the parabola, $TP = PV$.

13·72. EXAMPLE. Find the locus of the point of intersection of normals at the ends of a focal chord of the parabola $y^2 = 4ax$.

The tangents and normals at the ends of a focal chord $Q_1 Q_2$ form a rectangle $Q_1 T Q_2 N$.

By 13·71, $TP = PV$. Also TN is parallel to the axis. If N is (x, y), T is $(-a, y)$; hence V is $(\frac{1}{2}(x-a), y)$ and P is $(\frac{1}{4}(x-3a), y)$. But P lies on the parabola. Hence $y^2 = a(x-3a)$.

Thus the locus is a parabola with vertex $(3a, 0)$ and latus rectum a.

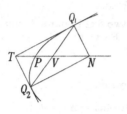 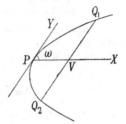

13·75. Oblique Axes

PX is the diameter through a fixed point P on the parabola, Q is a variable point on the curve and $Q_1 Q_2$ is the chord parallel to the tangent at P. $Q_1 Q_2$ is bisected by PX at V.

If the parameters of Q_1, Q_2 are t_1, t_2, the coordinates of V are $\frac{1}{2}a(t_1^2 + t_2^2)$, $a(t_1 + t_2)$ and those of P are $\frac{1}{4}a(t_1 + t_2)^2$, $a(t_1 + t_2)$. Also $t_1 + t_2 = 2 \cot \omega$.

Thus $\quad PV = \frac{1}{2}a(t_1^2 + t_2^2) - \frac{1}{4}a(t_1 + t_2)^2 = \frac{1}{4}a(t_1 - t_2)^2$

and $\quad QV = a(t_1 - t_2) \operatorname{cosec} \omega$,

$$QV^2 = 4a \operatorname{cosec}^2 \omega \, . \, PV = 4SP \, . \, PV.$$

Hence if P is taken as origin and PX, PY as oblique axes, the equation of the parabola is $y^2 = 4bx$, where $b = SP$.

With these axes $x : y : b = t^2 : 2t : 1$ are parametric equations, and the equations of the tangent and chord are the same as those in 13·2 with b substituted for a.

13·76. EXAMPLE. The diameter of a parabola through a point P on the curve meets a chord UV at L and meets the tangents at U, V in H, K. Prove that $PH \, . \, PK = PL^2$.

Taking the diameter and tangent at P as axes of coordinates, the equation of the parabola is $y^2 = 4bx$.

If U, V are $(bt_1{}^2, 2bt_1)$, $(bt_2{}^2, 2bt_2)$, the equations of UV and the tangents are

$$2x - (t_1 + t_2)\, y + 2bt_1 t_2 = 0,$$

$$x - t_1 y + bt_1{}^2 = 0,$$

$$x - t_2 y + bt_2{}^2 = 0.$$

$$\therefore \quad PL = -bt_1 t_2, \quad PH = -bt_1{}^2, \quad PK = -bt_2{}^2,$$

$$\therefore \quad PH \cdot PK = PL^2.$$

EXERCISE 13c

1. Find the distance from (x, y) to the curve $y^2 = 4ax$ measured parallel to the axis.

2. Tangents to a parabola at Q, Q' meet at T, and V is the mid-point of QQ'. Prove that the curve bisects TV.

3. Prove that the circle drawn on a focal chord as diameter touches the directrix.

4. What are the coordinates of the focus referred to the oblique axes of 13·75?

5. In 13·75, if QD is the perpendicular from Q to PX, prove that $QD^2 = 4aPV$.

6. Use the property $QV^2 = 4SP \cdot PV$ to write down the equation of the locus of the mid-point of a variable chord of $y^2 = 4ax$ of length l.

7. P, Q, R are points on a parabola. The diameters through P, Q meet RQ, RP at D, E. Prove that the tangents at P, Q meet at the mid-point of DE.

8. Q is a point on a chord PP' of a parabola such that $PQ \cdot QP' = k^2$, where k is constant. If the direction of PP' is fixed, find the locus of Q.

9. PQ is a chord of $y^2 = 4ax$. The lines through P, Q parallel to the tangents at Q, P meet at R. PQ varies but passes through the fixed point (f, g). Show that R lies on the curve $2y^2 - 2ax - 3gy + 6af = 0$.

10. With the figure of 13·75, the tangents at Q_1, Q_2 meet at T and the tangent at another point R meets the tangents at Q_1, Q_2, P in H, K, L. Prove that $TH : HQ_1 = KR : RH = Q_2K : KT$ and $KL = LH$.

11. A tangent to $y^2 + 4bx = 0$ meets $y^2 - 4ax = 0$ at P and Q. Prove that the mid-point of PQ lies on $y^2(2a + b) = 4a^2x$.

12. V is the mid-point and T is the pole of a chord QQ' of a parabola which meets the axis at E. The perpendicular at V to QQ' meets the axis at F. S is the focus and K is the foot of the perpendicular from T to the directrix. Prove that ST is parallel to EK and SK is parallel to FV.

13·8. Normals

13·81. By 13·26 the normal to $x:y:a = t^2:2t:1$ at the point t is
$$tx + y = at^3 + 2at.$$

This passes through (f, g) if
$$tf + g = at^3 + 2at,$$
i.e. $at^3 + (2a - f)t - g = 0.$

This is a cubic for the parameters of points the normals at which pass through (f, g). In G_3 it will not always have three roots.

If t_1, t_2, t_3 are the parameters of three points the normals at which pass through (f, g), since they are the roots of the cubic equation,
$$t_1 + t_2 + t_3 = 0, \quad t_2 t_3 + t_3 t_1 + t_1 t_2 = (2a - f)/a, \quad t_1 t_2 t_3 = g/a.$$

From $t_1 + t_2 + t_3 = 0$ it follows that the centroid of the feet of three concurrent normals lies on the axis of the parabola.

13·82. EXAMPLE. Find the point in which the normal at t_1 to $x:y:a = t^2:2t:1$ meets the curve again.

The normal is $\qquad t_1 x + y = at_1^3 + 2at_1.$

The point t_0 lies on it if
$$t_1 t_0^2 + 2t_0 = t_1^3 + 2t_1.$$

One root of this quadratic equation for t_0 is t_1, and the sum of the roots is $-2/t_1$. Hence the other is $-2/t_1 - t_1$. This is the parameter of the point in which the normal meets the curve again.

13·83. EXAMPLE. Find the locus of the point of intersection of normals at the ends of parallel chords of $y^2 = 4ax$.

1st method. The normals are
$$t_1 x + y = at_1^3 + 2at_1, \quad t_2 x + y = at_2^3 + 2at_2.$$

They meet where $x = a(2 + t_1^2 + t_1 t_2 + t_2^2)$, $y = -at_1 t_2 (t_1 + t_2)$. Since the chords are in a fixed direction, $t_1 + t_2$ is a constant, k.

Let $t_1 t_2 = h$. Then $x = a(2 + k^2 - h)$ and $y = -akh$.

Hence the locus is $kx = ak(2 + k^2) + y$, which is a line.

2nd method. Let t_3 be the parameter of the foot of the third normal through the point of intersection of the normals at t_1, t_2. Then by 13·81 $t_1 + t_2 + t_3 = 0$. But $t_1 + t_2 = k$. $\therefore t_3 = -k$.

Hence the locus is the normal at the point whose parameter is $-k$.

13·84. EXAMPLE. Find the locus of the foot of the perpendicular from $(a, 0)$ to a normal of $y^2 = 4ax$.

1st method. The normal is $tx + y = at^3 + 2at$. The perpendicular to it from $(a, 0)$ is $x - ty = a$. Elimination of t gives the required locus.

Multiply through the second equation by t and subtract from the first. Thus

$$y(1 + t^2) = at(1 + t^2). \qquad \therefore \quad y = at.$$

But $\qquad x - ty = a. \qquad \therefore \quad y^2 = a(x - a).$

2nd method. SQ is the perpendicular to the normal PG. Since $SG = SP$, Q is the mid-point of PG. Draw perpendiculars PN, QM to the axis. Then

$$NG = 2a. \qquad \therefore \quad MG = a.$$

But $\qquad QM^2 = SM . MG.$

$$\therefore \quad y^2 = (x - a)\, a.$$

In the first method, if the elimination of t is carried out by direct substitution for t from the second equation into the first, the locus is obtained in the form

$$\{y^2 - a(x - a)\}\{(x - a)^2 + y^2\} = 0.$$

In real geometry this represents the parabola and the point $(a, 0)$; but as this point lies on the parabola, the second factor adds nothing to the result. In complex geometry the locus is the parabola and the isotropic lines SI, SJ. These occur as part of the locus because they are tangents to the parabola $y^2 = 4ax$, and, as they are perpendicular to themselves, they are also normals. Thus if K is any point on SI or SJ, K should be regarded as a foot of a perpendicular from S to a normal. The portion SI, SJ of the locus can be found by the first method by considering the factor $1 + t^2$. It is not given by the second method, which is essentially a method of real geometry.

EXERCISE 13D

In Nos. 1–3, give the equation of the normal.

1. At $(9, -6)$ to $y^2 = 4x$.   **2.** At (x_1, y_1) to $y^2 = ax + by + c$.

3. To $y^2 = 4a(x+a)$ at the positive end of the latus rectum.

4. Prove that $2tx + y = 2kt^3 + kt$, where t is a parameter, is a normal to $y^2 = kx$.

5. Find the condition for $lx + my + n = 0$ to be a normal to $x^2 = 4ky$.

6. Find the normals to $y^2 = 4ax$ which pass through

$$\text{(i) } (5a, -18a), \qquad \text{(ii) } (10a, 3a).$$

7. Find the point of intersection of normals to $x : y : a = t^2 : 2t : 1$ at the points t_1, t_2.

8. Normals to $y^2 = 4ax$ make complementary angles with the axis. Prove that they meet on $y^2 - ax = -2a^2 \pm a^2$.

9. The normal at P to $y^2 = 4ax$ which makes an angle $\frac{1}{2}\pi + \theta$ with OX meets that axis at G and meets the curve again at Q. Find the lengths of PG and PQ.

10. Find the locus of the poles of normals to $y^2 = kx$.

11. Find the locus of the mid-points of normal chords of $y^2 = 4ax$.

12. Find the locus of a point Q taken on the normal at P to $y^2 = 4ax$ so that Q is twice as far from the axis as P.

13. PQ is a focal chord of a parabola $y^2 = 4ax$, and the normals at P, Q meet the curve again in P', Q'. Prove that $\mathbf{P'Q' = 3QP}$.

14. If the three normals from (f, g) to $y^2 = 4ax$ make angles $\theta_1, \theta_2, \theta_3$ with the axis, prove that $\tan(\theta_1 + \theta_2 + \theta_3) = g/(f-a)$.

15. If the normals to $y^2 = kx$ at (x_1, y_1), (x_2, y_2), (x_3, y_3) are concurrent, prove that $y_1 + y_2 + y_3 = 0$ and

$$x_1{}^2 + x_2{}^2 + x_3{}^2 = 2(x_2 x_3 + x_3 x_1 + x_1 x_2).$$

16. If two of the normals from (f, g) to $y^2 = 4ax$ are at right angles, prove that $g^2 = a(f - 3a)$.

17. If two of the normals from (f, g) to $y^2 = 4ax$ are coincident, prove that $27ag^2 = 4(f - 2a)^3$.

18. Prove that normals to $x:y:a = t^2:2t:1$ touch the curve

$$x-2a:y:a = 3t^2:-2t^3:1,$$

and sketch this curve.

19. Normals to $y^2 = 4ax$ at Q, R meet at $(at^2, 2at)$. Find the equation of QR and show that it meets the axis at the same point for all values of t.

13·9. Other Representations of the Parabola

13·91. When the properties of a single parabola are to be investigated, it is best to use the equations $y^2 = kx$ or $y^2 = 4ax$, or the parametric equations equivalent to them. But the curve may arise in the form

$$s \equiv ax^2 + 2hxy + by^2 + 2gx + 2fy + c = 0, \quad \text{where } ab = h^2,$$

and more general parametric equations may also be used.

13·92. EXAMPLE. What parabola is represented by

$$x^2 = 10x + 3y + 14?$$

The equation may be written $(x-5)^2 = 3y + 39 = 3(y+13)$. If PM, PN are the perpendiculars from $P(x, y)$ to the perpendicular lines $x - 5 = 0$ and $y + 13 = 0$, the equation expresses that $PM^2 = 3PN$.

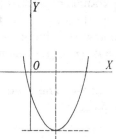

Hence the equation represents the parabola with axis $x - 5 = 0$, tangent at the vertex $y + 13 = 0$, and latus rectum 3.

Since the equation shows that $y + 13 > 0$, the curve lies above the line $y + 13 = 0$. Hence the focus is $(5, -12\frac{1}{4})$ and the directrix is $y = -13\frac{3}{4}$.

13·93. EXAMPLE. What curve is represented by

$$(3x + 4y)^2 = 16x - 62y - 6?$$

The equation can be written

$$(3x + 4y + k)^2 = (16 + 6k)x + (8k - 62)y + (k^2 - 6).$$

Choose k so that $3x + 4y + k = 0$

and $(16 + 6k)x + (8k - 62)y + (k^2 - 6) = 0$

are perpendicular lines, i.e. so that $3(16 + 6k) + 4(8k - 62) = 0$.

$$\therefore \quad k = 4.$$

The equation becomes $(3x + 4y + 4)^2 = 40x - 30y + 10$, or

$$\left(\frac{3x + 4y + 4}{5}\right)^2 = 2\left(\frac{4x - 3y + 1}{5}\right).$$

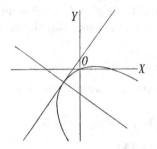 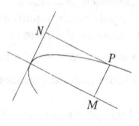

If PM, PN are the perpendiculars from $P(x, y)$ to the perpendicular lines $3x + 4y + 4 = 0$ and $4x - 3y + 1 = 0$, the equation expresses that the $PM^2 = 2PN$. Hence the curve is a parabola of latus rectum 2, having $3x + 4y + 4 = 0$ for axis and $4x - 3y + 1 = 0$ for tangent at the vertex. The parabola lies below the line $4x - 3y + 1 = 0$ because the equation requires that $4x - 3y + 1$ should be positive.

13·94. The method used in 13·93 can be applied to any equation

$ax^2 + 2hxy + by^2 + 2gx + 2fy + c = 0$, for which $ab - h^2 = 0$.

This condition makes $ax^2 + 2hxy + by^2 \equiv (px + qy)^2$. Hence the equation of the curve is

$$(px + qy + r)^2 = 2x(pr - g) + 2y(qr - f) + r^2 - c.$$

r may be chosen so that $p(pr - g) + q(qr - f) = 0$ unless

$$g : f = p : q.$$

The method of 13·93 breaks down in this exceptional case. The equation is however of the form

$$(px+qy)^2 + 2\lambda(px+qy)+c = 0, \quad \text{where } \lambda p = g, \quad \lambda q = f,$$

i.e. $$(px+qy+\lambda)^2 = \lambda^2 - c.$$

This represents parallel lines, coincident lines, or nothing according as $\lambda^2 >, =, < c$,

i.e. $\lambda^2 p^2 >, =, < cp^2$ (or $\lambda^2 q^2 >, =, < cq^2$),

i.e. $g^2 >, =, < ac$ (or $f^2 >, =, < bc$).

13·95. EXAMPLE. A curve has parametric equations

$$x = a_1 t^2 + 2a_2 t, \quad y = b_1 t^2 + 2b_2 t,$$

where $a_1 b_2 \neq a_2 b_1$. Show that it is a parabola.

1st method. $$a_1 t^2 + 2a_2 t - x = 0,$$
$$b_1 t^2 + 2b_2 t - y = 0,$$
$$t^2 : -2t : -1 = a_2 y - b_2 x : a_1 y - b_1 x : a_1 b_2 - a_2 b_1.$$

Elimination of t gives the locus equation

$$(a_1 y - b_1 x)^2 + 4(a_2 y - b_2 x)(a_1 b_2 - a_2 b_1) = 0,$$

which represents a parabola because the terms of the second degree form a square. When the equation is made homogeneous, the solution with $z = 0$ gives equal roots.

2nd method. $Xx + Yy + 1 = 0$ meets the curve where

$$(a_1 X + b_1 Y) t^2 + 2(a_2 X + b_2 Y) t + 1 = 0.$$

If $Xx + Yy + 1 = 0$ is a tangent, this quadratic for t has equal roots. Hence the envelope equation is

$$(a_2 X + b_2 Y)^2 = (a_1 X + b_1 Y).$$

But $(a_2 X + b_2 Y)^2 = (a_1 X + b_1 Y) Z$ is satisfied by $[0, 0, 1]$. Hence the curve is a parabola.

13·96. EXAMPLE. Find the directrix of the parabola

$$x = at^2 + 2bt, \quad y = ct^2 + 2dt.$$

1st method. The equations give

$$t^2 : 2t : 1 = by - dx : cx - ay : bc - ad.$$

Therefore the line

$$(by - dx) - (cx - ay)\, s + (bc - ad)\, s^2 = 0$$

meets the curve in points whose parameters are given by

$$t^2 - 2ts + s^2 = 0.$$

Hence it is the tangent at the point s. Regarded as a quadratic for s the equation has two roots s_1, s_2 which are the parameters of the points of contact of the tangents from (x, y); and

$$1 : s_1 + s_2 : s_1 s_2 = bc - ad : cx - ay : by - dx.$$

If (x_1, y_1) is on the directrix, the tangents from (x_1, y_1) are at right angles. Hence

$$(as_1 + b)(as_2 + b) + (cs_1 + d)(cs_2 + d) = 0,$$

$$(a^2 + c^2)(by - dx) + (ab + cd)(cx - ay) + (b^2 + d^2)(bc - ad) = 0,$$

$$ax + cy + b^2 + d^2 = 0.$$

2nd method. By 13·95 the envelope equation of the curve is

$$(bX + dY)^2 = aX + cY.$$

The pair of points at infinity on the tangents from (x, y) to $(bX + dY)^2 = (aX + cY)\, Z$ is, by 8·56,

$$(bX + dY)^2 + (aX + cY)(xX + yY) = 0.$$

Also if (x, y) is on the directrix, these points are in perpendicular directions. $\qquad \therefore \quad (b^2 + ax) + (d^2 + cy) = 0.$

The directrix is $ax + cy + b^2 + d^2 = 0$.

13·97. EXAMPLE. Prove that the envelope of a chord of a parabola which subtends a right angle at the vertex is a point.

1st method. Let the parabola be $x : y : a = t^2 : 2t : 1$ and let the ends of the chord $[X, Y]$ be t_1, t_2. Then, by 13·23,

$$X : Y : 1 = 2 : -t_1 - t_2 : 2at_1 t_2.$$

The lines joining the vertex to t_1, t_2 are at right angles if

$$t_1 t_2 + 4 = 0, \quad \text{i.e. } 4aX + 1 = 0.$$

Therefore the envelope of the chord is the point $(4a, 0)$.

2nd method. The pair of lines from the vertex to the points of intersection of $y^2 = 4ax$ with the chord

$$Xx + Yy + 1 = 0 \quad \text{are} \quad y^2 + 4ax(Xx + Yy) = 0.$$

These are at right angles if $4aX + 1 = 0$.

∴ the envelope of the chord is the point $(4a, 0)$.

This is the Frégier point (7·44) of the vertex.

13·98. EXAMPLE. Find the polar equation of the parabola referred to the focus as pole.

1st method. Moving the origin to the focus $(a, 0)$ by 2·51, the equation $y^2 = 4ax$ is replaced by $y^2 = 4a(x + a)$. Hence the polar equation is

$$r^2 \sin^2\theta = 4a(r \cos\theta + a).$$

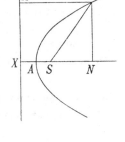

$$r^2 = r^2 \cos^2\theta + 4a(r \cos\theta + a)$$
$$= (r \cos\theta + 2a)^2,$$
$$\pm r = r \cos\theta + 2a,$$
$$r(1 - \cos\theta) = 2a \quad \text{or} \quad r(1 + \cos\theta) = -2a.$$

These equations are equivalent because the second is satisfied by $(-r_1, \theta + \pi)$ when the first is satisfied by (r_1, θ_1).

2nd method. Let P be any point (r, θ) on the curve. Then, by 13·4,

$$SP = PK = XN = XS + SN,$$
$$r = 2a + r \cos\theta, \qquad \therefore \quad r(1 - \cos\theta) = 2a.$$

EXERCISE 13 E

In Nos. 1–4, find the axis, tangent at the vertex, and the length of the latus rectum.

1. $x^2 + 3y = 0$.

2. $y^2 = 5x + 10$.

3. $y^2 = 4(y - x)$.

4. $x^2 + 2ax + 2by + c = 0$.

In Nos. 5–8, find the focus and directrix.

5. $(x + 1)^2 + y^2 = (x + 2)^2$.

6. $x^2 + 8y = 8$.

7. $y^2 = ax + b$.

8. $x = a + 2by + cy^2$.

In Nos. 9–14, find the axis and vertex and sketch the curve.

9. $(x + y)^2 = x - y$.

10. $(x - 2y)^2 = 2x + 6y - 1$.

11. $(x + 2y)^2 = 56x + 12y - 184$.

12. $4(5x + 12y)^2 + 8x + 53y - 34 = 0$.

13. $(7x - y)^2 = 276x + 32y - 531$.

14. $(7x - 24y)^2 = 82x + 76y - 1$.

15. Find the axis and vertex and the length of the latus rectum of the parabola $x = u \cos \alpha\, t$, $y = u \sin \alpha\, t - \tfrac{1}{2}gt^2$, where u, g, α are constants.

16. Find the new equation of the parabola

$$y = x \tan \alpha - \tfrac{1}{2}gu^{-2} \sec^2 \alpha\, x^2$$

when the origin is moved to the point $(u^2 \sin 2\alpha / 2g, \; u^2 \sin^2 \alpha / 2g)$.

17. Show that the parabola in No. 16 touches the parabola

$$x^2 = 4h(h - y),$$

where $h = u^2 / 2g$, for all values of α.

18. Prove that the equations $x = at^2 + 2bt + c$, $y = kt^2$ represent a parabola.

19. Find the tangent at t_1 to $x = at^2$, $y = b(t - 1)^2$.

20. Find the length of the chord $12x - 5y = 28a$ of

$$x = at^2, \quad y = 4a(t - 1)^2.$$

21. Find the tangent at t_1 to $x = a + bt^2$, $y = f + gt$ and the locus of the point of intersection of perpendicular tangents.

22. If the tangents at t_1, t_2 to $x = at + bt^2$, $y = ct + dt^2$ are at right angles, find the fixed point through which the chord $t_1 t_2$ passes.

23. Find the polar equation of $y^2 = 4ax$ with the vertex as pole, and of its inverse wo $r = 2a$. Sketch the inverse.

24. What is represented by $2a = r(1 + \cos \theta)$?

25. Find the inverse of a parabola wo the focus and sketch the curve.

EXERCISE 13F

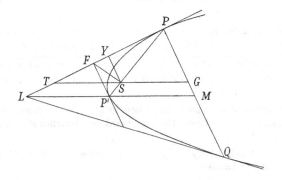

1. In the figure PSP' is a focal chord and PQ a normal chord of $y^2 = 4ax$. Obtain the lengths of SP, SP', PG, PQ in terms of a and θ $(=\angle STP)$.

2. Obtain as follows an alternative proof of 13·5. From TY produced cut off YP equal to TY. Prove that P lies on a fixed parabola by showing that $SP = PK$. Prove that TP is a tangent at P to this parabola.

3. A circle touches a given line and a given circle. Find the locus of its centre.

4. PSQ is a focal chord of a parabola with vertex A. Prove that $PS . SQ = AS . PQ$.

5. Prove that the semi latus rectum of a parabola is a harmonic mean between the segments of any focal chord.

6. If $PK, P'K'$ are perpendiculars to the directrix of a parabola from the ends of a focal chord, prove that PK', KP' meet at the vertex.

7. The normal at P to a parabola focus S meets the axis at G. GQ is drawn perpendicular to SP. Prove that $QP = 2a$.

8. From any point T on the tangent at P to a parabola focus S perpendiculars TM, TN are drawn to SP and the directrix. Prove that $SM = TN$ (Adams' Theorem).

9. T is the pole of a chord PQ of a parabola focus S. Prove that the triangles TSP, QST are similar and that $SP . SQ = ST^2$ and

$$TP^2 : TQ^2 = SP : SQ.$$

10. In the figure of No. 9 prove that TQ is a tangent to the circle PST, and if the circles PST, QST meet the axis again in U, V, prove that PU is parallel to TV.

11. Find the common tangents of $y^2 = 24x$ and $x^2 + y^2 = 25$.

12. Find the common tangents of $y^2 = 120x - 342y - 12441$ and $x^2 + y^2 + 280x + 342y = 0$.

13. If the distances of P_1, P_2 from the axis of a parabola are equal, prove that the distances of P_1, P_2 from the polars of P_2, P_1 are equal.

14. Prove that of all points on $y^2 = 4ax$ the vertex is the nearest to $(2a, 0)$, and that of other points on the curve the nearer to the vertex is the nearer to $(2a, 0)$.

15. A variable triangle is inscribed in $y^2 = 4ax$ so that two of the sides touch $y^2 = 4bx$. Prove that the third side touches $y^2 = 4cx$ where $(2a - b)^2 c = ab^2$.

16. Find the chord of $x^2 = 4ay$ which touches $y^2 = 4bx$ and subtends a right angle at the origin.

17. A variable chord of $y^2 = 4ax$ subtends a right angle at $(a, 2a)$. Prove that it passes through $(5a, -2a)$.

18. Chords of $y^2 = 4ax$ through $(a, -2a)$ make equal angles with the tangent at that point. Prove that their other ends are collinear with $(-3a, 2a)$.

19. Obtain the condition of perpendicularity of the chords $t_0 t_1$ and $t_0 t_2$ of the parabola $x : y : k = t^2 : t : 1$. Use it to prove that chords which subtend a right angle at the fixed point t_0 pass through the fixed point $(k + kt_0^2, -kt_0)$.

20. Show that the conic $\Sigma\{k_1(x - t_2 y + at_2^2)(x - t_3 y + at_3^2)\} = 0$ passes through the vertices of the triangle formed by the tangents at t_1, t_2, t_3 to the parabola $x : y : a = t^2 : 2t : 1$. Also find the ratios $k_1 : k_2 : k_3$ for which the conic is a circle and verify that the circle passes through $(a, 0)$.

21. The tangents at t_1, t_2, t_3 to the parabola $x : y : a = t^2 : 2t : 1$ form a triangle $Q_1 Q_2 Q_3$. Lines are drawn through Q_1 parallel to the chord $t_2 t_3$ and the tangent at t_1. Show that these lines and the similar lines drawn through Q_2 and Q_3 all touch the parabola

$$(y - 2at_1 - 2at_2 - 2at_3)^2 + 8a(x - at_2 t_3 - at_3 t_1 - at_1 t_2) = 0.$$

22. The tangent at P to $y^2 = 4ax$ is met in Q by the line through the vertex A perpendicular to AP. Z is the foot of the perpendicular from A to the tangent at P. If Z lies on $lx + my + na = 0$, prove that P has three possible positions and that the corresponding positions of Q lie on the line $(2l - n)x + 4my + 2na = 0$.

Chapter 14

THE ELLIPSE

14·1. In 6·41 the equation of the ellipse is obtained in the form

$$\frac{x^2}{a^2}+\frac{y^2}{b^2} = 1.$$

By 12·34 the polar of $P_1(x_1, y_1)$ wo this ellipse is

$$\frac{xx_1}{a^2}+\frac{yy_1}{b^2} = 1$$

and this is the tangent at P_1 if P_1 is a point on the curve.

By 12·45 the condition for P_1, P_2 to be conjugate points is

$$\frac{x_1 x_2}{a^2}+\frac{y_1 y_2}{b^2} = 1.$$

When $b = a$, the ellipse is a circle.

The polar of $P_1(x_1, 0)$ wo the circle $x^2+y^2 = a^2$ is $xx_1 = a^2$. This meets OP_1 at P_2 such that $OP_2 = a^2/x_1$.

$$\therefore \quad OP_1 . OP_2 = a^2.$$

Hence the polar of a point P_1 wo a circle centre O is the line perpendicular to OP_1 at the inverse of P_1.

By 12·32 the equation of the pair of tangents from P_1 to the ellipse $x^2/a^2+y^2/b^2 = 1$ is

$$\left(\frac{x^2}{a^2}+\frac{y^2}{b^2}-1\right)\left(\frac{x_1^2}{a^2}+\frac{y_1^2}{b^2}-1\right) = \left(\frac{xx_1}{a^2}+\frac{yy_1}{b^2}-1\right)^2.$$

14·2. Parametric Equations

14·21. In 6·43 parametric equations of $x^2/a^2+y^2/b^2 = 1$ are obtained in the form

$$1-\frac{x}{a} : \frac{y}{b} : 1+\frac{x}{a} = t^2 : t : 1$$

or

$$\frac{x}{a} : \frac{y}{b} : 1 = 1-t^2 : 2t : 1+t^2.$$

Every value of t determines just one point of the ellipse, but there is one point $(-a, 0)$ on the ellipse which is not given by any value of t. It can be obtained as a limit when $t \to \infty$.

In G_6 $x/a : y/b : z = s^2 - t^2 : 2st : s^2 + t^2$ is a proper parametric representation.

14·22. *Intersections of $lx + my + n = 0$ with the ellipse.* The point t of the ellipse lies on the line if

$$la(1 - t^2) + 2mbt + n(1 + t^2) = 0,$$

i.e. $$(n - la) t^2 + 2mbt + (n + la) = 0.$$

This is a quadratic for t whose roots are the parameters of the points of intersection of the line and curve.

14·23. *The Chord $t_1 t_2$.* The quadratic equation obtained by the method of 14·22 for the parameters of the points of intersection of the chord and curve must be $t^2 - t(t_1 + t_2) + t_1 t_2 = 0$. Therefore by the equations in 14·21 the chord must be

$$1 - \frac{x}{a} - \frac{y}{b}(t_1 + t_2) + \left(1 + \frac{x}{a}\right) t_1 t_2 = 0,$$

i.e. $$(1 - t_1 t_2) x/a + (t_1 + t_2) y/b = 1 + t_1 t_2.$$

14·24. *The Chord given by $ut^2 + 2vt + w = 0$.* By 14·21 the roots of $ut^2 + 2vt + w = 0$ are the parameters of the points of intersection of the ellipse with

$$u\left(1 - \frac{x}{a}\right) + 2v\frac{y}{b} + w\left(1 + \frac{x}{a}\right) = 0,$$

i.e. with $$(w - u) x/a + 2vy/b + (w + u) = 0.$$

Hence this is the chord joining the points whose parameters are given by $ut^2 + 2vt + w = 0$.

Alternatively, the result follows from 14·23 because

$$1 : t_1 + t_2 : t_1 t_2 = u : -2v : w.$$

14·25. *The Tangent at t_1.* This is found from the chord by making $t_2 \to t_1$. The equation is $(1 - t_1^2) x/a + 2t_1 y/b = 1 + t_1^2$. It can also be found by the method of 6·71.

14·26. *The Normal at t_1.* This is perpendicular to the tangent and passes through

$$x = a(1 - t_1^2)/(1 + t_1^2), \quad y = 2bt_1/(1 + t_1^2).$$

Hence the normal at t_1 is

$$2t_1 ax - (1 - t_1^2) by = 2c^2 t_1 (1 - t_1^2)/(1 + t_1^2),$$

where $c^2 = a^2 - b^2$.

14·27. *Envelope Equations.* By 14·25 parametric envelope equations are

$$aX : bY : -1 = 1 - t^2 : 2t : 1 + t^2.$$

Therefore the envelope equation is

$$a^2 X^2 + b^2 Y^2 = 1.$$

The results of 12·6 may be applied to this equation.

14·28. *Pole of a Chord.* If the chord is $[X_1, Y_1]$, by 12·64 its pole is

$$a^2 X X_1 + b^2 Y Y_1 = 1, \quad \text{i.e. } (-a^2 X_1, -b^2 Y_1).$$

If the chord joins the points t_1, t_2, and (x_1, y_1) is its pole, $xx_1/a^2 + yy_1/b^2 = 1$ is identical with the equation in 14·23. Hence

$$x_1 : y_1 : 1 = a(1 - t_1 t_2) : b(t_1 + t_2) : 1 + t_1 t_2. \tag{1}$$

If the ends of the chord have parameters given by

$$ut^2 + 2vt + w = 0,$$

this equation must be the same as

$$(1 - t^2) x_1/a + 2ty_1/b = 1 + t^2,$$

which expresses that the polar of (x_1, y_1) passes through the point t. Hence

$$\frac{x_1}{a} + 1 : -\frac{y_1}{b} : 1 - \frac{x_1}{a} = u : v : w.$$

$$\therefore \quad x_1 : y_1 : 1 = a(u - w) : -2bv : (u + w).$$

This may also be deduced from (1).

14·3. Eccentric Angle

14·31. In 6·41 the ellipse is represented by parametric equations
$$x = a\cos\phi, \quad y = b\sin\phi$$
and an indication is given of a method of obtaining the equation of the chord $\phi_1\phi_2$ by orthogonal projection. It is convenient to denote the eccentric angles by $\alpha - \beta, \alpha + \beta$. The chord is then

$$\begin{vmatrix} x/a & y/b & 1 \\ \cos(\alpha+\beta) & \sin(\alpha+\beta) & 1 \\ \cos(\alpha-\beta) & \sin(\alpha-\beta) & 1 \end{vmatrix} = 0,$$

i.e. $(x/a)\, 2\cos\alpha\sin\beta + (y/b)\, 2\sin\alpha\sin\beta = \sin\{(\alpha+\beta) - (\alpha-\beta)\}$,

i.e. $\qquad \dfrac{x}{a}\cos\alpha + \dfrac{y}{b}\sin\alpha = \cos\beta.$

14·32. Making $\beta \to 0$ the tangent is

$$\frac{x}{a}\cos\alpha + \frac{y}{b}\sin\alpha = 1.$$

The corresponding normal is
$$ax\sec\alpha - by\cosec\alpha = a^2 - b^2 = c^2,$$
and parametric envelope equations are
$$aX = -\cos\alpha, \quad bY = -\sin\alpha.$$

14·33. EXAMPLE. Find the pedal of $x^2/a^2 + y^2/b^2 = 1$ wo the origin.

Let any tangent to the ellipse be $x\cos\alpha + y\sin\alpha = p$; then, by 14·27, $\qquad a^2\cos^2\alpha + b^2\sin^2\alpha = p^2.$

But p and α are the polar coordinates of the foot of the perpendicular from the origin to the tangent. Hence the polar equation of the pedal is
$$a^2\cos^2\theta + b^2\sin^2\theta = r^2.$$

This curve can be traced by giving values to θ. Its cartesian equation is $a^2x^2 + b^2y^2 = (x^2 + y^2)^2$.

14·34. EXAMPLE. Find the envelope of the join of points on an ellipse whose eccentric angles differ by a constant.

Let the eccentric angles be $\phi - \alpha, \phi + \alpha$, so that α is constant. The equation of the join is

$$\frac{x}{a}\cos\phi + \frac{y}{b}\sin\phi = \cos\alpha$$

and so the line touches the ellipse $x^2/a_1{}^2 + y^2/b_1{}^2 = 1$, where $a_1 = a\cos\alpha$ and $b_1 = b\cos\alpha$.

14·35. EXAMPLE. Tangents to $x^2/c^2 + y^2/d^2 = 1$ are drawn from a point on $x^2/a^2 + y^2/b^2 = 1$. Find the envelope of the chord joining the other points where these tangents meet

$$x^2/a^2 + y^2/b^2 = 1.$$

1st method. Let either tangent from the point

$$x/a : y/b : 1 = 1 - t_1{}^2 : 2t_1 : 1 + t_1{}^2$$

meet the curve again in the point t.

Then $(1 - t_1 t)\, x/a + (t_1 + t)\, y/b = 1 + t_1 t$ is a tangent to

$$x^2/c^2 + y^2/d^2 = 1,$$

i.e. $[(1 - t_1 t)/a, \quad (t_1 + t)/b, \quad -1 - t_1 t]$

is a tangent to $c^2 X^2 + d^2 Y^2 = Z^2$.

i.e. $\therefore \quad \dfrac{c^2}{a^2}(1 - t_1 t)^2 + \dfrac{d^2}{b^2}(t_1 + t)^2 = (1 + t_1 t)^2,$

$$t^2\left(\frac{c^2}{a^2}t_1{}^2 + \frac{d^2}{b^2} - t_1{}^2\right) + 2tt_1\left(-\frac{c^2}{a^2} + \frac{d^2}{b^2} - 1\right) + \left(\frac{c^2}{a^2} + \frac{d^2}{b^2}t_1{}^2 - 1\right) = 0.$$

The roots of this quadratic in t are the parameters of the points where the tangents meet the curve again. Hence by 14·24 the chord joining these points is

$$\left(\frac{c^2}{a^2} - \frac{d^2}{b^2} - 1\right)(1 - t_1{}^2)\frac{x}{a} + 2\left(-\frac{c^2}{a^2} + \frac{d^2}{b^2} - 1\right)t_1\frac{y}{b}$$

$$+ \left(\frac{c^2}{a^2} + \frac{d^2}{b^2} - 1\right)(1 + t_1{}^2) = 0$$

or, say, $(1 - t_1{}^2)\, px/a + 2t_1 qy/b + (1 + t_1{}^2)\, r = 0.$

The homogeneous envelope coordinates of this line are

$$(1 - t_1{}^2)\,p/a, \quad 2t_1 q/b, \quad (1 + t_1{}^2)\,r$$

and these satisfy $a^2 X^2/p^2 + b^2 Y^2/q^2 = Z^2/r^2$. Therefore the envelope is $p^2 x^2/a^2 + q^2 y^2/b^2 = r^2$.

2nd method. Let either of the tangents drawn from $(a\cos\phi, b\sin\phi)$ meet the curve again in $(a\cos\phi_1, b\sin\phi_1)$.

The tangent is

$$\frac{x}{a}\cos\tfrac{1}{2}(\phi_1 + \phi) + \frac{y}{b}\sin\tfrac{1}{2}(\phi_1 + \phi) = \cos\tfrac{1}{2}(\phi_1 - \phi)$$

and its homogeneous envelope coordinates are

$$\frac{1}{a}\cos\tfrac{1}{2}(\phi_1 + \phi), \quad \frac{1}{b}\sin\tfrac{1}{2}(\phi_1 + \phi), \quad -\cos\tfrac{1}{2}(\phi_1 - \phi).$$

These satisfy $c^2 X^2 + d^2 Y^2 = Z^2$.

$$\therefore \quad \frac{c^2}{a^2}\cos^2\tfrac{1}{2}(\phi_1 + \phi) + \frac{d^2}{b^2}\sin^2\tfrac{1}{2}(\phi_1 + \phi) = \cos^2\tfrac{1}{2}(\phi_1 - \phi),$$

$$\frac{c^2}{a^2}(1 + \cos\phi_1 \cos\phi - \sin\phi_1 \sin\phi) + \frac{d^2}{b^2}(1 - \cos\phi_1 \cos\phi + \sin\phi_1 \sin\phi)$$

$$= 1 + \cos\phi_1 \cos\phi + \sin\phi_1 \sin\phi,$$

$$\therefore \quad \left(\frac{c^2}{a^2} - \frac{d^2}{b^2} - 1\right)\cos\phi_1 \cos\phi + \left(-\frac{c^2}{a^2} + \frac{d^2}{b^2} - 1\right)\sin\phi_1 \sin\phi$$

$$= -\frac{c^2}{a^2} - \frac{d^2}{b^2} + 1$$

or, say, $\qquad p\cos\phi_1 \cos\phi + q\sin\phi_1 \sin\phi = r.$

Hence $(a\cos\phi_1, b\sin\phi_1)$ lies on

$$(px/a)\cos\phi + (qy/b)\sin\phi = r,$$

which is therefore the equation of the chord. It is a tangent to $p^2 x^2/a^2 + q^2 y^2/b^2 = r^2$.

EXERCISE 14A

In Nos. 1–10, give the equation of the line or lines.

1. The tangent at $(3, -7)$ to $4x^2 + y^2 = 85$.

2. The tangents at the origin and $(-2, -1)$ to $x^2 + 4y^2 + 4x = 0$.

3. The tangents to $4x^2 + 5y^2 = 2$ perpendicular to $y = 3x$.

4. The polar of $(3, -8)$ wo $x^2 + 7y^2 = 3$.

5. The polar of the origin wo $(x-c)^2/a^2 + (y+d)^2/b^2 = 1$.

6. The polar of $X = k$ wo $a^2X^2 + b^2Y^2 = 1$.

7. The polar of $X + Y = 1$ wo $3x^2 + 4y^2 = 5$.

8. The polar of $(l, m, 0)$ wo $x^2/a^2 + y^2/b^2 = z^2$.

9. The pair of tangents from $(-3, 5)$ to $x^2 + 10y^2 = 1$.

10. The pair of tangents from $(2, 1)$ to $x^2 + 3y^2 + 2x - 5y = 1$.

In Nos. 11–14, give the coordinates of the point.

11. The pole of $4x - 3y + 5 = 0$ wo $7x^2 + 2y^2 = 1$.

12. The pole of OY wo $(x-p)^2 + qy^2 = r^2$.

13. The pole of $[5, 6]$ wo $3X^2 + 2Y^2 = 1$.

14. The pole of $[u, v, 0]$ wo $a^2X^2 + b^2Y^2 = Z^2$.

15. Give the equation of the pair of contacts of $X^2 + 3Y^2 = 1$ on $[1, 2]$.

16. Give the equation of the pair of contacts of $a^2X^2 + b^2Y^2 = Z^2$ on $[l, m, 0]$.

17. Show that no point of the ellipse $x^2/a^2 + y^2/b^2 = 1$ lies outside the rectangle formed by $x = \pm a, y = \pm b$.

18. Find the mid-point of the chord $lx + my + n = 0$ of
$$x^2/a^2 + y^2/b^2 = 1.$$

19. Find the equation of the pair of lines from the centre of
$$x^2/a^2 + y^2/b^2 = 1$$
to the ends of the chord $lx + my + n = 0$.

20. Find the equation of the circle which has the chord $x + y = 5$ of $2x^2 + 3y^2 = 35$ for a diameter.

21. Find the length of the perpendicular from the centre of an ellipse to a tangent which makes an angle ψ with the major axis.

22. Show that the sum of the squares of the reciprocals of two perpendicular diameters of an ellipse is constant.

23. What is the condition for $ax + by = 1$ to be a tangent to
$$4x^2 + y^2 = 9?$$

24. Prove that if $a^2l^2 + b^2m^2 = n^2$, then $lx + my + n = 0$ is a tangent to $x^2/a^2 + y^2/b^2 = 1$ and find the point of contact.

25. Two lines conjugate wo $x^2/a^2 + y^2/b^2 = 1$ pass one through $(a, 0)$ and the other through $(-a, 0)$. Prove that they meet on the ellipse $x^2/a^2 + 2y^2/b^2 = 1$.

26. If θ is the interior angle between the tangents from (x, y) to $x^2/a^2 + y^2/b^2 = 1$, prove that

$$(x^2 + y^2 - a^2 - b^2)\sin\theta = 2\cos\theta\sqrt{(a^2x^2 + b^2y^2 - a^2b^2)}.$$

Give the geometrical meanings of $x^2 + y^2 = a^2 + b^2$ and

$$a^2x^2 + b^2y^2 < a^2b^2.$$

27. Find the condition for the chord $t_1 t_2$ of the ellipse

$$x/a : y/b : 1 = 1 - t^2 : 2t : 1 + t^2$$

to pass through the origin.

28. State the condition for the chord $t_1 t_2$ of the ellipse

$$x^2/a^2 + y^2/b^2 = 1$$

to have a fixed direction.

29. State the condition for the chords $t_1 t_2$ and $t_3 t_4$ of the ellipse $x^2/a^2 + y^2/b^2 = 1$ to be at right angles.

30. Find the mid-point of the chord $t_1 t_2$ of $x^2/a^2 + y^2/b^2 = 1$.

31. State the relation between the eccentric angles of P_1, P_2 if the tangents at P_1, P_2 are (i) parallel, (ii) perpendicular.

32. State the condition for the chord joining the points whose eccentric angles are ϕ_1 and ϕ_2 to be in a fixed direction.

33. A tangent to the ellipse $x^2/a^2 + y^2/b^2 = 1$ meets the axes in P and Q. Find the locus of the mid-point of PQ.

34. Prove that the pole T and the mid-point V of a chord of an ellipse are collinear with the centre and that the ellipse separates T, V harmonically.

35. Write down the equations of the chords joining the vertices of $x^2/a^2 + y^2/b^2 = 1$ to the point whose eccentric angle is ϕ, and the coordinates of the poles of those lines.

36. If $\cos\theta + \cos\phi = 1$, prove that the pole of the chord $\theta\phi$ wo

$$x^2/a^2 + y^2/b^2 = 1$$

lies on $x^2/a^2 + y^2/b^2 = 2x/a$.

37. A is a fixed point and Q is a variable point on an ellipse centre C. The tangents at A, Q meet at P, and QR is the chord parallel to PC. Prove that R is fixed.

38. Prove that the parameters of concyclic points on

$$x/a : y/b : 1 = 1 - t^2 : 2t : 1 + t^2$$

satisfy $t_1 + t_2 + t_3 + t_4 = t_2 t_3 t_4 + t_1 t_3 t_4 + t_1 t_2 t_4 + t_1 t_2 t_3.$

39. Prove that the sum of the eccentric angles of concyclic points of an ellipse is $2n\pi$.

40. The tangents at P, Q to $x^2/a^2 + y^2/b^2 = 1$ are conjugate wo $x^2/c^2 + y^2/d^2 = 1$. Find the envelope of PQ.

14·4. Geometrical Properties of the Ellipse

14·41. The curve $x^2/a^2 + y^2/b^2 = 1$ is symmetrical about both axes.

It is assumed that $a > b > 0$. This involves no loss of generality. The vertices $A(a, 0)$ and $A'(-a, 0)$ and the major and minor axes and centre are defined in Chapter 6. Reference should be made to Exercise 6c, Nos. 12–15. In Chapter 6 the ellipse is obtained by orthogonal projection. Other methods of obtaining it are given in the present chapter.

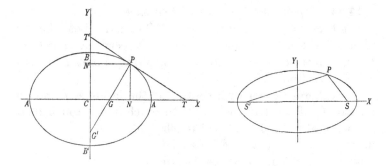

14·42. *The locus of P such that $SP + PS'$ is constant where S and S' are fixed points.*

Let S, S' be $(\pm c, 0)$ and denote the constant value of $SP + PS'$ by $2a$. Let $SP = h$, $PS' = h'$. Then

$$h'^2 - h^2 = (x+c)^2 - (x-c)^2 = 4cx$$

and $\qquad h' + h = 2a. \qquad \therefore \quad h' - h = 2cx/a,$

$$\therefore \quad a^2 + (cx/a)^2 = \tfrac{1}{2}(h^2 + h'^2) = CS^2 + CP^2 = c^2 + x^2 + y^2,$$

$$\therefore \quad x^2(1 - c^2/a^2) + y^2 = a^2 - c^2$$

or $x^2/a^2 + y^2/b^2 = 1$, where $b^2 = a^2 - c^2$. This shows that the ellipse $x^2/a^2 + y^2/b^2 = 1$ can be described by means of a string of length $2a$ with its ends fixed at a distance $2\sqrt{(a^2 - b^2)}$ apart.

14·43. *The locus of a point P fixed in a rod AB whose ends A, B move on fixed perpendicular lines.*

Let the perpendicular lines be taken as axes, and let $BP = a$, $PA = b$. Let P be an arbitrary point on the locus, and denote the angle OAP by t. Then $x = a \cos t$, $y = b \sin t$. These are parametric equations of an ellipse.

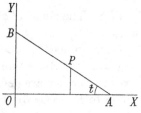

14·44. *A plane section of a right circular cone.*

Let V be the vertex of the cone, VC its axis. Let σ be the plane of section, α the plane through VC perpendicular to σ. Let α, σ meet in AA' and take an arbitrary point P on the curve of section.

The spheres drawn with centres at the incentre I and an ecentre I_1 of the triangle VAA' to touch σ will touch it at points S, S' on AA' and will touch the cone in circles EE', FF'. Let the generator VP cut these circles in H, K.

PS, PS' are tangents to the spheres.

$$\therefore \quad PS = PH, \qquad PS' = PK.$$

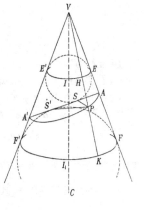

Thus $SP + PS' = HP + PK = EF$, a constant. Hence the locus of P is an ellipse (14·42).

It has been assumed that A and A' lie on the same side of V. The opposite case is considered in 15·44. If σ is parallel to a generator VE of the cone, the plane α through VE and the axis of the cone meets the surface of the cone also in another generator VE', and meets σ in a line $E'K$. The curve of section is the projection from V of the circle on EE' as diameter in the plane perpendicular to α.

With the notation of 9·2

$$VE = VE' = a, \quad EE' = b,$$

and the circle is $x^2 + y^2 = bx$. Hence by 9·21, the equation of the projection is

$$(bx')^2 + (ay')^2 = b^2 x'(x' + a),$$

i.e. $ay'^2 = b^2 x'$, a parabola.

The general case can also be investigated analytically in this way. See Exercise 9 A, No. 18.

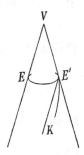

14·45. Foci

The points S, S' that occur in 14·42 and 14·44 are called the *foci* of the ellipse.

Since the constant value of $SP + PS'$ is $2a$, $SB = BS' = a$. Hence the foci can be constructed by drawing an arc centre B and radius a to cut the major axis.

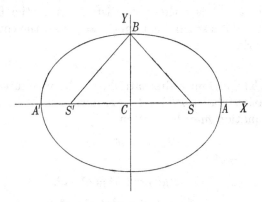

An ellipse may be obtained as the locus of a point P whose distance from a fixed point S is e times its perpendicular distance from a fixed line not through S, where e is a constant and $0 < e < 1$. e is called the *eccentricity*.

Take the fixed point S as origin and let the fixed line be $x = k$. Then the equation of the locus is

$$x^2 + y^2 = e^2(x - k)^2 \quad \text{or} \quad x^2(1 - e^2) + 2e^2 kx + y^2 = e^2 k^2.$$

By change of origin to $(-e^2k/(1-e^2), 0)$, this equation becomes

$$x^2(1-e^2)+y^2 = e^2k^2/(1-e^2).$$

This is the same as $x^2/a^2+y^2/b^2 = 1$ if

$$a^2 = \frac{e^2k^2}{(1-e^2)^2} \quad \text{and} \quad b^2 = \frac{e^2k^2}{1-e^2}.$$

Thus $\qquad\qquad 1-e^2 = \dfrac{b^2}{a^2}, \quad e^2 = \dfrac{a^2-b^2}{a^2}.$

But $\qquad\qquad a^2-b^2 = c^2. \qquad \therefore \quad e = c/a.$

Hence $x^2/a^2+y^2/b^2 = 1$ is the locus of a point whose distance from $(ae, 0)$ is e times its distance from

$$x = c+k = ae+a(1-e^2)/e = a/e,$$

where $e = c/a < 1$.

The line $x = a/e$ is called the directrix corresponding to the focus $(ae, 0)$.

The same curve is also the locus of a point whose distance from $S'(-ae, 0)$ is e times its distance from the fixed line $x = -a/e$. S' is a second focus and $x = -a/e$ is the corresponding directrix.

14·46. The foci may alternatively be obtained directly from the equation $x^2/a^2+y^2/b^2 = 1$ as follows.

The equation may be written

$$b^2x^2/a^2+y^2 = b^2,$$

or, using $c^2 = a^2-b^2$,

$$(1-c^2/a^2)\,x^2+y^2 = a^2-c^2,$$

$$x^2+c^2+y^2 = c^2x^2/a^2+a^2,$$

or, adding $\pm 2cx$ to each side,

$$(x \pm c)^2+y^2 = \frac{c^2}{a^2}\left(x \pm \frac{a^2}{c}\right)^2,$$

i.e. $SP^2 = e^2PK^2$ or $S'P^2 = e^2PK'^2$, where S, S' are $(\pm c, 0)$ or $(\pm \sqrt{(a^2-b^2)}, 0)$ and PK, PK' are the perpendiculars from P to the lines $x = \pm a^2/c = \pm a/e$.

$SP = ePK$ is called the focus-directrix property. No such property holds in G_3 for a point $(0, k)$ and line $y = k'$. See Exercise 14 B, No. 18.

14·47. *Latus Rectum.* The chord LSL' through a focus perpendicular to the major axis is called the *latus rectum.*

Its length $2l$ is found by substituting c, l for x, y in the equation $x^2/a^2 + y^2/b^2 = 1$.

Hence $l^2 = b^2(1 - c^2/a^2) = b^4/a^2, \quad l = b^2/a$.

This relation, together with $e^2 = (a^2 - b^2)/a^2$, determines a, b in terms of l, e.

14·48. *Focal Distances.* From $SP + PS' = 2a$, i.e. $r + r' = 2a$, by differentiation wo the arc s,

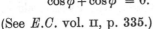

$$\frac{dr}{ds} + \frac{dr'}{ds} = 0,$$

$$\cos\phi + \cos\phi' = 0.$$

(See *E.C.* vol. II, p. 335.)

Hence $\phi + \phi' = \pi$, and SP, $S'P$ make supplementary angles with the tangent at P. Or: The tangent and normal at P are the angle-bisectors of SPS'.

14·49. From 14·48, if SY is the perpendicular from S to the tangent at P, and $S'P$ meets SY at V, the triangles SPY, VPY are congruent. Hence

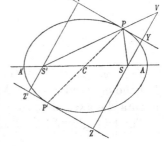

$$CY = \tfrac{1}{2}S'V = \tfrac{1}{2}(S'P + PS) = a.$$

Therefore Y lies on the auxiliary circle. Therefore the pedal of the ellipse wo a focus is the auxiliary circle.

14·50. Let Y, Y' be the feet of the perpendiculars from S, S' to the tangent at P, and let Z, Z' be the feet of the perpendiculars to the parallel tangent which touches the ellipse at the diametrically opposite point P' to P.

Then Z, Z' are the images of Y', Y through C. By 14·49, Y, Y', Z, Z' all lie on the auxiliary circle.

Also $SY . S'Y' = SY . ZS = A'S . SA = (c+a)(c-a)$.

$$\therefore \quad SY . S'Y' = b^2.$$

14·51. *Pedal Equation.* In the figure, triangles $SPY, S'PY'$ are similar.

$$\therefore \quad \frac{SY}{SP} = \frac{S'Y'}{S'P}, \qquad \text{i.e.} \quad \frac{p}{r} = \frac{b^2/p}{2a-r}.$$

$$\therefore \quad \frac{b^2}{p^2} = \frac{2a}{r} - 1.$$

This is the pedal equation wo the focus as pole.

14·52. *The Converse Pedal Property.* If S is a fixed point inside a fixed circle, and Y is a variable point on the circumference of the circle, the envelope of the perpendicular at Y to SY is an ellipse with focus S having the fixed circle as auxiliary circle.

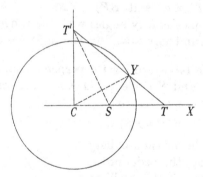

Take the centre C of the circle as origin and let S be $(c, 0)$. Let the perpendicular at Y to SY be $[X, Y]$ and suppose that it cuts the axes of coordinates in T, T'.

The triangles TYC, TST' are similar.

$$\therefore \quad CY . TT' = CT . ST',$$

$$\therefore \quad a\sqrt{\left(\frac{1}{X^2} + \frac{1}{Y^2}\right)} = -\frac{1}{X}\sqrt{\left(c^2 + \frac{1}{Y^2}\right)}.$$

$$\therefore \quad a^2(X^2 + Y^2) = c^2Y^2 + 1,$$

i.e. $\qquad\qquad\qquad a^2X^2 + b^2Y^2 = 1.$

The corresponding envelope when S is outside the circle is found in 15·52.

14·53. If a chord PP' meets the directrix corresponding to the focus S at F, then FS bisects an angle between SP, PS'.

For if $PK, P'K'$ are perpendiculars to the directrix,

$$SP = ePK, \qquad SP' = eP'K'.$$

$$\therefore \quad SP : S'P = PK : P'K' = PF : P'F.$$

Therefore FS bisects the angle at S.

By making $P' \to P$ it follows that if the tangent at P meets the directrix at R, then $\angle RSP$ is a right angle.

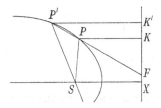 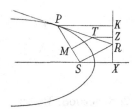

14·54. From the last result may be deduced Adams' Theorem:

If TM, TZ are perpendiculars to SP and the directrix, drawn from any point on the tangent at P, then $SM = eTZ$.

For $\qquad\qquad SM : SP = RT : RP = TZ : PK.$

But $\qquad\qquad SP = ePK \qquad \therefore \quad SM = eTZ.$

14·55. If tangents TP, TP' are drawn from a point T to an ellipse, TS bisects $\angle PSP'$.

For if TM, TM' are the perpendiculars to SP, SP', by 14·54, $SM = eTZ = SM'$. Hence TS bisects $\angle PSP'$.

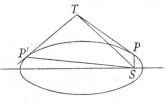

14·56. The proofs in 14·53–14·55 are applicable when $e = 1$. This shows that the results hold for the parabola.

14·57. The tangents TP, TP' are equally inclined to TS, TS'.

1st method. Let SY, $S'Y'$ be the perpendiculars from the foci to TP and $SZ, S'Z'$ the perpendiculars to TP'.

By 14·50, $SY . S'Y' = b^2 = SZ . S'Z'$. Hence it may be shown that the triangles SYZ, $S'Z'Y'$ are similar and so

$$\angle STP = \angle SZY = \angle S'Y'Z' = \angle S'TP'.$$

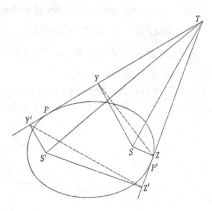

2nd method. By 12·38, the equation of the angle-bisectors of the lines through the origin parallel to TP, TP' is

$$\frac{x^2 - y^2}{g^2 - f^2 + a^2 - b^2} = -\frac{xy}{fg}.$$

But the lines through the origin parallel to TS, TS' are given by

$$\{xg - y(f - c)\}\{xg - y(f + c)\} = 0$$

and their angle-bisectors are given by the same equation as the bisectors of the parallels to TP, TP'. Hence TS, TS' are equally inclined to TP, TP'.

14·6. Locus of the Point of Intersection of Perpendicular Tangents

1st method. The tangents from (x_1, y_1) are $ss_{11} = s_1{}^2$,

i.e. $$\left(\frac{x^2}{a^2} + \frac{y^2}{b^2} - 1\right)\left(\frac{x_1{}^2}{a^2} + \frac{y_1{}^2}{b^2} - 1\right) = \left(\frac{xx_1}{a^2} + \frac{yy_1}{b^2} - 1\right)^2.$$

They are perpendicular if the sum of the coefficients of x^2 and y^2 in their equation is zero, i.e. if

$$\frac{1}{a^2}\left(\frac{y_1^2}{b^2}-1\right)+\frac{1}{b^2}\left(\frac{x_1^2}{a^2}-1\right)=0,$$

$$y_1^2-b^2+x_1^2-a^2=0.$$

Therefore (x_1, y_1) lies on $x^2+y^2=a^2+b^2$.

2nd method. By the envelope equation $a^2X^2+b^2Y^2=1$ it follows that an arbitrary tangent to the ellipse is

$$x\cos\alpha+y\sin\alpha=\sqrt{(a^2\cos^2\alpha+b^2\sin^2\alpha)}.$$

A tangent perpendicular to this is

$$-x\sin\alpha+y\cos\alpha=\sqrt{(a^2\sin^2\alpha+b^2\cos^2\alpha)}.$$

By squaring and adding it follows that the point of intersection of the tangents lies on

$$x^2+y^2=a^2+b^2.$$

This locus is called the *director circle* of the ellipse. It is discussed further in a later chapter. The corresponding locus for a parabola was proved in 13·61 to be the directrix of the parabola: in G_6 it would be the line-pair composed of the directrix and the line at infinity.

EXERCISE 14B

1. In the construction of 14·42 if the ends of the string are 3 in. apart, and the major axis is double the minor axis, what is the length of the string?

2. An ellipse of minor axis 3 in. is to be drawn by the construction of 14·42 with a string of length 5 in. Find the distance between the ends of the string.

3. Find the eccentricity of the ellipse drawn by means of a string 4 in. in length with ends $2\frac{1}{2}$ in. apart.

In Nos. 4–7, find the foci, directrices, and eccentricity.

4. $x^2+9y^2=36.$ **5.** $49x^2+9y^2=441.$

6. $3(x+y+1)^2+4(x-y-1)^2=12.$

7. $4(x^2+y^2)=(x\cos\alpha+y\sin\alpha-p)^2.$

In Nos. 8–11, find the eccentricity and the length of the latus rectum.

8. $x^2 + 2y^2 = 3$.   **9.** $4(x-1)^2 + 25(y+2)^2 = 100$.

10. $2x^2 + 4y^2 + 4x - 12y = 5$.   **11.** $81x^2 + 4y(y-2) = 320$.

12. Find the tangents to $9x^2 + 16y^2 = 144$ at the ends of the latera recta.

13. Find the condition that the chord $t_1 t_2$ of

$$x/a : y/b : 1 = 1 - t^2 : 2t : 1 + t^2$$

passes through a focus.

14. In the figure of 14·45 prove that $SB = CA$ and $AS \cdot SA' = b^2$.

15. Express the semi-axes of the ellipse in terms of the eccentricity and the latus rectum.

16. Show that the distances from the foci to the point (x, y) on $x^2/a^2 + y^2/b^2 = 1$ are $a \pm ex$.

17. Through any point P on $x^2/a^2 + y^2/b^2 = 1$ a line is drawn parallel to OX meeting OY at M and the circle on the minor axis as diameter at R. Prove that $MP : MR = a : b$.

18. Show that the equation $x^2/a^2 + y^2/b^2 = 1$ cannot be written in the form $x^2 + (y-p)^2 = (qy-r)^2$ if $b^2 < a^2$.

19. The adjacent corners A, B of a rectangular lamina $ABCD$ slide on two fixed perpendicular lines in the plane of the lamina. Prove that the locus of C is an ellipse of area πAD^2.

20. The normal at P meets the axis at G, and PK is drawn perpendicular to the directrix corresponding to the focus S. Prove that the triangles SPK and GSP are similar and deduce that $SG = eSP$.

21. If the tangent and normal at P meet the major axis at T and G, prove that $(S'GST) = -1$.

22. AA' is the major axis of an ellipse of which S, S' are foci. P is any point on the curve. $AR, A'R'$ parallel to $SP, S'P$ meet the tangent at P in R, R'. Prove that $AR + A'R' = AA'$.

23. The tangent to an ellipse at any point P meets a fixed tangent at T. At a focus S, a line is drawn perpendicular to ST to meet the tangent at P in Q. Prove that the locus of Q is a line which touches the ellipse.

24. T is any point on the tangent to an ellipse at an end of the minor axis. Prove that the other tangent from T also touches the circle through T and the foci.

25. Show that the envelope of the chords of contact of tangents to an ellipse of eccentricity e from a point on its director circle is an ellipse of eccentricity $e \sqrt{(2 - e^2)}$.

26. A tangent to $x^2/a^2 + y^2/b^2 = 1$ meets $x^2/a + y^2/b = a + b$ in P and Q. Prove that the tangents at P and Q are at right angles to one another.

27. Find the director circle of the ellipse $x^2/a^2 + y^2/b^2 = 2x/a$. If a and b increase without limit in such a way that $b^2/a \to 1$, find the limit of the ellipse and of its director circle and verify the result.

14·7. Diameters

14·71. It is proved in 12·34 that the harmonic conjugate P of a fixed point P_1 wo the ends Q, Q' of a variable chord through P_1 lies on a line $s_1 = 0$ called the polar of P_1, and that this line passes through the points of contact of tangents from P_1. Also if the tangents at Q, Q' meet at T, since the polar of T goes through P_1, that of P_1 goes through T. Thus T lies on HK. See the first figure.

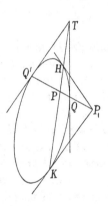 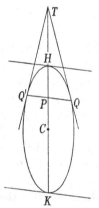

When the geometry is homogeneous and P_1 is a point at infinity $(l, m, 0)$, the polar of P_1 becomes the locus of the mid-points of chords parallel to $x/l = y/m$ and its equation is $xl/a^2 + ym/b^2 = 0$. This passes through the centre C and is therefore called a diameter of the ellipse. See the second figure.

The tangents at the ends H, K of the diameter

$$xl/a^2 + ym/b^2 = 0$$

are parallel to the chords bisected by that diameter, and the tangents at the ends Q, Q' of any one of the chords meet on HK.

The point at infinity on $xl/a^2 + ym/b^2 = 0$ is $(a^2m, -b^2l, 0)$ and this is the pole of $x/l = y/m$. Chords parallel to

$$xl/a^2 + ym/b^2 = 0$$

are bisected by $x/l = y/m$.

These two diameters, which are such that each bisects the chords parallel to the other, are called conjugate diameters. They are conjugate lines in the ordinary sense that each passes through the pole of the other.

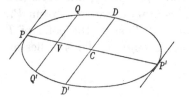 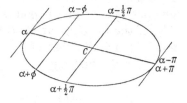

14·72. In the first figure $PVCP'$ and DCD' are conjugate diameters and QQ' is one of the chords bisected by PP'. The eccentric angles of the points can be taken as they are shown in the second figure. For (i) the sum of the eccentric angles of the ends of parallel chords is constant. Thus if P is the point α, Q, Q' must be $\alpha - \phi, \alpha + \phi$: also D, D' must be $\alpha - \beta, \alpha + \beta$, where $\beta = \frac{1}{2}\pi$ because DD' is a diameter. Or (ii) the ellipse can be regarded as the orthogonal projection of a circle: PP', DD' are then the projections of perpendicular diameters because the tangent at P is parallel to DD'; the eccentric angles of the points on the ellipse are equal to the vectorial angles of the corresponding points on the circle.

Since

$$CP^2 = a^2\cos^2\alpha + b^2\sin^2\alpha \quad \text{and} \quad CD^2 = a^2\sin^2\alpha + b^2\cos^2\alpha,$$

$$CP^2 + CD^2 = a^2 + b^2,$$

i.e. the sum of the squares of the lengths of conjugate diameters of an ellipse is constant.

Also, the parallelogram formed by the tangents at P, P', D, D' has constant area $4ab$, since it is the orthogonal projection of a square circumscribing a circle of radius a.

14·73. The conjugate diameters whose ends have eccentric angles $\frac{1}{4}\pi$, $\frac{3}{4}\pi$ are of equal length and are called the *equiconjugate* diameters of the ellipse. The equiconjugate diameters lie along the diagonals of the rectangle $x = \pm a$, $y = \pm b$.

The axes of the ellipse are one pair of conjugate diameters and they are the only perpendicular pair.

14·74. EXAMPLE. Find the chord of $x^2/a^2 + y^2/b^2 = z^2$ which is perpendicularly bisected by $lx + my = z$.

Let the chord be $x/l - y/m = kz$. The diameter which bisects it is the polar of $(l, m, 0)$, namely $xl/a^2 + ym/b^2 = 0$.

These two lines are concurrent with $lx + my = z$.

$$\therefore \quad \begin{vmatrix} 1/l & -1/m & k \\ l/a^2 & m/b^2 & 0 \\ l & m & 1 \end{vmatrix} = 0,$$

$$\therefore \quad k \begin{vmatrix} l/a^2 & m/b^2 \\ l & m \end{vmatrix} = - \begin{vmatrix} 1/l & -1/m \\ l/a^2 & m/b^2 \end{vmatrix}.$$

$$\therefore \quad k(a^2 - b^2) = a^2/l^2 + b^2/m^2.$$

Hence in G_3 the line perpendicularly bisected by $lx + my = 1$ is

$$(a^2 - b^2)(x/l - y/m) = a^2/l^2 + b^2/m^2.$$

14·75. EXAMPLE. Find the pedal equation of an ellipse wo the centre.

By 14·72, $r^2 + CD^2 = a^2 + b^2$

and $p \cdot CD = ab.$

Eliminating CD, $p^2 r^2 + a^2 b^2 = p^2(a^2 + b^2).$

14·76. EXAMPLE. Find the condition for

$$Ax^2 + 2Hxy + By^2 = 0$$

to represent conjugate diameters of $x^2/a^2 + y^2/b^2 = 1$.

If $Ax^2 + 2Hxy + By^2 = k\left(\dfrac{x}{l} - \dfrac{y}{m}\right)\left(\dfrac{xl}{a^2} + \dfrac{ym}{b^2}\right),$

$$A = \frac{k}{a^2}, \quad 2H = k\left(\frac{m}{lb^2} - \frac{l}{ma^2}\right), \quad B = -\frac{k}{b^2}.$$

Hence the condition is $a^2 A + b^2 B = 0$.

14·77. EXAMPLE. Find the least angle between conjugate diameters of $x^2/a^2 + y^2/b^2 = 1$.

Let the eccentric angles of P, D be $\phi, \phi - \tfrac{1}{2}\pi$.

Then $\tan \angle ACP = \dfrac{b \sin \phi}{a \cos \phi}$, $\tan \angle ACD = -\dfrac{b \cos \phi}{a \sin \phi}$,

$\tan \angle PCD = -\left(\dfrac{b \cos \phi}{a \sin \phi} + \dfrac{b \sin \phi}{a \cos \phi}\right)\bigg/\left(1 - \dfrac{b^2}{a^2}\right) = -\dfrac{2ab \operatorname{cosec} 2\phi}{a^2 - b^2}$.

Hence the acute angle between the conjugate diameters is

$$\tan^{-1}\{2ab \operatorname{cosec} 2\phi/(a^2 - b^2)\}.$$

This is least when $\phi = \tfrac{1}{4}\pi$ and is then $2\tan^{-1}(b/a)$.

14·78. If the figure of 14·72 is obtained by orthogonal projection from the corresponding figure for the circle, the ratio of lengths in the same direction is the same for both figures.

Hence QV^2/CD^2 and $PV.VP'/CP^2$ are the same for both. But in the circle

$$QV^2 = PV.VP' \quad \text{and} \quad CP^2 = CD^2.$$

Hence in the ellipse

$$QV^2 : PV.VP' = CD^2 : CP^2.$$

This is a special case of Newton's Theorem (14·79).

When CP, CD are taken as axes of coordinates and Q is the point (x, y), $PV.VP' = CP^2 - CV^2 = a_1^2 - x^2$,

where $CP = a_1$. Putting also $CD = b_1$,

$$y^2 : a_1^2 - x^2 = b_1^2 : a_1^2.$$

$$\therefore \quad x^2/a_1^2 + y^2/b_1^2 = 1.$$

This is the equation of the ellipse referred to conjugate diameters as oblique axes.

14·79. *Newton's Theorem.* If $P_1 Q R$ is a secant of the ellipse $x^2/a^2 + y^2/b^2 = 1$ through a fixed point P_1, and θ is the angle it makes with $y = 0$, then the point $(x_1 + r \cos \theta, y_1 + r \sin \theta)$ lies on the ellipse if $r = P_1 Q$ or $P_1 R$.

Hence $(x_1+r\cos\theta)^2/a^2+(y_1+r\sin\theta)^2/b^2 = 1$ is a quadratic in r whose roots are P_1Q and P_1R.

$$\therefore \quad P_1Q \cdot P_1R = \left(\frac{x_1^2}{a^2}+\frac{y_1^2}{b^2}-1\right)\bigg/\left(\frac{\cos^2\theta}{a^2}+\frac{\sin^2\theta}{b^2}\right).$$

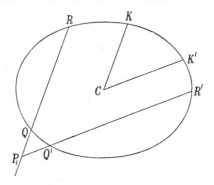

Similarly, for a secant $P_1Q'R'$ which makes an angle θ' with OX,

$$P_1Q' \cdot P_1R' = \left(\frac{x_1^2}{a^2}+\frac{y_1^2}{b^2}-1\right)\bigg/\left(\frac{\cos^2\theta'}{a^2}+\frac{\sin^2\theta'}{b^2}\right).$$

$$\therefore \quad \frac{P_1Q \cdot P_1R}{P_1Q' \cdot P_1R'}$$

is independent of (x_1, y_1) and depends only on a, b, θ, θ'.

Thus, for chords in fixed directions, $P_1Q \cdot P_1R : P_1Q' \cdot P_1R'$ is constant for all positions of P_1. This is known as Newton's Theorem.

The same method can be applied to the conic

$$ax^2+2hxy+by^2+2gx+2fy+c = 0,$$

and so Newton's Theorem holds for any conic.

For the ellipse, by taking one position of P_1 at the centre C,

$$\frac{P_1Q \cdot P_1R}{P_1Q' \cdot P_1R'} = \frac{CK^2}{CK'^2},$$

where CK, CK' are the semi-diameters parallel to P_1QR, $P_1Q'R'$.

14·8. Concyclic Points on the Ellipse

14·81. The chords of intersection of a circle and ellipse are inclined to the axis of the ellipse at supplementary angles.

1st method. Let $QR, Q'R'$ be the chords and let them meet at P. Then by a property of the circle $PQ \cdot PR = PQ' \cdot PR'$ and, by 14·79, $PQ \cdot PR : PQ' \cdot PR' = CK^2 : CK'^2$, where CK, CK' are the parallel semi-diameters.

Hence $CK = CK'$. Thus CK, CK', and therefore the chords, make supplementary angles with the axis.

2nd method. Let the circle be $x^2 + y^2 + 2gx + 2fy + c = 0$. The points t_1, t_2, t_3, t_4 of $x/a : y/b : 1 = 1 - t^2 : 2t : 1 + t^2$ lie on this circle if t_1, t_2, t_3, t_4 are the roots of

$$a^2(1 - t^2)^2 + 4b^2t^2 + 2ga(1 - t^4) + 4fbt(1 + t^2) + c(1 + t^2)^2 = 0.$$

Since the coefficients of t^3, t in this quartic equation are equal,

$$t_1 + t_2 + t_3 + t_4 = t_2 t_3 t_4 + t_1 t_3 t_4 + t_1 t_2 t_4 + t_1 t_2 t_3.$$

$\therefore (1 - t_1 t_2)(t_3 + t_4) + (1 - t_3 t_4)(t_1 + t_2) = 0$. This is a necessary and sufficient condition for the chords

$$(1 - t_1 t_2) x/a + (t_1 + t_2) y/b = 1 + t_1 t_2$$

and $\qquad (1 - t_3 t_4) x/a + (t_3 + t_4) y/b = 1 + t_3 t_4$

to be inclined at supplementary angles to OX. Another method is given in 16·82.

If ϕ is the eccentric angle of the point whose parameter is t, $t = \tan \frac{1}{2}\phi$. Hence the condition

$$(1 - t_1 t_2)(t_3 + t_4) + (1 - t_3 t_4)(t_1 + t_2) = 0$$

is equivalent to $\tan \frac{1}{2}(\phi_1 + \phi_2) + \tan \frac{1}{2}(\phi_3 + \phi_4) = 0$.

Hence the eccentric angles of four concyclic points are such that $\phi_1 + \phi_2 + \phi_3 + \phi_4 = 2n\pi$.

This is a necessary and sufficient condition for the points $\phi_1, \phi_2, \phi_3, \phi_4$ to be concyclic.

14·82. EXAMPLE. Find the equation of the locus of the point P which divides a chord of an ellipse of which the direction is fixed so that $QP^2 + PQ'^2$ is constant $(= 2k^2)$.

Take conjugate diameters one of which is parallel to QQ' as axes. Let P, Q be (x, y), (x, z). Then

$$(QP + PQ')^2 = 4z^2$$

and
$$(QP - PQ')^2 = 4y^2.$$

$$\therefore \quad k^2 = z^2 + y^2.$$

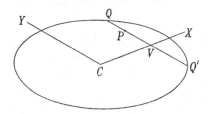

But
$$x^2/a_1{}^2 + z^2/b_1{}^2 = 1.$$

$$\therefore \quad x^2/a_1{}^2 - y^2/b_1{}^2 = 1 - k^2/b_1{}^2.$$

This is the equation of the locus.

EXERCISE 14c

1. What diameter of $x^2/a^2 + y^2/b^2 = 1$ bisects chords parallel to $y = tx$? What chords are bisected by $y = sx$?

2. Give the condition for $l_1 x + m_1 y = 0$ and $l_2 x + m_2 y = 0$ to be conjugate diameters of $ax^2 + by^2 = 1$.

3. CP, CD are conjugate semi-diameters of an ellipse and PN, DM are perpendicular to the major axis. Prove that

$$PN : CM = b : a = DM : CN.$$

4. Prove that $SP . PS' = CD^2$, where CD is conjugate to CP.

5. If PP' is a diameter and Q is any point on an ellipse, prove that QP, QP' are parallel to conjugate diameters.

6. Find the condition for the diameters of $ax^2 + by^2 = 1$ through its points of intersection with $lx + my + n = 0$ to be conjugate.

7. Prove that $ax^2 + 2hxy - by^2 = 0$ represents conjugate diameters of $ax^2 + by^2 = 1$.

8. Find the perpendicular to $lx + my + n = 0$ which is conjugate to it wo $x^2/a^2 + y^2/b^2 = 1$ and prove that the two lines meet the major axis in points which separate the foci harmonically.

9. The tangent at T to an ellipse meets a diameter CP at H, and the line through T parallel to the conjugate diameter meets CP at N. Prove that $CN \cdot CH = CP^2$.

10. Two conjugate diameters of an ellipse meet the tangent at one end of the major axis in Q and Q'. Prove that QQ' subtends supplementary angles at the foci.

11. Prove that two conjugate diameters and the directrix form a triangle whose orthocentre is the corresponding focus.

12. Prove that

$$ab(a^2 \sin^2 \theta - b^2 \cos^2 \theta)(x^2/a^2 - y^2/b^2) = 2xy(a^2 - b^2) \sin \theta \cos \theta$$

represents conjugate diameters of $x^2/a^2 + y^2/b^2 = 1$ and that their angle-bisectors are parallel to the tangent and normal at the point $(a \cos \theta, b \sin \theta)$.

13. Find the locus of the point of intersection of perpendiculars to the conjugate diameters CP, CD of an ellipse drawn at P and C respectively.

14. Prove that conjugate diameters of an ellipse cut the director circle at the ends of a chord which touches the ellipse.

15. Tangents to an ellipse at the ends of a fixed diameter QQ' are cut by a variable tangent at P in T, T'. Prove that $QT \cdot Q'T'$ is constant and that $TP : TQ = T'P : T'Q$.

16. Find the locus of the mid-points of focal chords of

$$x^2/a^2 + y^2/b^2 = 1.$$

17. Find the locus of the mid-points of chords of an ellipse which subtend a right angle at the centre.

18. Chords PKQ of $x^2/a^2 + y^2/b^2 = 1$ pass through a fixed point $K(k, 0)$ where $k^2 = a^2(a^2 - b^2)/(a^2 + b^2)$. Prove that $1/KP^2 + 1/KQ^2$ is constant.

19. Prove that the mid-point of a chord of length $2k$ of

$$x^2/a^2 + y^2/b^2 = 1$$

lies on the curve

$$a^2 b^2 (x^2/a^4 + y^2/b^4)(x^2/a^2 + y^2/b^2 - 1) + k^2(x^2/a^2 + y^2/b^2) = 0.$$

20. Prove that the common chords of an ellipse and a concentric circle are the sides and diagonals of a rectangle.

21. Given an ellipse drawn on paper, show how to find its axes and foci by geometrical construction.

22. A circle touches the ellipse $x = a \cos \phi, y = b \sin \phi$ at the point ϕ_1 and meets it again at the point ϕ_2. Find the parameter of the remaining point of intersection.

23. A circle has three-point contact with an ellipse at a given point whose eccentric angle is θ. Find the eccentric angle of the other point of intersection.

24. PCP' and DCD' are conjugate diameters of an ellipse and ϕ is the eccentric angle of D. Find the eccentric angle of the point where the circle PDP' meets the ellipse again.

14·9. Normals

14·91. The normal at (x_1, y_1) to $x^2/a^2 + y^2/b^2 = 1$ is

$$a^2 y_1 x - b^2 x_1 y = (a^2 - b^2) x_1 y_1.$$

This passes through a fixed point (f, g) if

$$a^2 y_1 f - b^2 x_1 g = (a^2 - b^2) x_1 y_1 = c^2 x_1 y_1,$$

i.e. if (x_1, y_1) lies on $a^2 fy - b^2 gx = (a^2 - b^2) xy$ as well as on

$$b^2 x^2 + a^2 y^2 = a^2 b^2.$$

These curves intersect in points whose ordinates satisfy

$$b^2 (a^2 fy)^2 = b^2 x^2 (c^2 y + b^2 g)^2 = a^2 (b^2 - y^2)(c^2 y + b^2 g)^2$$

and therefore, in G_6, in four points.

The equation $a^2 fy - b^2 gx = c^2 xy$ represents a rectangular hyperbola called the *hyperbola of Apollonius* which is discussed further in volume II.

14·92. The normal at the point ϕ to $x = a\cos\phi$, $y = b\sin\phi$ is

$$ax \sec\phi - by \operatorname{cosec}\phi = c^2.$$

This passes through the fixed point (f, g) if

$$af \sec\phi - bg \operatorname{cosec}\phi = c^2$$

or, if $t = \tan\tfrac{1}{2}\phi$,

$$2aft(1 + t^2) - bg(1 - t^4) = 2c^2 t(1 - t^2),$$

a quartic in t. The four roots of this equation satisfy

$$1 : t_1 + t_2 + t_3 + t_4 : \Sigma t_1 t_2 : \Sigma t_1 t_2 t_3 : t_1 t_2 t_3 t_4$$
$$= bg : -2(af + c^2) : 0 : -2(af - c^2) : -bg.$$

Hence $1 - \Sigma t_1 t_2 + t_1 t_2 t_3 t_4 = 0.$

$$\therefore \quad \cot\tfrac{1}{2}(\phi_1 + \phi_2 + \phi_3 + \phi_4) = 0,$$
$$\therefore \quad \phi_1 + \phi_2 + \phi_3 + \phi_4 = (2n + 1)\pi.$$

This is a necessary but not a sufficient condition for the concurrence of the normals at the points $\phi_1, \phi_2, \phi_3, \phi_4$. Two conditions are needed for the concurrence of four lines.

14·93. If the normals at $\phi_1, \phi_2, \phi_3, \phi_4$ are concurrent and ϕ_1' is the point diametrically opposite to ϕ_1, then since

$$\Sigma\phi = (2n+1)\pi \quad \text{and} \quad \phi_1' = \phi_1 + \pi,$$

it follows that $\phi_1' + \phi_2 + \phi_3 + \phi_4 = 0$ except for a multiple of 2π, and therefore by 14·81 the points $\phi_1', \phi_2, \phi_3, \phi_4$ are concyclic, i.e. the circle passing through the feet of three concurrent normals passes through the point diametrically opposite to the foot of the fourth normal.

14·94. EXAMPLE. Prove that the perpendiculars from the vertex of an ellipse to four concurrent normals meet the ellipse again in four concyclic points.

Let the perpendicular from A to the normal at P meet the ellipse again at P'. Let ϕ, ϕ' be the eccentric angles of P, P'. Since the tangent at P is parallel to AP' and the eccentric angle of A is zero, $\phi' = 2\phi$.

There are four positions of P for which the normals are concurrent, and their eccentric angles satisfy $\Sigma\phi = (2n+1)\pi$.

Hence for the corresponding points P',

$$\Sigma\phi' = 2(2n+1)\pi = 2m\pi.$$

Therefore the four points P' are concyclic.

14·95. EXAMPLE. Normals to an ellipse centre C at P_1, P_2, P_3, P_4 are concurrent. Prove that lines P_1Q_1, P_2Q_2, P_3Q_3, P_4Q_4 making with the major axis angles supplementary to those made by CP_1, CP_2, CP_3, CP_4 are also concurrent.

The equation of CP_1 is $xy_1 - yx_1 = 0$, and that of P_1Q_1 is $xy_1 + yx_1 = 2x_1y_1$. The normal at P_1 is $a^2y_1x - b^2x_1y = c^2x_1y_1$ and this passes through a point (f, g).

$$\therefore \quad a^2y_1f - b^2x_1g = c^2x_1y_1.$$

But this shows that $xy_1 + yx_1 = 2x_1y_1$ passes through

$$(2a^2f/c^2, \ -2b^2g/c^2).$$

Similarly, P_2Q_2, P_3Q_3, P_4Q_4 pass through this point.

14·96. EXAMPLE. Find the locus of the point of intersection of perpendicular normals to $x^2/a^2 + y^2/b^2 = 1$.

If the normals at U, U' meet at Q and the corresponding tangents meet at T, $TUQU'$ is a rectangle. The diagonal TQ bisects UU' at V and therefore passes through the centre C.

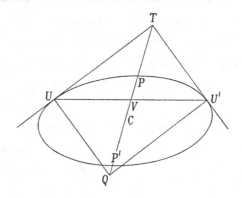

Since $$(TPVP') = -1,$$
$$CV \cdot CT = CP^2.$$
$$\therefore \tfrac{1}{2}(CT - QC) CT = CP^2.$$

But, by 14·6, $\quad CT = \sqrt{(a^2 + b^2)}$,

and if Q is the point (r, θ),

$$CQ = r \quad \text{and} \quad (CP \cos\theta)^2/a^2 + (CP \sin\theta)^2/b^2 = 1.$$

Hence

$$\tfrac{1}{2}\{\sqrt{(a^2 + b^2)} - r\} \sqrt{(a^2 + b^2)} = \frac{1}{(\cos^2\theta)/a^2 + (\sin^2\theta)/b^2}.$$

$$\therefore \tfrac{1}{2} r \sqrt{(a^2 + b^2)} = \frac{a^2 + b^2}{2} - \frac{a^2 b^2}{b^2 \cos^2\theta + a^2 \sin^2\theta},$$

$$\therefore r \sqrt{(a^2 + b^2)} = \frac{b^4 \cos^2\theta + a^4 \sin^2\theta - a^2 b^2}{b^2 \cos^2\theta + a^2 \sin^2\theta}.$$

This is the polar equation of the locus, and

$$(a^2 + b^2)(x^2 + y^2)(b^2 x^2 + a^2 y^2)^2 = \{b^4 x^2 + a^4 y^2 - a^2 b^2 (x^2 + y^2)\}^2$$
$$= (a^2 - b^2)^2 (a^2 y^2 - b^2 x^2)^2$$

is the cartesian equation.

14·97. EXAMPLE. If T is the pole of a chord PQ of

$$x^2/a^2 + y^2/b^2 = 1$$

and S is a focus, show that $SP \cdot SQ = ST^2 \cos^2\beta$, where 2β is the difference between the eccentric angles of P and Q.

Let the eccentric angles be $\alpha \pm \beta$; then T is the point

$$(a\cos\alpha\sec\beta,\ b\sin\alpha\sec\beta)$$

and $\qquad ST^2 = (ae - a\cos\alpha\sec\beta)^2 + (b\sin\alpha\sec\beta)^2.$

$$\begin{aligned}
\therefore\ ST^2\cos^2\beta &= a^2(e\cos\beta - \cos\alpha)^2 + b^2\sin^2\alpha \\
&= (a^2 - b^2)\cos^2\beta - 2a^2 e\cos\beta\cos\alpha \\
&\qquad\qquad + a^2\cos^2\alpha + b^2\sin^2\alpha \\
&= a^2\{1 + \cos(\alpha+\beta)\cos(\alpha-\beta)\} \\
&\qquad - b^2\{\cos(\alpha+\beta)\cos(\alpha-\beta)\} \\
&\qquad - a^2 e\{\cos(\alpha+\beta) + \cos(\alpha-\beta)\} \\
&= a^2\{1 - e\cos(\alpha+\beta)\}\{1 - e\cos(\alpha-\beta)\} = SP \cdot SQ.
\end{aligned}$$

14·98. EXAMPLE. Prove that chords of an ellipse which subtend a right angle at a fixed point on the curve are concurrent and determine the point of concurrence.

1st method. Let the ellipse be $x/a : y/b : 1 = 1 - t^2 : 2t : 1 + t^2$ and suppose that the chord $t_1 t_2$ subtends a right angle at the point t_0. Then the chords

$$(1 - t_0 t_1)\,x/a + (t_0 + t_1)\,y/b = 1 + t_0 t_1,$$

$$(1 - t_0 t_2)\,x/a + (t_0 + t_2)\,y/b = 1 + t_0 t_2$$

are at right angles.

$$\therefore\quad b^2(1 - t_0 t_1)(1 - t_0 t_2) + a^2(t_0 + t_1)(t_0 + t_2) = 0,$$

$$\therefore\quad (b^2 + a^2 t_0^2) - (b^2 - a^2)\,t_0(t_1 + t_2) + (a^2 + b^2 t_0^2)\,t_1 t_2 = 0.$$

This proves that the chord $t_1 t_2$ whose equation (14·23) is

$$\left(1 - \frac{x}{a}\right) - \frac{y}{b}(t_1 + t_2) + \left(1 + \frac{x}{a}\right)t_1 t_2 = 0$$

passes through the point (x, y) given by

$$\left(1 - \frac{x}{a}\right) : \frac{y}{b} : \left(1 + \frac{x}{a}\right) = b^2 + a^2 t_0^2 : (b^2 - a^2)\,t_0 : a^2 + b^2 t_0^2.$$

2nd method. See 7·44. The point of concurrence is on the normal and may be determined by taking the special case of chords OP, OQ parallel to the axes. If O is the point whose eccentric angle is ϕ, the equations of PQ and the normal are

$$bx \sin\phi + ay \cos\phi = 0$$

and $$ax \sin\phi - by \cos\phi = (a^2 - b^2) \sin\phi \cos\phi.$$

$$\therefore \quad x : y : a^2 - b^2 = a \cos\phi : -b \sin\phi : a^2 + b^2.$$

EXERCISE 14D

Prove the results in Nos. 1–8 in which P is a point on $x^2/a^2 + y^2/b^2 = 1$ the normal at which meets the axes in G, G', CF is drawn from the centre perpendicular to the normal, CD is conjugate to CP, and PN, PN' are perpendicular to the axes.

1. $CG = (c^2/a^2) CN.$ **2.** $CG' = (c^2/b^2) N'C.$

3. $PF \cdot PG = b^2.$ **4.** $PF \cdot PG' = a^2.$

5. $GN = (b^2/a^2) CN.$ **6.** $PF \cdot CD = ab.$

7. $PG : CD = b : a.$ **8.** $PG' : CD = a : b.$

9. Prove that $lx + my + n = 0$ is a normal to $x^2/a^2 + y^2/b^2 = 1$ if $a^2/l^2 + b^2/m^2 = (a^2 - b^2)^2/n^2.$

10. Prove that

$$2\lambda(\lambda^2 + 1) ax + (\lambda^4 - 1) by + 2\lambda(\lambda^2 - 1)(a^2 - b^2) = 0$$

is a normal to $x^2/a^2 + y^2/b^2 = 1$ for all values of $\lambda.$

11. The normal at P to $x^2/a^2 + y^2/b^2 = 1$ meets the axis at G. Find the locus of the mid-point of PG.

12. Prove that the locus of the poles of normal chords of

$$x^2/a^2 + y^2/b^2 = 1 \quad \text{is} \quad a^6/x^2 + b^6/y^2 = (a^2 - b^2)^2.$$

13. Prove that the product of the perpendiculars drawn to the normal at P of an ellipse, from the centre and from the pole of the normal, is equal to the product of the focal distances of P.

14. Find the locus of the mid-points of normal chords of

$$x = a \cos t, \quad y = b \sin t.$$

15. Points P, Q are taken on two similar and similarly situated ellipses centre O and $\angle POQ$ is bisected by the major axis. Prove that PQ is a normal to a fixed conic.

16. If α is the angle between normals to the ellipse $x = a\cos t$, $y = b\sin t$ at the points $t = \theta, t = \theta + \frac{1}{2}\pi$, prove that the eccentricity e is given by $e^2\sin 2\theta = 2\cot\alpha\sqrt{(1-e^2)}$.

17. Along the normal to $x^2/a^2 + y^2/b^2 = 1$ at the point P lengths PQ_1, PQ_2 are cut off equal to the conjugate semi-diameter CD. Show that Q_1, Q_2 lie on one of the circles $x^2 + y^2 = (a \pm b)^2$.

18. Prove that the locus of the point of intersection of the normal at P to $x^2/a^2 + y^2/b^2 = 1$ and the diameter conjugate to CP is

$$(b^2x^2 + a^2y^2)(x^2 + y^2)^2 = (a^2 - b^2)^2 x^2 y^2.$$

19. Prove that the length l of the normal chord of $x = a\cos t$, $y = b\sin t$ at a point P is given by $2a^2b^2/l = (a^2 + b^2)p - p^3$.

20. Use the result of No. 19 to find the least length of a normal chord of the ellipse when (i) $2b^2 \geqslant a^2$, (ii) $2b^2 < a^2$.

21. Prove that the normals to $x = a\cos\phi, y = b\sin\phi$ at the points ϕ_1, ϕ_2 meet in the point

$$ax: -by: c^2 = \cos\phi_1\cos\phi_2\cos\tfrac{1}{2}(\phi_1 + \phi_2)$$
$$: \sin\phi_1\sin\phi_2\sin\tfrac{1}{2}(\phi_1 + \phi_2): \cos\tfrac{1}{2}(\phi_1 - \phi_2).$$

22. Prove that the feet of the normals from (f, g) to $x^2/a^2 + y^2/b^2 = 1$ lie on the curve given by the parametric equations

$$x = ft/(t - b^2), \quad y = gt/(t - a^2).$$

23. Normals to an ellipse at P_1, P_2, P_3, P_4 are concurrent at $K(f, g)$. $P_1Q_1, P_2Q_2, P_3Q_3, P_4Q_4$ make with the major axis of the ellipse angles supplementary to those made by KP_1, KP_2, KP_3, KP_4. Prove that $P_1Q_1, P_2Q_2, P_3Q_3, P_4Q_4$ are concurrent and find the point of concurrence.

EXERCISE 14E

1. Find the equation of the ellipse the ends of whose axes are $(5, -3), (8, -1), (11, -3), (8, -5)$.

2. Find the equation of the ellipse of eccentricity $\frac{1}{2}$ whose vertices are $(2, 0)$ and $(0, 1)$.

3. Find the eccentricity and the length of the latus rectum of the ellipse $2(2x - 5y)^2 + 3(5x + 2y)^2 = 2(10y - 4x + 7)$.

4. Find the equation and the length of the chord of $2x^2 + 3y^2 = 35$ with mid-point $(3, 1)$, and find the equation of the circle on this chord as diameter.

5. Find the common tangents of $y^2 = 3x - 15$ and $3x^2 + 2y^2 = 11$.

6. Find the common tangents of

$$22X^2 + 33Y^2 = 6 \quad \text{and} \quad 20X^2 - 3Y^2 + 4X = 0.$$

7. Find the inverse of $x^2/a^2 + y^2/b^2 = 1$ wo $x^2 + y^2 = k^2$.

8. Find the inverse of $x^2/a^2 + y^2/b^2 = 2x/a$ wo $x^2 + y^2 = k^2$ and sketch the curve.

9. Use the property $SP = ePK$ to show that the polar equation of an ellipse with a focus as pole is of the form

$$r\{1 \pm e \cos(\theta - \alpha)\} = k$$

and give the geometrical meanings of k and α.

10. Find the locus of the poles of tangents of an ellipse wo a circle centre a focus.

11. The normal to an ellipse at P meets the major axis at G and GK is drawn perpendicular to the focal distance SP. Prove that PK is equal to the semi-latus rectum.

12. M is the mid-point of a focal chord PP' of an ellipse and the normals at P, P' meet at N. Prove that MN is parallel to the major axis.

13. A circle whose centre is on the major axis touches the ellipse at P and Q and passes through a focus S. Prove that SP is equal to the latus rectum.

14. The polar of T wo $x^2/a^2 + y^2/b^2 = 1$ goes through a focus. Prove that the product of the perpendiculars from T and the centre to this polar is equal to the square of the semi-minor axis.

15. A chord of fixed direction of an ellipse centre C meets the major axis at X and the perpendicular bisector of the chord meets that axis at Y. Prove that $CX : CY$ is constant.

16. Prove that conjugate lines through a focus of an ellipse are at right angles.

17. A variable line l passes through a fixed point (h, k). Prove that the envelope of the line which is perpendicular to l and conjugate to it wo $x^2/a^2 + y^2/b^2 = 1$ is a parabola.

18. The tangents to $x^2/c^2 + y^2/d^2 = 1$ at P and Q are conjugate wo $x^2/a^2 + y^2/b^2 = 1$. Prove that PQ touches

$$x^2/\{c^4(b^2 + d^2)\} + y^2/\{d^4(a^2 + c^2)\} = 1/(a^2d^2 + b^2c^2).$$

19. The lines joining a point P of $x^2/a^2 + y^2/b^2 = 1$ to $(\pm k, 0)$ meet the ellipse again in Q and R, and T is the pole of QR. Prove that PT is bisected by the minor axis.

20. Find the equation of the ellipse centre the origin with semi-axes a_1 and b_1 and a major axis which makes an angle ϕ with $OX, (a_1 > b_1)$.

21. If two concentric ellipses touch, prove that the angle between their major axes is given by

$$(a_1^2 - b_2^2)^2 (a_2^2 - b_1^2) \tan^2 \theta = (a_1^2 - a_2^2)(b_1^2 - b_2^2),$$

where a_1, b_1 and a_2, b_2 are the semi-axes.

22. Prove that the locus of a point two of the normals from which to $x^2/a^2 + y^2/b^2 = 1$ are coincident is

$$(a^2x^2 + b^2y^2 - c^4)^3 + 27a^2b^2c^4x^2y^2 = 0,$$

where $c^2 = a^2 - b^2$.

23. Find the locus of the point of intersection of normals to

$$x^2/a^2 + y^2/b^2$$

at the ends of chords making an angle γ with OX.

24. If $e^2 > 2(\sqrt{2} - 1)$, prove that there are eight normal chords of

$$x^2/a^2 + y^2/b^2 = 1$$

which meet the ellipse at the ends of conjugate diameters, and that the length k of such a chord satisfies $k^4 - (a^2 + b^2)k^2 + 2a^2b^2 = 0$.

25. From the ends of chords of $x^2/a^2 + y^2/b^2 = 1$ which touch $x^2/c^2 + y^2/d^2 = 1$ the other tangents to this second ellipse are drawn. Find the locus of their point of intersection.

26. Can the locus in No. 25 coincide with the first ellipse? If one triangle can be inscribed in $x^2/a^2 + y^2/b^2 = 1$ and circumscribed about $x^2/c^2 + y^2/d^2 = 1$, prove that an infinity of such triangles can be drawn.

Chapter 15

THE HYPERBOLA

15·1. In 6·51 the equation of a hyperbola is given in the form

$$\frac{x^2}{a^2} - \frac{y^2}{b^2} = 1.$$

By 12·3 and 12·4 formulae for the hyperbola corresponding to the results for the ellipse in 14·1 can be obtained. Such formulae differ from those for the ellipse only in containing $-b^2$ in place of b^2.

In complex geometry there is no distinction between the two curves. The present chapter is concerned mainly with G_3, but it is sometimes convenient to use homogeneous geometry G_5.

15·2. Parametric Equations

15·21. The equation $\dfrac{x^2}{a^2} - \dfrac{y^2}{b^2} = 1$ may be written

$$\frac{\dfrac{x}{a} - 1}{\dfrac{y}{b}} = \frac{\dfrac{y}{b}}{\dfrac{x}{a} + 1}, \ = t, \text{ say,}$$

$$\therefore \quad \frac{x}{a} - 1 : \frac{y}{b} : \frac{x}{a} + 1 = t^2 : t : 1$$

or $\qquad x/a : y/b : 1 = 1 + t^2 : 2t : 1 - t^2.$

This parametric representation is complete except that the point $(-a, 0)$ is only obtained as a limit by making $t \to \infty$. Each value of t, except ± 1, gives a point of the curve.

The following results for the hyperbola

$$x/a : y/b : 1 = 1 + t^2 : 2t : 1 - t^2$$

correspond to those with the same numbers in 14·2.

15·22. $lx + my + n = 0$ meets the hyperbola in points whose parameters are the roots of $(la - n)t^2 + 2mbt + (la + n) = 0$.

15·23. The chord $t_1 t_2$ is $(1 + t_1 t_2)x/a - (t_1 + t_2)y/b = 1 - t_1 t_2$.

15·24. The chord given by $ut^2 + 2vt + w = 0$ is
$$(w + u)x/a + 2vy/b + (w - u) = 0.$$

15·25. The tangent at t_1 is $(1 + t_1{}^2)x/a - 2t_1 y/b = 1 - t_1{}^2$.

15·26. The normal at t_1 is
$$2t_1 ax + (1 + t_1{}^2)by = 2c^2 t_1(1 + t_1{}^2)/(1 - t_1{}^2),$$
where $c^2 = a^2 + b^2$.

15·27. The envelope equations are
$$-aX : bY : 1 = 1 + t^2 : 2t : 1 - t^2$$
and $a^2 X^2 - b^2 Y^2 = 1$.

The results of 12·6 can be applied to this envelope equation.

15·28. The pole of $[X_1, Y_1]$ is $a^2 XX_1 - b^2 YY_1 = 1$, i.e. $(-a^2 X_1, b^2 Y_1)$.

The pole of the chord $t_1 t_2$ is given by
$$x : y : 1 = a(1 + t_1 t_2) : b(t_1 + t_2) : 1 - t_1 t_2.$$

The pole of the chord given by $ut^2 + 2vt + w = 0$ is
$$x_1 : y_1 : 1 = a(u + w) : -2bv : (u - w).$$

15·29. EXAMPLE. Give the condition for $3x - 4y + c = 0$ to touch $3x^2 - 2y^2 + 7 = 0$.

In G_5 the corresponding line and curve are
$$[3, -4, c] \quad \text{and} \quad \tfrac{1}{3}X^2 - \tfrac{1}{2}Y^2 + \tfrac{1}{7}Z^2 = 0.$$

Hence the condition is $3 - 8 + \tfrac{1}{7}c^2 = 0$, $c = \pm \sqrt{35}$.

15·3. Other representations of the hyperbola are given in 6·52, 6·54, 6·55. Results corresponding to those in 15·2 can be found for the hyperbola $x = \tfrac{1}{2}a\left(t + \dfrac{1}{t}\right)$, $y = \tfrac{1}{2}b\left(t - \dfrac{1}{t}\right)$. See Exercise 15A, Nos. 17–21.

In G_5 or G_6 the curve corresponding to

$$x/a : y/b : 1 = 1 + t^2 : 2t : 1 - t^2$$

has the proper parametric representation

$$x/a : y/b : z = s^2 + t^2 : 2st : s^2 - t^2,$$

where (s, t) is the parameter. The point $(-a, 0, 1)$ is given by $s = 0, t = 1$; and the values $(1, \pm 1)$ of (s, t) give the points at infinity. This should be compared with 15·21.

In G_5 and G_6 $x/a : y/b : z = s^2 + t^2 : 2st : s^2 - t^2$ represents a central conic. A simpler representation is given in 15·81.

EXERCISE 15A

In Nos. 1–8, give the equation of:

1. The polar of $(2, -3)$ wo $(x - 2y)(x + 2y) = 1$.

2. The tangents at $(\pm a, b, 0)$ to $x^2/a^2 - y^2/b^2 = z^2$.

3. The polar of $(0, 0, 1)$ wo $x^2/a^2 - y^2/b^2 = z^2$.

4. The pole of $[1, k, 0]$ wo $a^2X^2 - b^2Y^2 = Z^2$.

5. The polar of the origin wo $a^2X^2 - b^2Y^2 = Z^2$.

6. The tangent from $(3, 6)$ to $y^2 = 4(x^2 - 1)$.

7. The tangents from $(3, 6, 1)$ to $y^2 = 4(x^2 - z^2)$.

8. The pair of contacts of $a^2X^2 - b^2Y^2 = Z^2$ on $[0, 0, 1]$.

9. Give the equation of the pair of tangents from (a, k) to

$$x^2/a^2 - y^2/b^2 = 1$$

and deduce the separate equations.

10. Find the constant value of the sum of the squares of the reciprocals of perpendicular diameters of $x^2/a^2 - y^2/b^2 = 1$. Explain the result when $b^2 < a^2$.

11. Show how the parameter t changes when the point

$$(a(1 + t^2)/(1 - t^2),\ 2bt/(1 - t^2))$$

describes the hyperbola. What is the condition for the points t_1, t_2 to be on the same branch?

In Nos. 12–16, the hyperbola is $x/a : y/b : 1 = 1 + t^2 : 2t : 1 - t^2$.

12. Give the condition for the chord $t_1 t_2$ to pass through the origin.

13. Give the condition for the chord $t_1 t_2$ to have a fixed direction.

14. Give the condition of perpendicularity for the chords $t_1 t_2$, $t_3 t_4$.

15. Find the mid-point of the chord $t_1 t_2$.

16. Find the locus of the point of intersection of perpendicular tangents.

In Nos. 17–21, the hyperbola is $x = \tfrac{1}{2}a\left(t+\dfrac{1}{t}\right), y = \tfrac{1}{2}b\left(t-\dfrac{1}{t}\right)$.

17. Give the equation of the chord $t_1 t_2$ and deduce the equations of the tangent and normal at t_1.

18. Find the equation for the parameters of the points of intersection of the hyperbola with $lx + my + n = 0$.

19. Find the equation of the chord joining the points whose parameters are the roots of $ut^2 + 2vt + w = 0$.

20. Find the pole of the chord $t_1 t_2$.

21. Find the pole of the chord in No. 19.

22. Prove that the chord of $x = a \operatorname{ch} \phi, y = b \operatorname{sh} \phi$ joining the points $\alpha + \beta, \alpha - \beta$ is

$$\frac{x}{a} \operatorname{ch} \alpha - \frac{y}{b} \operatorname{sh} \alpha = \operatorname{ch} \beta,$$

and deduce the equation of the tangent at the point α.

23. Find the equation of the chord $\phi_1 \phi_2$ of the hyperbola

$$x = a \sec \phi, \; y = b \tan \phi,$$

and deduce the equation of the tangent at ϕ.

24. Find the pedal of $x^2 - y^2 = a^2$ wo the origin and sketch the curve.

25. A tangent to $x^2/a^2 - y^2/b^2 = 1$ meets the axes at T, T', and the rectangle $TOT'K$ is completed. Find the locus of K.

26. The tangents to a hyperbola at the ends of a chord QQ' whose mid-point is V meet at T Prove that VT passes through the centre C, and if it meets the curve at P, prove that $CV \cdot CT = CP^2$.

15·4. Geometrical Properties of the Hyperbola

15·41. The curve $x^2/a^2 - y^2/b^2 = 1$ is symmetrical about both axes. a^2 may be greater than, equal to, or less than b^2.

The vertices $A(a, 0)$, $A'(-a, 0)$ and the transverse and conjugate axes and centre are defined in 6·56. The circle $x^2 + y^2 = a^2$ is called the auxiliary circle.

Reference should be made to Exercise 6 D, Nos. 7–10.

We give here various ways of obtaining the hyperbola.

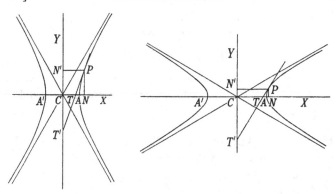

15·42. The locus of P such that $SP \sim PS'$ is constant, where S and S' are fixed points.

Take S, S' to be $(\pm c, 0)$ and denote the constant value of $SP \sim PS'$ by $2a$. Then the algebra of 14·42 gives, as before,

$$x^2(1 - c^2/a^2) + y^2 = a^2 - c^2.$$

Since a is now less than c, this equation is written

$$x^2/a^2 - y^2/b^2 = 1, \quad \text{where} \quad b^2 = c^2 - a^2.$$

15·44. The hyperbola may also be obtained as a section of a right circular cone. The method of 14·44 is used, but A, A' are on opposite sides of V.

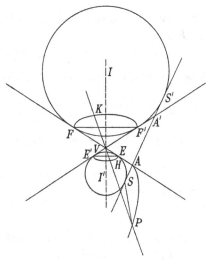

When P is on one branch,
$$PS' = PK, \quad PS = PH.$$
$$\therefore \quad SP - PS' = -HK = -EF.$$

When P is on the other branch, $SP - PS' = +EF$, and $SP - PS' = \pm EF$ leads to the equation in 15·42, with $AA' = 2a$ and $SS' = 2c$.

15·45. Foci. The points S, S' which occur in 15·42 and 15·44 are called the *foci* of the hyperbola.

If B is the point $(0, b)$, the foci may be constructed by making
$$S'C = CS = AB,$$
since $c = \sqrt{(a^2 + b^2)}$ by 15·42.

This point B does not lie on the hyperbola.

15·46. By the method of 14·46 the equation of the hyperbola may be written
$$(x \pm c)^2 + y^2 = \frac{c^2}{a^2}\left(x \pm \frac{a^2}{c}\right)^2,$$

and therefore the hyperbola is the locus of a point P such that
$$SP = ePK \quad \text{or} \quad S'P = ePK',$$

where PK, PK' are perpendiculars to the fixed lines $x = \pm a^2/c$.

Also $e^2 = \dfrac{c^2}{a^2} = \dfrac{a^2 + b^2}{a^2}$, and so the *eccentricity* (e) of a hyperbola is greater than 1.

The foci are $(\pm c, 0)$, $(\pm \sqrt{(a^2 + b^2)}, 0)$, or $(\pm ae, 0)$, and the corresponding directrices are $x = \pm a^2/c = \pm a/e$.

The locus of P such that $SP = ePK$ is an ellipse, parabola, or hyperbola according as $e < 1, e = 1, e > 1$. A parabola is said to have eccentricity 1.

15·47. The latus rectum LSL' is of length $2l$ found by putting $x = c$, $y = l$ in $x^2/a^2 - y^2/b^2 = 1$.

Hence $l^2 = b^2(c^2/a^2 - 1) = b^4/a^2$, $\therefore l = b^2/a$.

This relation, together with $e^2 = (a^2 + b^2)/a^2$, determines a, b in terms of l, e.

15·48. The work of 14·48, 14·49, 14·5, 14·6 is applicable to the hyperbola with slight modifications.

The director circle of $x^2/a^2 - y^2/b^2 = 1$ is $x^2 + y^2 = a^2 - b^2$. When $b^2 \geqslant a^2$, the hyperbola has no perpendicular tangents.

The properties of normals and of concyclic points are similar to those for the ellipse. See, for example, Exercise 15B, Nos. 26–31.

15·51. *Pedal Equation.* As in 14·51 it is proved that $\dfrac{p}{r} = \dfrac{b^2/p}{S'P}$. If P is on the branch nearer to S, $r' - r = 2a$, and if P is on the farther branch, $r - r' = 2a$. Hence the pedal equation is

$$\frac{b^2}{p^2} = 1 \pm \frac{2a}{r}.$$

15·52. *The Converse Pedal Property.* If S is a fixed point outside a fixed circle, and Y is a variable point on the circle, the envelope of the perpendicular at Y to SY is a hyperbola with focus S having the fixed circle as auxiliary circle. The method of 14·52 leads to the envelope equation $a^2X^2 - b^2Y^2 = 1$.

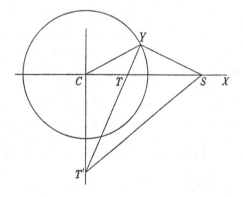

EXERCISE 15B

Prove the results of Nos. 1–6 with the notation of Exercise 6D, No. 7. The normal at P meets the axes in G, G' and CF is a perpendicular to the normal.

1. $CG \cdot CT = a^2 + b^2$.   **2.** $CG = (c^2/a^2) CN$.

3. $CG' = (c^2/b^2) CN'$.   **4.** $PF \cdot PG = -b^2$.

5. $PF \cdot PG' = a^2$.   **6.** $(STS'G) = -1$.

7. A rod APB is movable about the end A and a string BPC is tied to the other end B and to a fixed point C. Show that if the string is kept stretched in two straight parts BP and PC, the locus of P is one branch of a hyperbola. State the exception.

8. A shot is fired from A so as to strike a target at B, and the sound of the firing is heard at P one second before the sound of the destruction of the target. Find the locus of P.

9. Show, as in Exercise 14B, No. 18, that the hyperbola

$$(c^2 - a^2) x^2 - a^2 y^2 = a^2(c^2 - a^2), \quad c^2 > a^2,$$

has no foci on the y-axis in real geometry.

10. AB is the arc of a sector of a circle centre C, and a hyperbola of eccentricity 2 is drawn with focus A and directrix BC. If the hyperbola meets the arc AB at D, prove that CD trisects $\angle ACB$.

11. Show that the distances from the foci to the point (x_1, y_1) on $x^2/a^2 - y^2/b^2 = 1$ are $ex_1 \pm a$.

In Nos. 12–15, find the foci, directrices, and eccentricity.

12. $x^2 = 4(y^2 + 1)$.   **13.** $9x^2 - 4y^2 = 36$.

14. $25x^2 - 144y^2 = 3600$.   **15.** $(x-1)^2 - (y+2)^2 = 1$.

In Nos. 16–19, find the length of the latus rectum and the eccentricity.

16. $3x^2 - 4y^2 = 12$.   **17.** $x^2 - y^2 + 6x + 8y - 9 = 0$.

18. $4y^2 - 9x^2 = 36$.   **19.** $x^2 - 4(y+1)^2 = 1$.

20. Find the product of the perpendiculars from $(a\,\mathrm{ch}\,\phi, b\,\mathrm{sh}\,\phi)$ to $x/a \pm y/b = 0$, showing that it is independent of ϕ.

21. Find the locus of the mid-points of normal chords of $x^2 - y^2 = a^2$.

22. Find the locus of the poles of normal chords of $x^2/a^2 - y^2/b^2 = 1$.

23. The normal to $x^2/a^2 - y^2/b^2 = 1$ at P meets the transverse axis in G. GQ is perpendicular to that axis and equal to GP. Find the locus of P.

24. Find the Frégier's Point of $(a\,\mathrm{ch}\,\phi, b\,\mathrm{sh}\,\phi)$ wo the hyperbola $x^2/a^2 - y^2/b^2 = 1$.

25. The chords $t_0 t_1$ and $t_0 t_2$ of $x:y:a = 1+t^2:2t:1-t^2$ are equally inclined to the tangent at t_0. Prove that the chord $t_1 t_2$ meets the tangent at t_0 in a fixed point.

26. If the points $(a\,\mathrm{ch}\,\phi_n, b\,\mathrm{sh}\,\phi_n)$, $n = 1$ to 4, are concyclic, prove that $\Sigma\phi = 0$.

27. If the points t_1, t_2, t_3, t_4 of $x/a:y/b:1 = 1+t^2:2t:1-t^2$ are concyclic, prove that $\Sigma t + \Sigma t_1 t_2 t_3 = 0$.

28. Prove that the feet of the normals to $x^2/a^2 - y^2/b^2 = 1$ from (f, g) lie on the curve $a^2 y(f-x) = b^2 x(y-g)$.

29. Prove that the feet of four concurrent normals to a hyperbola cannot lie on the same branch of the curve.

30. Prove that the normals to $x/a:y/b:1 = 1+t^2:2t:1-t^2$ at t_1, t_2, t_3, t_4 are concurrent if $\Sigma t_1 t_2 = 0$ and $t_1 t_2 t_3 t_4 = -1$.

31. Use the converses of Nos. 27, 30 to prove that if normals at P, Q, R, S to a hyperbola are concurrent and P' is the point diametrically opposite to P, then P', Q, R, S are concyclic.

15·6. Asymptotes

15·61. The lines $\dfrac{x}{a} \pm \dfrac{y}{b} = 0$ are called asymptotes of

$$\frac{x^2}{a^2} - \frac{y^2}{b^2} = 1.$$

See 6·56. The joint equation of the asymptotes is $\dfrac{x^2}{a^2} - \dfrac{y^2}{b^2} = 0$.

When $a = b$, the asymptotes are at right angles and the curve is a rectangular hyperbola. Since $e^2 = (a^2 + b^2)/a^2$, the eccentricity of a rectangular hyperbola is $\sqrt{2}$.

15·62. Any line parallel to the asymptote $x/a + y/b = 0$ is $x/a + y/b = k$ and this meets the hyperbola

$$x/a:y/b:1 = 1+t^2:2t:1-t^2,$$

where
$$(1+t^2) + 2t = k(1-t^2);$$

$$1+t = k(1-t),$$

$$t = \frac{k-1}{k+1}.$$

Hence a line parallel to an asymptote meets the curve in one point only. When $k \to 0$, $t \to -1$ and there is no point of the curve corresponding to $t = -1$. The asymptote itself does not meet the curve. Similar results hold for lines $\dfrac{x}{a} - \dfrac{y}{b} = k$. In G_5 the asymptotes of $\dfrac{x^2}{a^2} - \dfrac{y^2}{b^2} = z^2$ are $\dfrac{x^2}{a^2} - \dfrac{y^2}{b^2} = 0$ and they are the tangents at infinity. Certain properties of the asymptotes can be found by treating them as tangents.

15·63. Conics $x^2/a^2 - y^2/b^2 = k$ have the same asymptotes for all values of k.

$x^2/a^2 - y^2/b^2 = k_1$ is transformed into $x'^2/a^2 - y'^2/b^2 = k_2$ by the substitution $x' = px$, $y' = py$, where $p^2 = k_2/k_1$. The conics may therefore be called similar. In G_3 the transformation does not exist when k_1 and k_2 have opposite signs and the conics are not then similar in the ordinary sense.

EXERCISE 15c

1. Find the limits when $t \to 1$ and $t \to -1$ of the tangent at t to the hyperbola $x/a : y/b : 1 = 1 + t^2 : 2t : 1 - t^2$.

2. Find the tangents at $(a, \pm b, 0)$ to
$$x/a : y/b : z = s^2 + t^2 : 2st : s^2 - t^2.$$

3. Write down the equation of the tangents from (a, b) to
$$x^2/a^2 - y^2/b^2 = 1$$
and simplify it.

In Nos. 4–7, find the asymptotes of the hyperbola.

4. $x^2 - 9y^2 = 9$. **5.** $4x^2 - y^2 + 4 = 0$.

6. $(x+6)^2 - (y-5)^2 = 2$. **7.** $y^2 - x^2 = 2y - 4x$.

8. Find the rectangular hyperbola with vertices $(5, 4)$ and $(-3, -2)$.

9. Find the hyperbolas with asymptotes $x \pm 2y = 0$ passing through $(6, 2\frac{1}{2})$ and $(-5, -3)$. What are their eccentricities?

10. What is the relation between the eccentricities of
$$x^2/a^2 - y^2/b^2 = \pm 1?$$

11. SY is the perpendicular from a focus to an asymptote. Prove that $CY = a$ and $SY = b$.

12. Prove that the foot of the perpendicular from a focus to an asymptote of a hyperbola lies on the directrix.

13. TQ is a tangent to a hyperbola from a point T on an asymptote. Prove that $\angle TSQ = \angle CTS$.

14. TQ is the tangent to a hyperbola from a point T on an asymptote. Prove that $\angle STQ = \angle CTS'$.

15. Prove a property of a hyperbola corresponding to Adams' Theorem (14·54) when T lies on an asymptote.

15·7. Diameters

15·71. The properties of diameters regarded as the polars of points at infinity are given in 14·7. These apply to the hyperbola.

The locus of the mid-points of chords of $x^2/a^2 - y^2/b^2 = 1$ which are parallel to $x/l = y/m$ is $xl/a^2 = ym/b^2$. These lines are conjugate diameters.

The diameter $x/l = y/m$ meets $x^2/a^2 - y^2/b^2 = 1$ in points $(\pm \rho l, \pm \rho m)$ if $\rho^2(l^2/a^2 - m^2/b^2) = 1$.

The conjugate diameter $xl/a^2 = ym/b^2$ meets $x^2/a^2 - y^2/b^2 = 1$ in points $(\pm \sigma a^2/l, \pm \sigma b^2/m)$ if $\sigma^2(a^2/l^2 - b^2/m^2) = 1$.

The intersections exist in real geometry for $x/l = y/m$ if $l^2/a^2 > m^2/b^2$, and for $xl/a^2 = ym/b^2$ if $l^2/a^2 < m^2/b^2$. Hence, of two conjugate diameters, one and only one meets the hyperbola.

15·72. If PCP' is a diameter which meets the hyperbola and if it is represented by $x/l = y/m$, then the conjugate diameter $xl/a^2 = ym/b^2$ meets the hyperbola $x^2/a^2 - y^2/b^2 = 1$ of G_4 in points D, D' whose coordinates $\pm \sigma a^2 l, \pm \sigma b^2 m$ are given by $\sigma^2(b^2/m^2 - a^2/l^2) = -1$ and

$$CP^2 + CD^2 = \rho^2(l^2 + m^2) + \sigma^2(a^4/l^2 + b^4/m^2)$$
$$= \frac{a^2b^2(l^2 + m^2)}{b^2l^2 - a^2m^2} - \frac{a^4m^2 + b^4l^2}{b^2l^2 - a^2m^2}$$
$$= a^2 - b^2.$$

In complex geometry, there is no distinction between the ellipse and the hyperbola. The results of Chapter 15 differ from those of Chapter 14 by having $-b^2$ in place of b^2. In 14, $b^2 = a^2 - c^2$, and, in 15, $b^2 = c^2 - a^2$.

In real geometry, it is convenient to use the points Δ, Δ' given by

$$\sigma^2(b^2/m^2 - a^2/l^2) = +1.$$

These points exist in real geometry, but they do not lie on the hyperbola $x^2/a^2 - y^2/b^2 = 1$.

$$C\Delta^2 = -CD^2.$$

Hence $CP^2 - C\Delta^2 = a^2 - b^2.$

The points Δ, Δ' lie on the hyperbola $x^2/a^2 - y^2/b^2 = -1$, and this is called the *conjugate hyperbola*. It has the same asymptotes as $x^2/a^2 - y^2/b^2 = 1$. Δ, Δ' are sometimes called the ends of the diameter conjugate to PCP' although they do not lie on the original curve. In the special case when PCP' is the transverse axis ACA' ($y = 0$), the conjugate diameter ($x = 0$) meets the conjugate hyperbola in the points B, B' whose coordinates are $0, \pm b$.

Some other properties of conjugate diameters are given in Exercise 15 D.

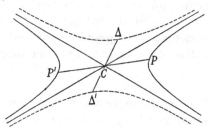

15·78. When CP and $C\Delta$ are taken as coordinate axes, the equation of the hyperbola, found from $x^2/a^2 - y^2/b^2 = 1$ by a substitution of the form

$$x = lx' + my', \qquad y = l'x' + m'y',$$

is of the form $Px'^2 + Qx'y' + Ry'^2 = 1$.

Since the values of x' for a given value of y' are equal and opposite, $Q = 0$. Also $y = 0$ when $x = a_1 = CP$. $\therefore Pa_1^2 = 1$.

The same substitution transforms $x^2/a^2 - y^2/b^2 = -1$ into $Px'^2 + Ry'^2 = -1$, and for this curve $x' = 0$ when $y' = b_1 = C\Delta$. $\therefore Rb_1^2 = -1$.

Hence the equation of the original hyperbola referred to the oblique axes $CP, C\Delta$ is

$$\frac{x'^2}{a_1^2} - \frac{y'^2}{b_1^2} = 1.$$

15·79. EXAMPLE. The tangent to a hyperbola at a point P meets an asymptote at Q, and a chord UV parallel to QP meets the same asymptote at K. Prove that $KU.KV = PQ^2$.

Let the axes be CP and $C\Delta$. Then the equation of the curve is

$$\frac{x^2}{a_1^2} - \frac{y^2}{b_1^2} = 1, \text{ where } a_1 = CP, b_1 = C\Delta.$$

Also the equation of the asymptotes is

$$\frac{x^2}{a_1^2} - \frac{y^2}{b_1^2} = 0.$$

Hence if M is the mid-point of UV and $CM = x$,

$$\frac{MU^2}{b_1^2} = \frac{x^2}{a_1^2} - 1.$$

Also $\dfrac{PQ^2}{b_1^2} = \dfrac{CP^2}{a_1^2} = 1.$ $\therefore PQ = b_1$ $\therefore MK = \dfrac{b_1 x}{a_1}.$

Thus $\qquad KU.KV \equiv MK^2 - MU^2 = b_1^2 = PQ^2.$

EXERCISE 15D

1. If the diameters of $x/a : y/b : 1 = 1 + t^2 : 2t : 1 - t^2$ through the points t_1 and t_2 are conjugate, prove that $(1 + t_1^2)(1 + t_2^2) = 4t_1 t_2$.

2. P is the point ϕ_1 on $x = a\operatorname{ch}\phi$, $y = b\operatorname{sh}\phi$. Prove that the diameter conjugate to CP cannot pass through any point ϕ_2, and find where it meets $x^2/a^2 - y^2/b^2 = -1$.

3. If the normal at P to a hyperbola meets the axes at G, G', prove that $PG : C\Delta = b : a = C\Delta : PG'$.

4. Prove that $SP.PS' = C\Delta^2$.

5. Prove that the parallelogram which has P, Δ, P', Δ' for the mid-points of its sides has area $4ab$.

6. Prove that the mid-points of the sides of the parallelogram $P\Delta P'\Delta'$ lie on the asymptotes.

7. For the hyperbola $x^2 - y^2 = a^2$, prove that the triangle $CP\Delta$ is isosceles and has constant area.

8. Prove that conjugate diameters of a rectangular hyperbola make complementary angles with its axes.

9. Find the condition for $Ax^2 + 2Hxy + By^2 = 0$ to be conjugate diameters of $x^2/a^2 - y^2/b^2 = 1$.

10. Show that an asymptote of a hyperbola is its own conjugate diameter, and that the two asymptotes are separated by any pair of conjugate diameters harmonically.

15·8. Asymptotes as Axes of Coordinates

15·81. It is proved in 6·61 that the equation of a hyperbola referred to its asymptotes as axes is $xy = k^2$. The value of the constant k may be found by taking the particular position A of P.

Thus

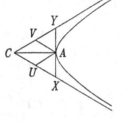

$$k^2 = AU \cdot AV = \tfrac{1}{4}CY^2 = \tfrac{1}{4}(a^2 + b^2).$$

The formulae of Chapter 12 can be applied to $xy = k^2$. For example, the polar of (x_1, y_1) is $xy_1 + yx_1 = 2k^2$. The curve may be represented parametrically by

$$x : y : k = t^2 : 1 : t$$

but no point is given by $t = 0$. The same equations represent a central conic in G_4. For the corresponding curve $xy = k^2z^2$ in G_6 a proper parametric representation is

$$x : y : kz = t^2 : s^2 : ts,$$

and this is a standard form of equation of a central conic in G_6. In G_5 it represents a hyperbola.

Results corresponding to those in 14·2 and 15·2 are given below for $xy = k^2$.

15·82. The line $lx + my + n = 0$ meets $x : y : k = t^2 : 1 : t$ in points given by

$$lt^2 + nt/k + m = 0.$$

15·83. The chord $t_1 t_2$ is

$$x + t_1 t_2 y = (t_1 + t_2)k.$$

15·84. The chord given by $ut^2 + 2vt + w = 0$ is

$$ux + wy + 2vk = 0.$$

15·85. The tangent at t_1 is

$$x + t_1^2 y = 2t_1 k.$$

15·86. If the hyperbola is rectangular, the normal at t_1 is

$$t_1^2 x - y = k(t_1^4 - 1)/t_1.$$

15·87. The envelope equations, found from 15·85, are

$$X : Y : 1/k = 1 : t^2 : -2t$$

and $$4k^2 XY = 1.$$

The results of 12·6 apply to this equation.

15·88. The pole of $[X_1, Y_1]$ is $2k^2(XY_1 + YX_1) = 1$, i.e.

$$(-2k^2 Y_1, -2k^2 X_1).$$

The pole of the chord $t_1 t_2$ is

$$x_1 : y_1 : 2k = t_1 t_2 : 1 : t_1 + t_2.$$

The pole of the chord given by $ut^2 + 2vt + w = 0$ is

$$x_1 : y_1 : 2k = w : u : -2v.$$

15·89. Any equation of the form

$$xy = ax + by + c$$

represents a hyperbola. For the equation can be written

$$(x - b)(y - a) = ab + c,$$

or, by a change of origin,

$$xy = ab + c.$$

15·90. Example. E, F, P are points on $x : y : k = t^2 : 1 : t$, of which E and F are fixed and P varies. PE, PF meet an asymptote in E', F'. Prove that $E'F'$ is constant.

Let the parameters of E, F, P be a, b, t. Then PE is

$$x + yat = k(a + t),$$

and E' is $(k(a + t), 0)$.

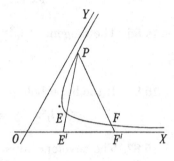

Similarly F' is $(k(b + t), 0)$.

Therefore

$$E'F' = k(a \sim b),$$

which is independent of t.

15·91. EXAMPLE. $\{\omega\}$. Find the equation of the normal at t_1 to $x : y : k = t^2 : 1 : t$.

If the normal is $ux + vy = k$, the condition of perpendicularity with $x + t_1^2 y = 2kt_1$ is, by 3·31,

$$u + vt_1^2 = (ut_1^2 + v) \cos \omega.$$

Also, since the normal passes through t_1, $ut_1^2 + v = t_1$. Hence by elimination of u, v, the normal is

$$\begin{vmatrix} x & y & k \\ 1 - t_1^2 \cos \omega & t_1^2 - \cos \omega & 0 \\ t_1^2 & 1 & t_1 \end{vmatrix} = 0,$$

i.e. $\qquad xt_1(t_1^2 - \cos \omega) - yt_1(1 - t_1^2 \cos \omega) = k(t_1^4 - 1).$

15·92. EXAMPLE. Prove that the orthocentre of a triangle inscribed in a rectangular hyperbola lies on the curve.

Let the hyperbola be $x : y : k = t^2 : 1 : t$. Then the chords $t_1 t_2$ and $t_3 t_4$ are

$$x + t_1 t_2 y = k(t_1 + t_2)$$

and $\qquad x + t_3 t_4 y = k(t_3 + t_4),$

and they are at right angles if

$$1 + t_1 t_2 t_3 t_4 = 0.$$

Hence if $t_1 t_2 t_3$ is the given triangle there is a point t_4 on the curve such that $t_1 t_2$ is perpendicular to $t_3 t_4$, and by the symmetry of the relation $1 + t_1 t_2 t_3 t_4 = 0$ also $t_1 t_3$ is perpendicular to $t_2 t_4$ and $t_2 t_3$ to $t_1 t_4$. Thus the point t_4, $= -1/(t_1 t_2 t_3)$, is the orthocentre.

15·93. EXAMPLE. Prove that an angle between the tangent at A to a rectangular hyperbola and a chord AP is equal to the angle subtended by AP at the point B of the curve diametrically opposite to A.

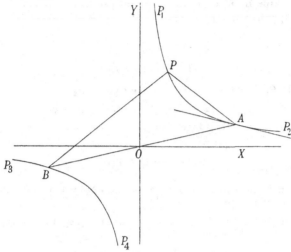

Let the hyperbola be $x:y:k = t^2:1:t$, and let the parameters of A, P be a, p; then the parameter of B is $-a$.

By 15·83, AP and BP are parallel to $x + apy = 0$ and $x - apy = 0$. Hence they make equal angles with OY on opposite sides.

By 15·85, the tangent at A is $x + a^2y = 2ka$; also BA is $x - a^2y = 0$. Hence these lines also make equal angles with OY on opposite sides. Therefore an angle between AP and the tangent at A is equal to the angle between BP and BA.

15·94. EXAMPLE. If the tangents at P, Q, R to an ellipse are parallel to QR, RP, PQ, prove that the centroid of P, Q, R is the centre of the ellipse.

In G_4 any central conic is $x:y:k = t^2:1:t$.

Let p, q, r be the parameters of P, Q, R. Then, by 15·83 and 15·85,
$$p^2 = qr, \quad q^2 = pr, \quad \text{and so} \quad q^2 = p^3/q.$$
$$\therefore \quad q = \omega p, \quad r = \omega^2 p, \quad \text{where} \quad \omega^3 = 1, \quad \omega \neq 1.$$

Hence the centroid, whose coordinates are

$$\tfrac{1}{3}(p+q+r), \ \tfrac{1}{3}(1/p+1/q+1/r).$$

is the origin, i.e. the centre of the conic.

The same must be true for an ellipse in G_3. No such points exist on a hyperbola in G_3.

EXERCISE 15E

1. If P_1, P_2 are on $xy = k^2$, prove that $P_1 P_2$ is

$$(y_1 + y_2)x + (x_1 + x_2)y = (x_1 + x_2)(y_1 + y_2).$$

2. Prove that the tangent at (x_1, y_1) to $xy = k^2$ is $x/x_1 + y/y_1 = 2$.

3. Find the condition of conjugacy for the diameters of

through t_1 and t_2. $x : y : k = t^2 : 1 : t$

4. Find the condition for $ax^2 + 2hxy + by^2 = 0$ to be conjugate diameters of $xy = k^2$.

5. Find the equation of the normal at (x_1, y_1) to the hyperbola $xy = k^2$ of eccentricity e.

In Nos. 6–9, find the asymptotes of the hyperbola.

6. $xy - 3x - 2y + 2 = 0$.   7. $12xy = 8x - 3y$.

8. $x(2y - 3) = 3y + 2$.   9. $y = (ax + b)/(cx + d)$.

10. Express the distances of the foci of $xy = k^2$ from a point (x, y) on the curve as rational functions of x, y, k, e.

11. A line $ABCD$ meets a hyperbola in B, C and the asymptotes in A, D. Prove that $AB = CD$.

12. A chord PQ of $xy = k^2$ passes through a fixed point (f, g). Prove that the locus of the remaining vertices of the parallelogram formed by drawing lines through P, Q parallel to the asymptotes is

$$xy + k^2 = gx + fy.$$

13. The tangents at P, Q to a hyperbola meet one asymptote in L, M. Prove that PQ meets the asymptote at the mid-point of LM.

14. A tangent to a hyperbola meets the asymptotes OX, OY at Q, R. The lines through Q, R parallel to OY, OX meet the curve in U, V. Find the envelope of UV.

15. Through a fixed point A of a fixed hyperbola, lines AM, AN are drawn parallel to the asymptotes, and PN, PM are drawn parallel to them through a variable point P on the curve. Show that MN passes through the centre.

16. Find the locus of the mid-point of a chord of

$$axy + bx + cy + d = 0$$

through a fixed point (e, f).

17. The line $PABQ$ meets $xy = k^2$ in A, B and its asymptotes in P, Q. DCD' is the diameter of $xy = -k^2$ parallel to the line. Prove that $PA . AQ = CD^2$.

18. PP' is a diameter of $xy = k^2$ and the tangent at P meets the asymptotes in Q, R. Prove that $P'Q$ and $P'R$ touch $3xy + k^2 = 0$.

19. A circle through the foci of a hyperbola meets the asymptotes in P and Q. Prove that PQ touches the hyperbola or is parallel to the transverse axis.

20. Given the base of a triangle and the difference between the base angles, prove that the vertex lies on one of two rectangular hyperbolas.

21. Prove that if $abc \neq 0$ two triangles exist in G_4 which are inscribed in $xy = k^2$ and have sides parallel to $y = ax$, $y = bx$, $y = cx$. Find the condition for the existence of the triangles in G_3.

22. Prove the property of Exercise 14 c, No. 9, for any central conic by using the equation $xy = k^2$.

23. EE' is a fixed diameter of a central conic and the tangents at E, E' meet a variable tangent in T, T'. Prove that $ET . E'T'$ is constant.

In Nos. 24–39, the curve is a rectangular hyperbola.

24. Prove that no two tangents are at right angles.

25. The normal at P meets the transverse axis at G. Prove that the triangle POG is isosceles, where O is the centre.

26. The normal at P to $xy = k^2$ meets the curve again at Q. Prove that $PQ = OP^3/k^2$, where O is the centre.

27. Prove that the diameter of $xy = k^2$ conjugate to $x - t^2y = 0$ is $x + t^2y = 0$.

28. Show that the asymptotes bisect the angles between any two conjugate diameters.

29. Show that I and J are conjugate points wo any rectangular hyperbola.

30. A circle cuts $xy = 1$ at P_1, P_2, P_3, P_4. Prove that

$$x_1 x_2 x_3 x_4 = 1 = y_1 y_2 y_3 y_4.$$

31. A circle of radius a meets a rectangular hyperbola centre C at P, Q, R, S; prove that $CP^2 + CQ^2 + CR^2 + CS^2 = 4a^2$.

32. Each of two perpendicular chords of $xy = k^2$ subtends a right angle at P and the chords meet at Q. Prove that P and Q are conjugate points.

33. Prove that chords of $xy = k^2$ which subtend a right angle at (f, g) envelop the parabola $k^2(X^2 + Y^2) + (f^2 + g^2)XY + gX + fY = 0$.

34. Prove that the feet of the normals from (f, g) to $xy = k^2$ lie on the curve $x : y : 1 = gt - ft^2 : g - ft : 1 - t^2$.

35. If the normals at P_1, P_2, P_3, P_4 to $xy = k^2$ are concurrent, prove that they meet at $(x_1 + x_2 + x_3 + x_4, \ y_1 + y_2 + y_3 + y_4)$ and prove that $x_1 x_2 x_3 x_4 = y_1 y_2 y_3 y_4$.

36. Find the locus of the point of intersection of normals to $xy = k^2$ at the ends of chords parallel to $y = mx$.

37. In the figure of 15·93, if P describes the curve from P_1 to P_2 and then from P_3 to P_4, discuss the changes in the value of $\angle PAB - \angle PBA$.

38. If A, B, C are the points $(t, 1/t)$ where $t = a, b, c$, prove that the circumcentre of ABC is

$$(\tfrac{1}{2}(a + b + c + 1/abc), \ \tfrac{1}{2}(1/a + 1/b + 1/c + abc)).$$

39. With the notation of No. 38 if D is given by $t = d$, and

$$a^2 b^2 c^2 d^2 \neq 1,$$

prove that the centre of the rectangular hyperbola through the circumcentres of the triangles BCD, ACD, ABD, ABC is

$$(\tfrac{1}{2}(a + b + c + d), \ \tfrac{1}{2}(1/a + 1/b + 1/c + 1/d)).$$

40. Show that the conic through $(a, 1/a), (b, 1/b), (c, 1/c), (d, 1/d)$ and $(0, 0)$ is $x^2 + p_4 y^2 + p_2 xy = p_1 x + p_3 y$, where $p_1 = a + b + c + d$, $p_2 = \Sigma ab$, $p_3 = \Sigma abc$, $p_4 = abcd$. Find the Frégier's point F of the origin wo this conic, and prove that the circles BCD, ACD, ABD, ABC subtend equal angles at F.

Chapter 16

$$s = \kappa s'$$

16·1. Form of an Equation

16·11. It is sometimes stated that whereas the procedure in pure geometry is tentative so that each problem calls for the exercise of some degree of ingenuity, the procedure in analytical geometry is almost automatic.

This is only true in a limited sense. Consider, for example, the problem of Frégier's point (7·44). A is a given point on a conic, AB and AC are perpendicular chords and the envelope of BC is required. It might be thought that the natural procedure would be to start with A, AB, AC and then to find BC, and lastly to seek its envelope. The simple analytical solution proceeds almost in the reverse order, beginning with BC which is taken to be the line $[X, Y]$. And this is a typical rather than an isolated example. There is a technique to be learnt in analytical geometry; automatic processes too often lead to laborious algebra.

An important idea, which often saves clumsy algebra and makes an analytical solution as simple as a solution by pure geometry, is the idea of the *form* of an equation. This idea has already been used in this book in numerous examples. See for example 4·6, 4·7, 4·8, 4·9, 5·15, 5·61, 7·44.

The present chapter is concerned mostly with the form $s = \kappa s'$ where $s = 0$ and $s' = 0$ are often, but not always, the equations of conics. The importance of this form of equation in elementary geometry is easily recognised. But it remains true that a simple problem of elementary geometry usually possesses a simple solution by the methods of pure geometry. This consideration may appear to discount the value of analytical geometry as a tool. It is to be remembered that it also has an important role in establishing the foundations of pure geometry, which can no longer be regarded as a subject apart. In connexion with the foundations of geometry the ideas of Chapter 8 are essential.

16·12. *The equation* $s = \kappa s'$. If $s = 0$ and $s' = 0$ are two conics and κ is a constant, then $s = \kappa s'$ is a conic through the points of intersection of $s = 0$ and $s' = 0$; for the coordinates of any one of these points satisfy $s = 0$ and $s' = 0$ and therefore satisfy $s = \kappa s'$.

If $s = 0$ and $s' = 0$ meet in the four points A, B, C, D, then any conic σ which passes through A, B, C, D has an equation of the form $s = \kappa s'$. For if P_1 is a point on σ, κ can be chosen so that $s = \kappa s'$ passes through P_1, the value being s_{11}/s_{11}'. It is assumed that P_1 is not on $s' = 0$. For the value s_{11}/s_{11}' of κ, the conic $s = \kappa s'$ has five points in common with σ, and therefore coincides with σ. Thus:

$s = \kappa s'$ is a conic through the four points of intersection of $s = 0$ and $s' = 0$ and every conic through those four points has an equation of that form.

16·13. For its complete validity, this theorem must be regarded as belonging to G_6. The proof shows that $s = \kappa s'$ passes through any points of intersection that exist, but it is only in G_6 that there are necessarily four points of intersection. For example the conics $s \equiv xy + 1 = 0$ and $s' \equiv xy + 2 = 0$ in G_3 have no points of intersection. The corresponding conics in G_6 are $s \equiv xy + z^2 = 0$ and $s' \equiv xy + 2z^2 = 0$ and these have two pairs of coincident points of intersection at $(1, 0, 0)$ and $(0, 1, 0)$. $s = \kappa s'$ also passes through these points and the tangents at the points are the same as for $s = 0$ and $s' = 0$; hence, by the convention, it is said to pass through the two pairs of coincident points.

It is assumed in 16·12 that two conics $\sigma = 0$ and $s = \kappa s'$ coincide when they have five common points A, B, C, D, P_1. Possible exceptions might arise when points such as A, B, P_1 or A, B, C are collinear. But, in 16·12, P_1 is an arbitrary point on $s = \kappa s'$ and so A, B, P_1 need not be collinear. If A, B, C are collinear, s and s' must be of the forms $\alpha\beta$ and $\alpha\gamma$, where $\alpha = 0$ is ABC, and $\beta = 0$, $\gamma = 0$ are lines through D. Every conic through A, B, C, D then consists of $\alpha = 0$ and a line $\beta = \kappa\gamma$ through D, and so

$$\sigma \equiv \alpha(\beta - \kappa\gamma) \equiv s - \kappa s'.$$

Hence the result of 16·12 is valid even in this exceptional case.

16·14. There is one conic through the points of intersection A, B, C, D of $s = 0$ and $s' = 0$ which does not have an equation $s = \kappa s'$. This is the conic $s' = 0$. Usually this exception leads to no inconvenience. It can be avoided by the use of $\kappa s = \kappa' s'$ where (κ, κ') is a homogeneous parameter. Theoretically the best form of equation for conics through four points is

$$\lambda s + \lambda' s' = 0,$$

but in particular examples the form $s = \kappa s'$ is often convenient. See 4·73.

16·15. There are ∞^1 conics of the system $s = \kappa s'$. The five degrees of freedom possessed by a conic are reduced to one degree by the conditions for passing through A, B, C, D.

A unique conic of a system $s = \kappa s'$ passes through a point P_1. The name "pencil" was first given to a system of concurrent lines, but it is also applied to any system which has the property that just one member passes through an arbitrary point. Thus a system, $s = \kappa s'$, of conics is called a *pencil of conics*.

16·16. Three conics of a pencil $s = \kappa s'$ are degenerate. If two conics of the system meet in A, B, C, D, the degenerate conics are the line-pairs AB, CD, and AC, BD, and AD, BC. The condition for $\lambda s + \lambda' s' = 0$ to be a line-pair is

$$\begin{vmatrix} \lambda a + \lambda' a' & \lambda h + \lambda' h' & \lambda g + \lambda' g' \\ \lambda h + \lambda' h' & \lambda b + \lambda' b' & \lambda f + \lambda' f' \\ \lambda g + \lambda' g' & \lambda f + \lambda' f' & \lambda c + \lambda' c' \end{vmatrix} = 0.$$

This is of the form

$$\delta \lambda^3 + \theta \lambda^2 \lambda' + \theta' \lambda \lambda'^2 + \delta' \lambda'^3 = 0$$

and is a cubic for $\lambda : \lambda'$ whose three roots correspond to the three line-pairs.

The values of the coefficients in the cubic equation are given by

$$\delta = abc + 2fgh - af^2 - bg^2 - ch^2,$$
$$\theta = Aa' + Bb' + Cc' + 2Ff' + 2Gg' + 2Hh',$$
$$\theta' = A'a + B'b + C'c + 2F'f + 2G'g + 2H'h,$$
$$\delta' = a'b'c' + 2f'g'h' - a'f'^2 - b'g'^2 - c'h'^2,$$

where the notation of 0·6 is used.

16·17. The condition for $\lambda s + \lambda' s' = 0$ to be a parabola is

$$\begin{vmatrix} \lambda a + \lambda' a' & \lambda h + \lambda' h' \\ \lambda h + \lambda' h' & \lambda b + \lambda' b' \end{vmatrix} = 0$$

or $\qquad C\lambda^2 + (ab' + a'b - 2hh')\lambda\lambda' + C'\lambda'^2 = 0.$

This is a quadratic for $\lambda : \lambda'$. Hence two conics of the system are parabolas. Either or both of these may be a pair of parallel lines.

16·18. The condition for $\lambda s + \lambda' s' = 0$ to be a rectangular hyperbola is
$$\lambda(a + b) + \lambda'(a' + b') = 0.$$

If $a + b = 0$ and $a' + b' = 0$, this equation is true for all values of λ, λ'. Otherwise it is true for just one value of $\lambda : \lambda'$. Hence there is in general just one rectangular hyperbola through the points of intersection of two conics. But if the conics are both rectangular hyperbolas, then every conic through their four points of intersection is also a rectangular hyperbola.

A pair of perpendicular lines is a special case of a rectangular hyperbola. If then the perpendiculars from A to BC and from B to AC meet in D, the four points A, B, C, D are the points of intersection of the two rectangular hyperbolas AD, BC and AC, BD. Hence also AB, CD is also a rectangular hyperbola, i.e. AB is at right angles to CD. Thus:

The perpendiculars from the vertices of a triangle to the opposite sides meet in a point (the orthocentre).

Also

Every conic through the vertices of a triangle and the orthocentre is a rectangular hyperbola.

16·19. In order that $\lambda s + \lambda' s' = 0$ may represent a circle, two conditions have to be satisfied by $\lambda : \lambda'$, namely

$$\lambda(a - b) + \lambda'(a' - b') = 0$$

and $\qquad\qquad\qquad\qquad \lambda h + \lambda' h' = 0.$

This is usually impossible. In general no conic of the pencil $s = \kappa s'$ is a circle.

16·20. EXAMPLE. Find the equation of the conic through $(4, 0)$, $(0, 3)$, $(-2, 0)$, $(0, -1)$, $(12, 12)$.

Two of the conics through the first four of the points are the lines $(4, 0)$ $(0, 3)$ and $(-2, 0)$ $(0, -1)$ and the lines $(4, 0)$ $(-2, 0)$ and $(0, 3)$ $(0, -1)$. Hence an arbitrary conic through the four points is

$$(x/4 + y/3 - 1)(x/2 + y/1 + 1) = kxy.$$

This passes also through $(12, 12)$ if

$$6.19 = k.12.12. \qquad \therefore \quad k = \tfrac{19}{24}.$$

Thus the required conic is $(3x + 4y - 12)(x + 2y + 2) = 19xy$.

16·21. EXAMPLE. Find the condition for the points of intersection of $ax^2 + 2hxy + by^2 = d$ and $px^2 + 2qxy + ry^2 = u$ to be concyclic.

Any conic through the four points is

$$ax^2 + 2hxy + by^2 - d = k(px^2 + 2qxy + ry^2 - u).$$

This is a circle if

$$a - kp = b - kr \quad \text{and} \quad h = kq.$$

Usually it is impossible to choose k so as to satisfy these two equations. But they are satisfied by

$$k = h/q = (a - b)/(p - r)$$

provided that these fractions are equal. The condition is

$$h(p - r) = q(a - b).$$

The case $q = 0, p = r$ is not exceptional. This can be seen by using $ks = s'$ or $ks = k's'$ instead of $s = ks'$.

16·22. EXAMPLE. P, P' are points on an ellipse whose foci are S, S', and $SP, S'P'$ are in the same direction. Find the envelope of PP'.

Let $PS, P'S'$ meet the ellipse $x^2/a^2 + y^2/b^2 = 1$ again in Q, Q'. Then if PP' is $Xx + Yy + 1 = 0$, QQ' is $Xx + Yy - 1 = 0$, and hence the parallel lines $PS, P'S'$ are

$$(Xx + Yy)^2 - 1 = \kappa(x^2/a^2 + y^2/b^2 - 1).$$

Since they are parallel, $(X^2 - \kappa/a^2)(Y^2 - \kappa/b^2) = X^2Y^2$. Since they pass through the foci, $a^2e^2X^2 - 1 = \kappa(e^2 - 1)$.

Eliminating κ,

$$(a^2e^2X^2 - 1)/(a^2b^2) = (X^2/b^2 + Y^2/a^2)(e^2 - 1).$$

$$\therefore \quad X^2/b^2 + Y^2b^2/a^4 = 1/(a^2b^2),$$

$$\therefore \quad a^4X^2 + b^4Y^2 = a^2.$$

This is the envelope equation of an ellipse with semi-axes a and b^2/a along OX and OY.

16·23. Example. Find the circumcircle of the triangle formed by

$$ax^2 + 2hxy + by^2 = 0 \quad \text{and} \quad lx + my + n = 0.$$

Let $px + qy = 0$ be the tangent at the origin. Then the circle is

$$ax^2 + 2hxy + by^2 = (lx + my + n)(px + qy),$$

where $\quad a - b = lp - mq, \quad 2h = mp + lq.$

Therefore the circle is

$$\begin{vmatrix} ax^2 + 2hxy + by^2 & (lx + my + n)x & (lx + my + n)y \\ a - b & l & -m \\ 2h & m & l \end{vmatrix} = 0.$$

16·3. The Equation S = kS′

16·32. If $S = 0$, $S' = 0$ are the envelope equations of two conics, $S = kS'$ represents a conic touching their four common tangents; and every conic touching these four lines has an equation of the form $S = kS'$. This is proved by the dual of the arguments in 16·12.

16·35. There are ∞^1 conics of the system $S = kS'$. A unique conic of the system touches a given line.

16·36. Three conics of the system are point-pairs. The condition for $\lambda S + \lambda' S' = 0$ to be a point-pair is

$$\begin{vmatrix} \lambda A + \lambda'A' & \lambda H + \lambda'H' & \lambda G + \lambda'G' \\ \lambda H + \lambda'H' & \lambda B + \lambda'B' & \lambda F + \lambda'F' \\ \lambda G + \lambda'G' & \lambda F + \lambda'F' & \lambda C + \lambda'C' \end{vmatrix} = 0,$$

and this is a cubic equation for $\lambda : \lambda'$.

16·37. $\lambda S + \lambda' S' = 0$ is a parabola for one value of $\lambda : \lambda'$ namely that given by $\lambda C + \lambda' C' = 0$ which is the condition of tangency for $[0, 0, 1]$.

16·38. EXAMPLE. Find the parabola which touches the lines $[1, 2]$, $[2, 6]$, $[-3, 6]$, $[-1, 0]$.

The points of intersection of $[1, 2][2, 6]$ and $[-3, 6][-1, 0]$ are given by

$$\begin{vmatrix} X & Y & 1 \\ 1 & 2 & 1 \\ 2 & 6 & 1 \end{vmatrix} \begin{vmatrix} X & Y & 1 \\ -3 & 6 & 1 \\ -1 & 0 & 1 \end{vmatrix} = 0,$$

i.e. $$(4X - Y - 2)(3X + Y + 3) = 0.$$

The points of intersection of $[1, 2][-1, 0]$ and $[2, 6][-3, 6]$ are given by $(X - Y + 1)(Y - 6) = 0$. Hence any conic which touches the four lines is

$$(4X - Y - 2)(3X + Y + 3) = k(X - Y + 1)(Y - 6)$$

The constant term is zero when $k = 1$. Hence the parabola is

$$(4X - Y - 2)(3X + Y + 3) = (X - Y + 1)(Y - 6),$$

i.e. $$X^2 + X = Y.$$

EXERCISE 16A

In Nos. 1, 2, find the conics through:

1. $(5, 0)$, $(7, 11)$, $(0, -3)$, $(-2, 0)$, $(0, 4)$.

2. $(2, -2)$, $(2, 0)$, $(4, -3)$, $(6, -3)$, $(4, -2)$.

In Nos. 3, 4, find the parabolas through:

3. $(1, 3)$, $(2, 2)$, $(1, 6)$, $(10, 2)$.

4. $(-2, 3)$, $(6, 7)$, $(22, 11)$, $(-2, -1)$.

5. Find the rectangular hyperbola through $(3, -1)$, $(5, -1)$, $(1, 0)$, $(1, 4)$.

6. Find the rectangular hyperbola through the points of intersection of $x^2 + y^2 = 1$ and $x^2 + 16y^2 = 4$.

7. Find the common chords of $x^2 + y^2 = 25$ and $x^2 + xy + y^2 = 36$.

8. Find the conic through (r, r) and the points of intersection of $x^2 + y^2 = r^2$ and $ax^2 + 2hxy + by^2 = 1$.

9. Find the conic through the origin and the points of intersection of $y^2 = 4a(x - a)$ and $(x - 4a)^2 + y^2 = 9a^2$.

10. Find the parabolas through the points of intersection of $x^2 + y^2 = 4$ and $3x^2 + 2xy + 5y^2 = 15$.

11. Find the rectangular hyperbola through the points of

$$8y^2 = a(x + 2a)$$

which are at distance a from the origin.

12. Show that two pairs of common chords of $x^2 + y^2 = 4$ and $2x^2 + 7xy + 2y^2 + 14x + 14y + 20 = 0$ coincide with $x + y + 2 = 0$. Find also the other pair.

13. Prove that all rectangular hyperbolas through the vertices of a triangle pass through its orthocentre.

14. Find the conic through $(5, -6)$ and the points where $xy = 2$ meets $3x^2 + y^2 = 7$.

15. Give the general equation of the conic which meets

$$x^2/a^2 + y^2/b^2 = 1$$

at the ends of its axes. Can the conic be a parabola?

16. The equation of AB, AD is $bx^2 + ay^2 = 0$ and the equation of CB, CD is $ax^2 + 2hxy + by^2 + 2gx + 2fy + c = 0$. Prove that the diagonals not through A of the quadrilateral formed by the four lines are at right angles if

$$(a + b)(f^2 + g^2) = 2fgh + c(a^2 + b^2 - h^2).$$

In Nos. 17–19, find the conic which touches the lines.

17. $[1, 3], [0, 1], [-2, 3], [-3, 7], [2, 7]$.

18. $[a, b], [a, -b], [-a, b], [-a, -b], [c, 0]$.

19. $[1, 1, 1], [3, 5, 1], [-6, 8, 1], [-2, 0, 1]$, and the line at infinity.

In Nos. 20, 21, find the point-pairs of the system $S = kS'$.

20. $S \equiv a^2 X^2 + b^2 Y^2 - 1$, $S' \equiv b^2 X^2 + a^2 Y^2 - 1$.

21. $S \equiv X^2 + Y^2 - 6XY - 1$, $S' \equiv X^2 + Y^2 + 6XY - 1$.

22. Find the value of k for which

$$(2X^2 + 3XY + 4Y^2 + 2X + 1) = k(X^2 - 2XY + 3Y^2 + 2Y + 1)$$

touches $x + y + 1 = 0$.

23. Find the parabola of the system in No. 22.

24. Find the value of λ such that the conic

$$\lambda(a^2 X^2 + b^2 Y^2 - Z^2) = X^2 + Y^2$$

touches $[X_1, Y_1, Z_1]$. Can this conic be a parabola?

16·4. The theorem of 16·12 applies when one or both of the conics $s = 0$, $s' = 0$ is degenerate, and the particular cases which arise from degeneracy are important.

16·41. $s = k\alpha\beta$ is a conic through the points where the lines $\alpha = 0$ and $\beta = 0$ meet $s = 0$.

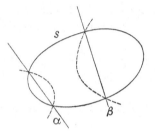 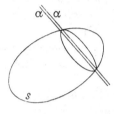

16·42. $s = k\alpha^2$ arises from $s = k\alpha\beta$ when $\beta \to \alpha$. It is a conic which touches $s = 0$ at each of two points on $\alpha = 0$. It is said to have *double contact* with $s = 0$.

16·43. $s = k\alpha\tau$, where $\tau = 0$ is a tangent to $s = 0$, meets $s = 0$ at two points on $\alpha = 0$ and touches it at the contact of $\tau = 0$.

16·44. $s = k\alpha\beta$, when $\alpha = 0$ and $\beta = 0$ meet on $s = 0$, gives a conic like that in 16·43.

16·45. $s = k\alpha\tau$, where $\tau = 0$ is a tangent to $s = 0$, and $\alpha = 0$ passes through the contact of $\tau = 0$, represents a conic which meets $s = 0$ actually in two points only, but the contact of $\tau = 0$ counts as three points of intersection, and $s = k\alpha\tau$ is said to have *three-point contact* with $s = 0$.

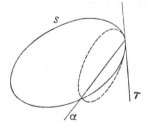 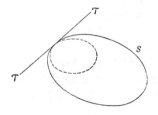

16·46. $s = k\tau^2$, where $\tau = 0$ is a tangent to $s = 0$, has *four-point contact* with $s = 0$.

16·47. $s = k\alpha$ passes through the two points where $\alpha = 0$ meets $s = 0$. But ∞^3 conics have this property whereas $s = k\alpha$ contains only ∞^1 members. If the geometry is homogeneous, the equation $s = k\alpha$ is replaced by $s = k\alpha z$. This has the same points at infinity as $s = 0$: the asymptotes of the two conics are parallel.

The equations $s = 0$ and $s = k\alpha$ have the same second degree terms $ax^2 + 2hxy + by^2$. They are similar conics. See 15·63.

16·48. $\alpha\beta = k\gamma\delta$ is a conic through the four points of intersection of the lines α, β with the lines γ, δ.

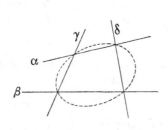 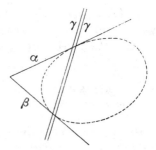

16·49. $\alpha\beta = k\gamma^2$ arises from $\alpha\beta = k\gamma\delta$ when $\delta \to \gamma$. It is a conic touching $\alpha = 0, \beta = 0$ at their points of intersection with $\gamma = 0$. This very important form gives the equation of a conic which touches two sides of a triangle at the ends of the third side.

16·50. $\alpha^2 = k\gamma^2$ is a conic composed of two lines $\alpha = \gamma\sqrt{k}$ and $\alpha = -\gamma\sqrt{k}$.

The examples in 16·41–16·50 illustrate the cases in which $s = 0$ and $s' = 0$ have points of intersection $ABCD, AABC, AABB, AAAB, AAAA$.

16·51. EXAMPLE. If $P_1 N_1, P_2 N_2$ are ordinates of a parabola and the circle on $N_1 N_2$ as diameter touches the parabola, prove that $N_1 P_1 + N_2 P_2 = N_1 N_2$.

Let the parabola be $y^2 = 4ax$ and let the chord of contact be $x = b$. Then the equation of the circle is

$$y^2 - 4ax = k(x-b)^2, \quad \text{where } k = -1.$$

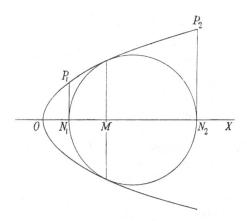

Hence at the points of intersection N_1, N_2 of the circle with the axis $y = 0$, $4ax = (x-b)^2$.

$$\therefore \quad \sqrt{(4ax)} = \pm (x-b).$$

If M is $(b, 0)$, $\quad N_1 P_1 = b - x_1 = N_1 M$,

and $\qquad\qquad N_2 P_2 = x_2 - b = M N_2$.

$$\therefore \quad N_1 P_1 + N_2 P_2 = N_1 N_2.$$

16·52. EXAMPLE. Determine the circle which has four-point contact with $x^2/a^2 + y^2/b^2 = 1$ at $(a, 0)$.

Since $x = a$ is the tangent to the ellipse at $(a, 0)$, the equation of the circle is
$$x^2/a^2 + y^2/b^2 - 1 = k(x-a)^2.$$

The coefficients of x^2, y^2 must be equal.

Hence $k = 1/a^2 - 1/b^2$ and the circle is

$$x^2/a^2 + y^2/b^2 - 1 + (a^2 - b^2)(x-a)^2/a^2b^2 = 0,$$

i.e. $\qquad x^2 + y^2 - 2(a^2 - b^2)\, x/a + a^2 - 2b^2 = 0,$

or $\qquad \left(x - \dfrac{a^2 - b^2}{a}\right)^2 + y^2 = \left(\dfrac{b^2}{a}\right)^2.$

16·53. EXAMPLE. Interpret the equations $y^2 = 4ax$ and $xy = k^2$ as examples of $s = \kappa s'$.

It is necessary to consider the corresponding equations in G_6.

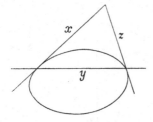 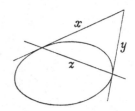

$y^2 = 4axz$ is an example of $\alpha\beta = k\gamma^2$. It touches $x = 0$ and $z = 0$ at their points of intersection with $y = 0$, i.e. at the vertex of the parabola and at the point at infinity on the axis.

$xy = k^2z^2$ is also an example of $\alpha\beta = k\gamma^2$. It touches $x = 0$, $y = 0$ at their points of intersection with $z = 0$: the hyperbola touches its asymptotes at infinity.

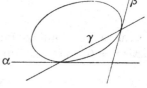

16·6. Special cases of $S = kS'$ dual to the examples in 16·4 are as follows.

16·61. $S = kAB$. This is a conic touching the tangents to $S = 0$ from the points $A = 0$ and $B = 0$.

16·62. $S = kA^2$ arises from $S = kAB$ when $B \to A$. It is a conic which touches $S = 0$ twice, the tangents passing through $A = 0$. This gives the same figure as 16·42.

16·63. $S = kAT$, where $T = 0$ is a contact of $S = 0$, is a conic touching the tangents from $A = 0$ to $S = 0$ and also touching $S = 0$ at $T = 0$.

16·64. $S = kAB$, where $A = 0$ and $B = 0$ lie on the same tangent of $S = 0$, gives the same figure as 16·63.

16·65. $S = k\mathrm{AT}$, where $\mathrm{T} = 0$ is a contact of $S = 0$ and $\mathrm{A} = 0$ lies on the tangent which has that contact, is a conic which actually has only two tangents in common with $S = 0$. But conventionally the tangent $A T$ counts as three, where A and T are the points $\mathrm{A} = 0$, $\mathrm{B} = 0$. $S = k\mathrm{AT}$ is therefore said to have *three-line contact* with $S = 0$.

16·66. $S = k\mathrm{T}^2$ has *four-line contact* with $S = 0$. The tangent whose contact is $\mathrm{T} = 0$ counts as four common tangents of $S = 0, S = k\mathrm{T}^2$.

16·67. $S = k\mathrm{AZ}$ is a conic touching the tangents to $S = 0$ from $\mathrm{A} = 0$ and the tangents from the origin $(\mathrm{Z} = 0)$.

16·68. $\mathrm{AB} = k\Gamma\Delta$ is a conic touching the lines AC, AD, BC, BD, where A, B, C, D are the points $\mathrm{A} = 0$, $\mathrm{B} = 0$, $\Gamma = 0$, $\Delta = 0$.

16·69. $\mathrm{AB} = k\Gamma^2$ arises from 16·68 when $\Delta \to \Gamma$. It is a conic touching AC at A and BC at B. The figure is the same as that of 16·49.

16·70. $\mathrm{A}^2 = k\Gamma^2$ is a conic composed of the two points $\mathrm{A} = \Gamma\sqrt{k}$, $\mathrm{A} = -\Gamma\sqrt{k}$.

The examples in 16·61–16·70 illustrate the cases in which $S = 0$ and $S' = 0$ have common tangents $abcd$, $aabc$, $aabb$, $aaab$, $aaaa$.

16·71. EXAMPLE. If $\alpha \equiv x$, $\beta \equiv y$, $\gamma \equiv 2x - 3y + 1$, find the equation of the conic through $(-3, 1)$ touching $\alpha = 0$ and $\beta = 0$ where $\gamma = 0$ meets them.

By 16·49, the equation is

$$xy = k(2x - 3y + 1)^2,$$

where $\qquad\qquad -3 = k(-6 - 3 + 1)^2 = 64k,$

i.e. $\qquad\qquad 64xy + 3(2x - 3y + 1)^2 = 0.$

16·72. EXAMPLE. If $\alpha \equiv x$, $\beta \equiv y$, $\gamma \equiv 2x - 3y + 1$, find the equation of the conic touching $\alpha = 0$ and $\beta = 0$ where $\gamma = 0$ meets them and also touching $4x + 5y = 1$.

Using G_5 or G_6, the vertices of the triangle formed by $\alpha = 0, \beta = 0, \gamma = 0$ are $Z = 0, Y + 3Z = 0, X - 2Z = 0$. Also the coordinates of $4x + 5y = z$ are $[4, 5, -1]$.

By 16·69, the equation is

$$(Y + 3Z)(X - 2Z) = kZ^2,$$

where $\qquad (5 - 3)(4 + 2) = k,$

i.e. $\qquad (Y + 3Z)(X - 2Z) = 12Z^2,$

or, returning to G_3, $\quad (Y + 3)(X - 2) = 12$

or $\qquad XY + 3X - 2Y - 18 = 0.$

By 12·7 this is the same conic as

$$4x^2 - 60xy + 9y^2 + 4x - 6y + 1 = 0.$$

16·73. EXAMPLE. Prove that if a conic touches four given tangents to a given conic and the chord of contact of one pair of them, it also touches the chord of contact of the other pair.

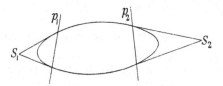

Let $S = 0$ be the given conic, and $[X_1, Y_1]$, $[X_2, Y_2]$ the chords of contact. The poles of these chords are $S_1 = 0$ and $S_2 = 0$, and these form a point-pair conic whose common tangents with $S = 0$ are the given tangents. Hence any conic which touches these tangents is

$$S = kS_1 S_2.$$

This also touches $[X_1, Y_1]$ if $S_{11} = kS_{11}S_{12}$, i.e. if $k = 1/S_{12}$. For this value of k, by symmetry, the conic also touches $[X_2, Y_2]$. The equation of the conic is $S_{12}S = S_1 S_2$.

EXERCISE 16B

1. Give the equation of an arbitrary conic circumscribing the convex quadrilateral whose sides are $\pm x/a + y/b = 1$, $\pm x/a - y/c = 1$. If this conic can be a circle, what relation do a, b, c satisfy and what is the equation of the circle?

2. Give the envelope equation of an arbitrary conic inscribed in the square formed by the lines $[\pm 1, \pm 1]$ and deduce the locus equation.

3. Find the parabolas through the ends of the chords $lx \pm my = 1$ of $ax^2 + by^2 = 1$.

4. Find the parabola which touches the axes and the tangents from $(\frac{1}{2}k, \frac{1}{2}k)$ to $4k^2 XY = 1$.

5. If a circle has double contact with $ax^2 + by^2 = c$, prove that the chord of contact is parallel to $x = 0$ or $y = 0$.

6. QT is a tangent from a point Q on an ellipse of eccentricity e to a circle which has double contact with the ellipse, the chord of contact being parallel to the minor axis. QL is drawn perpendicular to the chord. Prove that $QT = eQL$.

7. Give the general locus and envelope equations of the conics which have (i) three-point contact, (ii) four-point contact, with the parabola $x : y : a = t^2 : 2t : 1$ at the point t.

8. Show that a circle cannot have four-point contact with a parabola except at the vertex.

9. Find the circle which has three-point contact with the rectangular hyperbola $xy = k^2$ at $(kt, k/t)$.

10. Find the rectangular hyperbola which has four-point contact with $y^2 = 4ax$ at $(at^2, 2at)$.

11. Find the rectangular hyperbola which touches $3x + y = 2$ and $2x + 3y = 1$ at $(\frac{1}{5}, \frac{7}{5})$ and $(-7, 5)$.

12. If $s = 0$ and $s' = 0$ are two conics and $s - s'$ is the product of linear factors α and β, prove that the conic $k^2 \alpha^2 + \beta^2 = 2k(s + s')$ has double contact with both $s = 0$ and $s' = 0$.

13. If $\alpha\beta = 0$ is a line-pair of the system $s = ks'$, prove that there is a conic which has double contact with $s = 0$ along $\alpha = \lambda\beta$ and with $s' = 0$ along $\alpha = -\lambda\beta$.

14. State the relation of the conics

$$x^2/a^2 + y^2/b^2 - 1 = \lambda(x/a + y/b - 1)(x/a - y/b + 1),$$

$$x^2/a^2 + y^2/b^2 - 1 = \kappa y(y - b)$$

to the ellipse $x^2/a^2 + y^2/b^2 - 1 = 0$, and give the value of λ in terms of κ, b for which they are identical.

15. State the relation of the conic $\kappa(aY^2 - X) = (aX - 1)^2$ to the parabola $aY^2 = X$ and discuss the special case $\kappa = 2a$.

In Nos. 16–24, interpret the equation.

16. $x^2 + y^2 = \kappa^2(x\cos\alpha + y\sin\alpha - a)^2$.

17. $(y^2 + 2kx)(y_1 y_2 + kx_1 + kx_2) = (yy_1 + kx + kx_1)(yy_2 + kx + kx_2)$, where $y_1^2 + 2kx_1 = 0 = y_2^2 + 2kx_2$.

18. $xy - k^2 = \lambda(x + yt^2 - 2kt)^2$. **19.** $xy - k^2 = (\lambda x + \mu y)y$.

20. $4k^2XY - 1 = \lambda Y^2$. **21.** $aY^2 - X = \kappa(aX - 2aY + 1)^2$.

22. $aY^2 - XZ = Z^2/4a$. **23.** $a^2X^2 + b^2Y^2 - 1 = \lambda X^2$.

24. $a^2X^2 + b^2Y^2 - 1 = \lambda(X - Y)^2$.

16·8. Concyclic Points

16·81. A pencil of conics can be drawn to pass through four given concyclic points A, B, C, D. If the axes of one conic of the system are taken as axes of coordinates, the equation of that conic is
$$ax^2 + by^2 + c = 0.$$

If the equation of the circle $ABCD$ is
$$x^2 + y^2 + 2px + 2qy + r = 0,$$
an arbitrary conic of the system is
$$ax^2 + by^2 + c = k(x^2 + y^2 + 2px + 2qy + r).$$

The asymptotes of this conic are parallel to
$$ax^2 + by^2 = k(x^2 + y^2)$$
and therefore the axes, which bisect the angles between the asymptotes, are parallel to the axes of coordinates. Thus:

All conics through four concyclic points have parallel axes.

16·82. The common chords of $x^2 + y^2 + 2px + 2qy + r = 0$ and $ax^2 + by^2 + c = 0$ are degenerate conics of the system
$$ax^2 + by^2 + c = k(x^2 + y^2 + 2px + 2qy + r)$$
given by three special values of k (see 16·16). The axes of the line-pairs are their angle-bisectors, which are parallel to the axes of coordinates. Thus:

The common chords of a circle and a conic taken in pairs make supplementary angles with the axes of the conic.

16·83. The argument of 16·81 and 16·82 is applicable when $ax^2 + by^2 + c = 0$ is replaced by $y^2 = 4ax$. There are two parabolas of the system, and their axes are parallel to the axes of the other conics.

16·84. *Eccentric Angles of Concyclic Points of an Ellipse.* The chords of $x^2/a^2 + y^2/b^2 = 1$ joining points whose eccentric angles are ϕ_1, ϕ_2 and ϕ_3, ϕ_4 are (by 14·31) $\alpha = 0$ and $\beta = 0$, where

$$\alpha \equiv \frac{x}{a} \cos \tfrac{1}{2}(\phi_1 + \phi_2) + \frac{y}{b} \sin \tfrac{1}{2}(\phi_1 + \phi_2) - \cos \tfrac{1}{2}(\phi_1 - \phi_2)$$

and $\quad \beta \equiv \dfrac{x}{a} \cos \tfrac{1}{2}(\phi_3 + \phi_4) + \dfrac{y}{b} \sin \tfrac{1}{2}(\phi_3 + \phi_4) - \cos \tfrac{1}{2}(\phi_3 - \phi_4).$

The four points $\phi_1, \phi_2, \phi_3, \phi_4$ are concyclic if and only if one conic of the system $x^2/a^2 + y^2/b^2 - 1 = k\alpha\beta$ is a circle. The conditions for the conic to be a circle are that the coefficients of x^2 and y^2 should be equal, and that the coefficient of xy should be zero. It is not usually possible to choose k to satisfy these two conditions. But the coefficient of xy is zero if

$$\cos \tfrac{1}{2}(\phi_1 + \phi_2) \sin \tfrac{1}{2}(\phi_3 + \phi_4) + \cos \tfrac{1}{2}(\phi_3 + \phi_4) \sin \tfrac{1}{2}(\phi_1 + \phi_2) = 0,$$

i.e. $\quad\quad\quad\quad \sin \tfrac{1}{2}(\phi_1 + \phi_2 + \phi_3 + \phi_4) = 0,$

i.e. $\quad\quad\quad\quad\quad \phi_1 + \phi_2 + \phi_3 + \phi_4 = 2n\pi.$

The coefficients of x^2 and y^2 are then equal if

$$\frac{1}{a^2} - \frac{1}{b^2} = k \left\{ \frac{1}{a^2} \cos \tfrac{1}{2}(\phi_1 + \phi_2) \cos \tfrac{1}{2}(\phi_3 + \phi_4) \right.$$

$$\left. - \frac{1}{b^2} \sin \tfrac{1}{2}(\phi_1 + \phi_2) \sin \tfrac{1}{2}(\phi_3 + \phi_4) \right\}$$

$$= k \left\{ \frac{1}{a^2} \cos^2 \tfrac{1}{2}(\phi_1 + \phi_2) + \frac{1}{b^2} \sin^2 \tfrac{1}{2}(\phi_1 + \phi_2) \right\}$$

in which the coefficient of k is not zero. Hence if the condition $\Sigma\phi = 2n\pi$ holds, k can be chosen to satisfy the second condition. Therefore $\Sigma\phi = 2n\pi$ is a necessary and sufficient condition for the points $\phi_1, \phi_2, \phi_3, \phi_4$ to be concyclic.

16·85. EXAMPLE. On the chord $\phi_1 \phi_2$ of the ellipse $x^2/a^2 + y^2/b^2 = 1$ as diameter, a circle is described. Find the equation of the chord joining its other points of intersection with the ellipse.

In the equation $x^2/a^2 + y^2/b^2 - 1 = k\alpha\beta$ of 16·84, since the coefficient of xy is zero, $k\beta$ can be replaced by

$$\lambda \left\{ \frac{x}{a} \cos \tfrac{1}{2}(\phi_1 + \phi_2) - \frac{y}{b} \sin \tfrac{1}{2}(\phi_1 + \phi_2) - l \cos \tfrac{1}{2}(\phi_1 - \phi_2) \right\}.$$

The centre of the circle is the mid-point of the chord $\phi_1 \phi_2$, given by

$$x_1 = a \cos \tfrac{1}{2}(\phi_1 + \phi_2) \cos \tfrac{1}{2}(\phi_1 - \phi_2),$$

$$y_1 = b \sin \tfrac{1}{2}(\phi_1 + \phi_2) \cos \tfrac{1}{2}(\phi_1 - \phi_2).$$

But x_1/y_1 is also equal to the ratio of the coefficients of x and y in the equation of the circle.

$$\therefore \quad \frac{a \cos \tfrac{1}{2}(\phi_1 + \phi_2)}{b \sin \tfrac{1}{2}(\phi_1 + \phi_2)} = \frac{x_1}{y_1} = \frac{b(l+1) \cos \tfrac{1}{2}(\phi_1 + \phi_2)}{a(l-1) \sin \tfrac{1}{2}(\phi_1 + \phi_2)},$$

$$\therefore \quad \frac{l+1}{l-1} = \frac{a^2}{b^2}, \qquad l = \frac{a^2 + b^2}{a^2 - b^2},$$

$\therefore$ the chord is

$$\frac{x}{a} \cos \tfrac{1}{2}(\phi_1 + \phi_2) - \frac{y}{b} \sin \tfrac{1}{2}(\phi_1 + \phi_2) = \frac{a^2 + b^2}{a^2 - b^2} \cos \tfrac{1}{2}(\phi_1 - \phi_2).$$

16·86. EXAMPLE. A circle meets $x^2/a^2 + y^2/b^2 = 1$ at the ends of a diameter and at the ends of a chord through (f, g). Find the locus of the centre of the circle.

Let the circle be $(x - x_1)^2 + (y - y_1)^2 = r_1^2$, so that the centre is (x_1, y_1). The chords of intersection are

$$(px - qy)\{p(x - f) + q(y - g)\} = 0.$$

This must be of the form

$$(x - x_1)^2 + (y - y_1)^2 - r_1^2 + \lambda \left(\frac{x^2}{a^2} + \frac{y^2}{b^2} - 1 \right) = 0.$$

$$\therefore \quad \frac{p(pf + qg)}{2x_1} = \frac{q(pf + qg)}{-2y_1} = \frac{p^2}{1 + \lambda/a^2} = \frac{-q^2}{1 + \lambda/b^2}.$$

$$\therefore \quad = \frac{a^2p^2 + b^2q^2}{a^2 - b^2},$$

$$\therefore \quad \frac{fx_1 - gy_1}{2} = \frac{a^2x_1{}^2 + b^2y_1{}^2}{a^2 - b^2},$$

$\therefore$ the locus is $a^2x^2 + b^2y^2 = \frac{1}{2}(a^2 - b^2)(fx - gy)$.

16·87. EXAMPLE. Find the locus of the centre of a circle which has three-point contact with $y^2 = 4ax$.

The equation of such a circle is of the form

$$y^2 - 4ax = \kappa \alpha \tau,$$

where $\tau = 0$ is a tangent to $y^2 = 4ax$, and $\alpha = 0$ is a line through the contact of $\tau = 0$.

Also, by 16·82, $\alpha = 0$ and $\tau = 0$ make supplementary angles with OX. Taking $\tau \equiv x - ty + at^2$, $\alpha \equiv x + ty - 3at^2$ and the circle is

$$y^2 - 4ax = \kappa(x + ty - 3at^2)(x - ty + at^2).$$

Equating coefficients of x^2 and y^2, $-1 = \kappa(1 + t^2)$. The centre is found by partial differentiation wo x and wo y. (See 12·52.) Hence it is given by

$$-4a = \kappa(2x - 2at^2), \quad +2y = \kappa t(-2ty + 4at^2).$$

$$\therefore \quad x = 2a + 3at^2, \quad y = -2at^3.$$

These are parametric equations of the locus of the centre. By change of origin to the point $(2a, 0)$, the locus of the centre becomes $27ay^2 = 4x^3$, a semi-cubical parabola.

EXERCISE 16c

1. Prove that central conics with parallel axes meet in concyclic points.

2. Prove that a central conic with an axis parallel to the axis of a parabola meets the parabola in concyclic points.

3. Prove that the two parabolas through four concyclic points have perpendicular axes.

4. Prove that the points of intersection of

$$x^2 - 4xy + 4y^2 + 42x - 9y - 109 = 0$$

and $\qquad 17x^2 + 12xy + 8y^2 - 46x - 9y - 13 = 0$

are concyclic.

5. $ABCD$ is a cyclic quadrangle; AB, DC are produced to meet at E, and BC, AD are produced to meet at F. Prove that the bisectors of $\angle CEB$ and $\angle CFD$ are at right angles.

6. Two parabolas are drawn through the points of intersection of a circle and a rectangular hyperbola. Prove that the tangents at any one of the four points to the conics form a harmonic pencil.

7. Given a conic drawn on paper, state how to find the positions of its axes.

8. $\{\omega\}$. If $(x/a + y/b - 1)^2 = 2kxy$ is a circle, prove that $a^2 = b^2$ and find k.

9. $\{\omega\}$. Give the general equation of a conic which meets the axes at $(a, 0), (a', 0), (0, b), (0, b')$, and determine whether the conic can be a circle.

16·9. Equations of More General Form

16·91. If $f(x, y) = 0$ and $g(x, y) = 0$ are two curves, the coordinates of every point of intersection of the curves satisfy both equations. They therefore satisfy every equation which can be derived algebraically from $f(x, y) = 0$ and $g(x, y) = 0$. Thus the derived equation represents a curve passing through all the points of intersection.

In the special case of 16·12, $f = 0$ and $g = 0$ were conics and the derived curve $f = \kappa g$ was also a conic. In general, the curves $f = 0$ and $g = 0$ need not be curves of the same type, and the derived equation need not be of the form $f = \kappa g$. When it is of that form, κ may be a function of x, y instead of being a constant. But the derived curve always passes through the common points of $f = 0$ and $g = 0$.

In 4·8 an equation is derived from $\alpha \equiv lx + my + n = 0$ and $s = 0$ which represents the pair of lines joining the origin to the common points of $\alpha = 0$ and $s = 0$. This is an example of the general process of combining the two equations $f = 0$ and $g = 0$, but it is not an example of $f = \kappa g$.

Sometimes the process of combining the equations is modified so that the deductions are not applicable to all the points of intersection. In 16·94 the derived curve passes through three of the common points of the two conics, but not through the fourth point.

16·92. EXAMPLE. Interpret the equation

$$a\alpha^2 = p\beta\gamma + q\gamma\alpha + r\alpha\beta,$$

where α, β, γ are linear in x, y, z and a, p, q, r are constants.

Consider the three lines $\alpha = 0$, $\beta = 0$, $\gamma = 0$ and let their points of intersection be A, B, C. Then the coordinates of B satisfy $\alpha = 0$ and $\gamma = 0$; hence they satisfy the given equation. The same is true of the coordinates of C, but is not true of the coordinates of A. The equation is of the second degree. It represents a conic through B and C but not through A.

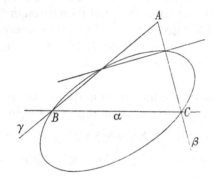

Since the equation can be written

$$p\beta\gamma = \alpha(a\alpha - r\beta - q\gamma),$$

it passes through the point of intersection of

$$\beta = 0, \quad a\alpha = r\beta + q\gamma,$$

and through the point of intersection of

$$\gamma = 0, \quad a\alpha = r\beta + q\gamma.$$

Thus $a\alpha = r\beta + q\gamma$ is the line joining the points other than B and C in which the conic meets AB, AC.

16·93. EXAMPLE. Find the equation of the circle through the points of intersection of $y^2 = x$ and $(x - b)y = c$.

The coordinates of the points of intersection satisfy

$$xy^2 - by^2 = cy$$

and therefore satisfy

$$x^2 - by^2 = cy$$

and

$$x^2 + y^2 - (b + 1)y^2 = cy$$

and

$$x^2 + y^2 = (b + 1)x + cy.$$

This is the equation of the circle.

It should be noticed that this is not an example of $s = \kappa s'$. This is because the circle does not pass through four points of intersection of the given conics. In G_6 the corresponding circle does not pass through the point $(1, 0, 0)$ of intersection of $y^2 = xz$ and $(x - bz)y = cz^2$.

16·94. EXAMPLE. Find the equation of the conics through the points of intersection other than the origin of

$$s \equiv ax^2 + 2hxy + by^2 + 2gx + 2fy = 0$$

and

$$s' \equiv a'x^2 + 2h'xy + b'y^2 + 2g'x + 2f'y = 0,$$

where $fg' \neq gf'$.

The given equations are

$$x(ax + 2hy + 2g) = -y(by + 2f),$$

$$x(a'x + 2h'y + 2g') = -y(b'y + 2f').$$

Coordinates other than $(0, 0)$ which satisfy these equations also satisfy

$$\sigma \equiv (ax + 2hy + 2g)(b'y + 2f') - (a'x + 2h'y + 2g')(by + 2f) = 0.$$

This is therefore the equation of one conic through the three points and it does not pass through the origin. Evidently $\sigma + ks + k's' = 0$ also passes through the three points for all values of k, k'. Since k, k' can be chosen to make

$$\sigma + ks + k's' = 0$$

pass through two other points, any given conic through the three points has an equation of this form.

A system $\sigma + ks + k's' = 0$ of conics is called a *net of conics*.

16·95. *Pascal's Theorem.* If A, B, C, D, E, F are six points on a conic, then AB, DE and BC, EF and CD, FA meet in collinear points.

Let the equations of AB, BC, CD, DE, EF, FA be $\alpha = 0$, $\alpha' = 0$, $\beta = 0$, $\beta' = 0$, $\gamma = 0$, $\gamma' = 0$. Then $\alpha\beta\gamma = k\alpha'\beta'\gamma'$ is a cubic curve, and it passes through the nine points of intersection

$$\alpha = \alpha' = 0, \qquad \alpha = \beta' = 0, \qquad \alpha = \gamma' = 0,$$

$$\beta = \alpha' = 0, \qquad \beta = \beta' = 0, \qquad \beta = \gamma' = 0,$$

$$\gamma = \alpha' = 0, \qquad \gamma = \beta' = 0, \qquad \gamma = \gamma' = 0,$$

since the coordinates of these points make both sides of

$$\alpha\beta\gamma = k\alpha'\beta'\gamma'$$

zero. In the figure the nine points are $B, X, A, C, D, Z, Y, E, F$.

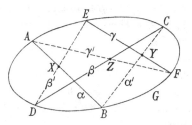

Let k be chosen so that the coordinates of an arbitrarily chosen point G on the conic also satisfy $\alpha\beta\gamma = k\alpha'\beta'\gamma'$.

For this special value of k, the conic and cubic have 10 points in common. It follows from 7·61 that the cubic is degenerate and contains the conic as part of itself; the rest of the cubic is a line. But X, Y, Z lie on the cubic and not on the conic. Hence X, Y, Z are collinear.

The dual of Pascal's Theorem can be proved by a dual method, in which a curve of class 3 is used.

16·96. *The Double Contact Theorem.* If two conics both have double contact with a third conic, then the chords of contact are concurrent with a pair of the common chords of the two conics and separate these chords harmonically.

Let $s = 0$ be the third conic, and let $\alpha = 0$ and $\beta = 0$ be the chords of contact. Then the other conics have equations

$$s = k\alpha^2, \quad s = l\beta^2,$$

and these can be taken to be

$$s = \alpha^2, \quad s = \beta^2,$$

by alteration of α, β by constant factors.

One conic through the points of intersection of $s = \alpha^2$, $s = \beta^2$ is
$$s - \alpha^2 = s - \beta^2,$$
i.e. $(\alpha + \beta)(\alpha - \beta) = 0.$

This is a line pair, and is one of the pairs of common chords. The two lines $\alpha + \beta = 0$, $\alpha - \beta = 0$ separate $\alpha = 0$ and $\beta = 0$ harmonically. Thus the theorem is proved. Its dual can be proved by the dual method.

16·97. Parabola Referred to Tangents as Axes

Using homogeneous geometry, let $A(a, 0, 1)$ and $B(0, b, 1)$ be the contacts of a parabola with the axes of coordinates. The envelope equations of A, B, O are

$$aX + Z = 0, \quad bY + Z = 0, \quad Z = 0.$$

Hence the envelope equation of the parabola is

$$(aX + Z)(bY + Z) = kZ^2.$$

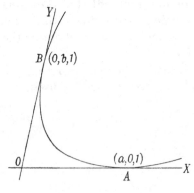

Also $k = 1$, since the equation must be satisfied by $[0, 0, 1]$. Hence the equation is

$$-abXY = Z(aX + bY),$$

or, parametrically,

$$aX : bY : Z = 1 - T : T : T^2 - T.$$

If the locus equation is required, the equations of AB, OB, OA are $x/a + y/b = 1$, $x = 0$, $y = 0$. Hence the equation of the parabola is $(x/a + y/b - 1)^2 = kxy$.

Also $k = 4/ab$, to make the second degree terms a perfect square.

Hence the equation is

$$\frac{x^2}{a^2} - \frac{2xy}{ab} + \frac{y^2}{b^2} - \frac{2x}{a} - \frac{2y}{b} + 1 = 0, \quad \text{or} \quad \left(\frac{x}{a}\right)^{\frac{1}{2}} + \left(\frac{y}{b}\right)^{\frac{1}{2}} = 1,$$

or, parametrically, $x = at^2$, $y = b(1 - t)^2$.

EXERCISE 16D

1. Find the equation of the parabola through the origin with its axis parallel to $x = 0$ and passing through the points of intersection of $y = k$ and $ax^2 + 2hxy + by^2 = 0$.

2. Find the equation of the lines through the origin and the points of intersection of $(x^2 + y^2)^2 = 9(x^2 - y^2)$ and $x^2 + 4y^2 = 4$.

3. Find the equation of the circle through the origin and the points of intersection of $x - y - 2 = 0$ and $x^2 + y + 1 = 0$.

4. Find the equation of the circle through the points of intersection of $y = x^2$ and $xy = x + y$.

5. Show that $xy = 2gx - fy$ passes through the points of contact of tangents from (f, g) to $y^2 = 4ax$.

6. Find the circle through the points of intersection other than the origin of $y = 2x^2 + 3xy$ and $x = 4xy + y^2$.

7. Prove that $xy + (2a - f)y = 2ag$ meets $y^2 = 4ax$ in points P, Q, R concyclic with the vertex. Also show that P, Q, R lie on

$$x^2 + (2a - f)x = \tfrac{1}{2}gy.$$

8. Find the circumcentre, centroid, and orthocentre of the triangle PQR in No. 7.

9. Examine the proof of Pascal's Theorem given in 16·95 when A, E, C lie on one line, and B, D, F on another.

10. $AA'BB'CC'$ is a hexagon circumscribing a conic and the envelope equations of the vertices are $A = 0$, $A' = 0$, $B = 0$, $B' = 0$, $\Gamma = 0$, $\Gamma' = 0$. Show that, if k is properly chosen, the curve

$$A B \Gamma = k A' B' \Gamma',$$

of class 3, has seven common tangents with the conic, and from the fact that it must be degenerate deduce that $A B'$, $C A'$, $B C'$ are concurrent. This is Brianchon's Theorem. It is the dual of 16·95.

11. If two conics both have double contact with a third, prove that the poles of the chords of contact separate harmonically two of the points of intersection of the common tangents of the conics.

12. If an ellipse touches both asymptotes of a hyperbola, prove that two of the common chords are parallel.

13. If each of two intersecting circles has double contact with a parabola, prove that the common chord of the circles is midway between the chords of contact.

In Nos. 14–19, prove the properties of the parabola

$$x = at^2, \quad y = b(1-t)^2.$$

14. The tangents perpendicular to the axes are

$$x + y \cos \omega = ab \cos \omega/(a + b \cos \omega), \quad y + x \cos \omega = ab \cos \omega/(b + a \cos \omega).$$

15. The directrix is $(a + b \cos \omega) x + (b + a \cos \omega) y = ab \cos \omega$.

16. The tangent at the vertex is

$$x/(b + a \cos \omega) + y/(a + b \cos \omega) = ab/(a^2 + 2ab \cos \omega + b^2).$$

17. The focus is $x : y : 1 = ab^2 : a^2 b : a^2 + 2ab \cos \omega + b^2$.

18. The axis is $bx - ay = ab(b^2 - a^2)/(a^2 + 2ab \cos \omega + b^2)$.

19. The latus rectum is $4a^2 b^2 \sin^2 \omega/(a^2 + 2ab \cos \omega + b^2)^{\frac{3}{2}}$.

In Nos. 20–28, x, y, z are cartesian coordinates in three dimensions, $\alpha = 0$ and $\beta = 0$ are (linear) equations of planes, $s = 0$ and $s' = 0$ are (second degree) equations of conicoids, $\sigma = 0$ and $\sigma' = 0$ are equations of spheres of the form $x^2 + y^2 + z^2 + 2ux + 2vy + 2wz + d = 0$. Interpret the given equation.

20. $\alpha = k\beta$.   **21.** $\sigma = k\sigma'$.   **22.** $\sigma = k\alpha$.

23. $\sigma = k$.   **24.** $s = ks'$.   **25.** $s = k\alpha\beta$.

26. $s = k\alpha^2$.   **27.** $\alpha^2 = k\beta^2$.   **28.** $s^2 = k^2 s'^2$.

EXERCISE 16ᴇ

1. Segments LM of length c and $L'M'$ of length c' of the oblique axes are such that LL' and MM' are parallel. Prove that LM' and ML' envelop a parabola which touches the axes.

2. ABC is a triangle right-angled at C. $CA = b$, $CB = a$. Write down the equation referred to axes CA and CB of an arbitrary conic through A, B, C such that its normal at C bisects AB. Find the locus of the centre of this conic.

3. The lines joining $(c, 0)$ to the points where $[X, Y]$ meets $y^2 = 4ax$ meet the curve again in two points. Prove that these points lie on $[X', Y']$ such that $c^2XX' = 1$ and $Y' = -cYX'$. Also if $[X, Y]$ passes through one fixed point, prove that $[X', Y']$ passes through another.

4. If (f_1, g_1), $(-f_1, -g_1)$ and (f_2, g_2), $(-f_2, -g_2)$ are the ends of conjugate diameters of an ellipse, prove that the ellipse is

$$(g_1 x - f_1 y)^2 + (g_2 x - f_2 y)^2 = (g_1 f_2 - f_1 g_2)^2.$$

Also use the properties $CP^2 + CD^2 = a^2 + b^2$ and $p \cdot CD = ab$ to show that the squares of the sum and difference of the semi-axes are

$$(f_1 \pm g_2)^2 + (g_1 \mp f_2)^2.$$

5. A conic through $A(0, 4)$ and $B(7, -3)$ has double contact with $2x^2 + xy + 4y^2 + 6x + 4y + 1 = 0$. Prove that the chord of contact passes through one of two fixed points and find their coordinates. Verify that these points separate A and B harmonically.

6. Find the equation of the circle through the points of intersection other than the origin of $2(x^2 + xy + y^2) = x$ and $6x^2 + 5y^2 = y$.

7. A conic is drawn to touch two pairs of tangents to another conic. Prove that if it touches one chord of contact, it also touches the other.

8. $\{\omega\}$. Find the equation of the rectangular hyperbola which touches the axes at $(a, 0)$ and $(0, b)$.

9. Find the contacts of $pXY + qX + rY + s = 0$ with the axes.

10. Interpret the equations

$$(aX + Z)(bY + Z) = k(cX + dY + Z) Z,$$

$$aX + bY + 2Z = k(cX + dY + 2Z),$$

and use them to find the locus of the centres of conics touching four given lines.

11. Interpret the equation

$$(1 + t^2)(y^2 - 4ax) + (x - ty + at^2)(x + ty + \lambda) = 0,$$

and find the value of λ for which it has three-point contact with $y^2 = 4ax$.

12. Interpret the equation

$$(x_1 X + y_1 Y + z_1 Z)(x_2 X + y_2 Y + z_2 Z) = \lambda(X^2 + Y^2).$$

What is the envelope of a line the product of whose distances from two given points is constant?

13. Show that the feet of the normals from (h, k) to $ax^2 + by^2 = 1$ lie on the hyperbola $(a - b)xy - akx + bhy = 0$.

14. If the normals to $ax^2 + by^2 = 1$ at the ends of the chords which are the polars of $(x_1, y_1), (x_2, y_2)$ meet at (h, k), prove that

$$ax_1 x_2 = -1, \quad by_1 y_2 = -1$$

and $\qquad h : k : b - a = y_1 + y_2 : -x_1 - x_2 : ab(x_1 y_2 + x_2 y_1).$

15. If the normals to $x^2/a^2 + y^2/b^2 = 1$ at the points whose eccentric angles are $\phi_1, \phi_2, \phi_3, \phi_4$ meet in a point (h, k), prove, by comparing the equations

$$(a^2 - b^2)xy + b^2 kx - a^2 hy = 0$$

and $\qquad \dfrac{x^2}{a^2} + \dfrac{y^2}{b^2} - 1 = k\alpha\beta,$

where $\qquad \alpha \equiv \dfrac{x}{a}\cos\tfrac{1}{2}(\phi_1 + \phi_2) + \dfrac{y}{b}\sin\tfrac{1}{2}(\phi_1 + \phi_2) - \cos\tfrac{1}{2}(\phi_1 - \phi_2),$

$$\beta \equiv \dfrac{x}{a}\cos\tfrac{1}{2}(\phi_3 + \phi_4) + \dfrac{y}{b}\sin\tfrac{1}{2}(\phi_3 + \phi_4) - \cos\tfrac{1}{2}(\phi_3 - \phi_4),$$

that $\qquad \phi_1 + \phi_2 + \phi_3 + \phi_4 = (2n + 1)\pi.$

16. In No. 15 show that $k\beta$ can be replaced by

$$\dfrac{x}{a}\sec\tfrac{1}{2}(\phi_1 + \phi_2) + \dfrac{y}{b}\operatorname{cosec}\tfrac{1}{2}(\phi_1 + \phi_2) + \sec\tfrac{1}{2}(\phi_1 - \phi_2) = 0.$$

Deduce that if the normals at ϕ_1, ϕ_2, ϕ_3 are concurrent, then

$$\sin(\phi_2 + \phi_3) + \sin(\phi_3 + \phi_1) + \sin(\phi_1 + \phi_2) = 0.$$

MISCELLANEOUS EXERCISE B

[These are arranged in sets of seven]

1. The joins of the fixed points $(a, 0)$ $(b, 0)$ and $(0, c)$ $(0, d)$ are divided in the same ratio at P and Q. Find the locus of the mid-point of PQ.

2. Two parallel tangents to $(x-a)^2 + y^2 = 0$ form a square with two parallel tangents to $(x+a)^2 + y^2 = b^2$. Prove that each diagonal of the square passes through a fixed point.

3. Circles of constant radius k are drawn through the ends of a variable diameter of $x^2/a^2 + y^2/b^2 = 1$. Prove that the locus of their centres is
$$(x^2 + y^2)(a^2x^2 + b^2y^2 + a^2b^2) = k^2(a^2x^2 + b^2y^2).$$

4. A, B, C, D are concyclic points on a rectangular hyperbola and E is the orthocentre of ABC. Prove that D and E are opposite ends of a diameter of the hyperbola.

5. S is a fixed point and s is a fixed circle. Prove that the envelope of chords of s which subtend a right angle at S is a conic with one focus at S.

6. If a triangle is self-polar wo a parabola, prove that the joins of the mid-points of the sides are tangents to the parabola.

7. Q is any point on $x^2 + y^2 = 2ax$ and OQ meets $x = a$ at R. The lines through Q, R parallel to OY, OX meet at P. Show that the locus of P has parametric equations
$$x : y : a = 2 : t + t^3 : 1 + t^2.$$
Find the points of inflexion and sketch the curve. Also find the locus of the point of intersection of tangents at the ends of chords through $(2a, 0)$.

8. Fixed finite lines AB, CD are divided in the same ratio at P, Q. Find the envelope of PQ.

9. Two circles meet at A and B. A line through A meets one circle at P and the parallel line through B meets the other circle at Q. Show that the locus of the mid-point of PQ is a circle.

10. Prove that the circles through the focus of a parabola touching the curve at the ends of a focal chord cut orthogonally.

11. Find the envelope of the polars of the fixed point (h, k) wo the conic $x^2/a^2 + y^2/b^2 = 1$ when a and b vary in such a way that $a^2 - b^2$ is constant.

12. Lines joining a point P on $x^2/a^2 + y^2/b^2 = 1$ to $(\lambda a, 0), (\lambda' a, 0)$ meet the ellipse again in Q, R. Prove that the envelope of QR is

$$x^2/a^2 + (1 - \lambda\lambda')^2 y^2/\{(1 - \lambda^2)(1 - \lambda'^2) b^2\} = 1$$

and discuss the case $\lambda\lambda' = 1$.

13. A circle has double contact with a hyperbola. From a point P on the hyperbola a line PQ is drawn parallel to one asymptote meeting the chord of contact at Q. Prove that PQ is equal to the length of the tangent from P to the circle.

14. Show that the inverse of a central conic wo a vertex is a cubic curve through I and J.

15. If $\lambda(ax^2 + 2hxy + by^2 + 2x) + \mu(a'x^2 + 2h'xy + b'y^2 + 2y) + 2vxy = 0$ is a line-pair, prove that one line goes through the origin and that the other envelops $(ax + b'y + 2)^2 = 4a'bxy$.

16. Prove that any four concyclic points A, B, C, D can be taken to be $(t_1, 1/t_1), (t_2, 1/t_2), (t_3, 1/t_3), (t_4, 1/t_4)$, where $t_1 t_2 t_3 t_4 = 1$, and verify that $BC . AD \pm CA . BD \pm AB . CD = 0$.

17. If PSQ and $P'SQ'$ are perpendicular focal chords of an ellipse, prove that $(PS . SQ)^{-1} + (P'S . SQ')^{-1}$ is constant.

18. Given the position of the vertex A of a triangle ABC, the size of the angle A, and the line along which BC lies, prove that the circumcentre lies on a hyperbola focus A. Prove the same for the incentre, and find the locus of the nine-point centre.

19. A given circle has centre $(c, 0)$ and radius r. The power of a point $P(x, y)$ wo the circle is $e^2 x^2$ where e is constant. Prove that the locus of P is a conic of eccentricity e, and that the directrices of the conic are given by

$$e^2(1 - e^2) x^2 - 2e^2 cx = r^2.$$

20. Find the equation of the conic through (a, b) touching the axes at $(h, 0)$ and $(0, k)$.

21. Obtain the parametric equations $x : y : 1 = a + bt^2 : at + bt^3 : 1 + t^2$ of the curve $x(x^2 + y^2) = ax^2 + by^2$.

Find the relation between the parameters of two points which are collinear with $(a, 0)$ and show that the circle through two such points and the origin touches the x-axis.

22. A point P has the sum of the squares of its distances from m fixed points equal to the sum of the squares of its distances from n fixed lines. Find the relation between m and n so that the locus of P may be a rectangular hyperbola.

23. Find the length of the common chord of the circles

$$x^2 + y^2 - 2g_1 x - 2f_1 y = 0 \quad \text{and} \quad x^2 + y^2 - 2g_2 x - 2f_2 y = 0.$$

24. The joins of the vertices A, A' of an ellipse to a variable point P of the curve meet the tangents at A', A in Q, R. Find the envelope of QR.

25. AB is a fixed diameter of a rectangular hyperbola. Prove that the circles which touch the curve and pass one through A and the other through B are equal.

26. Prove that the points on $y^2 = 4ax$ nearest to $(3a, 0)$ are the ends of the latus rectum, and find the points nearest to $(4a, 0)$ and $(\frac{1}{2}a, 0)$.

27. The line joining two points conjugate wo a given conic passes through a fixed point, and one of the two points lies on a fixed line. Prove that the other point lies on a conic through the fixed point and through the pole of the fixed line and through the points of intersection of the fixed line and the given conic.

28. A chord of $a^2y = x^3$ passes through a fixed point on the y-axis. Find the locus of the mid-point of the chord.

29. A line parallel to $y = ax$ meets $y = bx$ at B and $y = cx$ at C. Find the equation of the locus of the point P in BC such that BP/PC is a constant, d.

30. Sketch the curves $r = a\cos 3\theta$ and $r = a\cos 5\theta$, and find the degrees of their cartesian equations.

31. Prove that the locus of the circumcentre of the triangle formed by two fixed tangents and a variable tangent of a parabola is a line.

32. Find the eccentricity of the hyperbola

$$(x+y)^2 - (x-y-1)^2 = 5$$

and the length of its latus rectum.

33. Conics are drawn with given vertices. Show that the points of contact of tangents from a given point lie on a hyperbola through the given vertices.

34. Chords of $(px+qy)^2 + 2gx + 2fy = 0$ through the origin make equal angles with the tangent at the origin. Prove that the join of their other ends goes through the point

$$x : y : f^2 + g^2 = f : -g : (qg - pf)(qf + pg).$$

35. The tangent at P to $x(x^2+y^2) = ay^2$ meets the curve at Q and OY at R. The tangent at Q meets OY at S. Prove that $RO = 8OS$.

36. Prove that the envelope of the radical axis of equal circles taken one from each of the systems

$$x^2 + y^2 + 2\lambda x + c^2 = 0$$

and $$x^2 + y^2 - 2\mu y - c^2 = 0$$

is a rectangular hyperbola.

37. Find the angle between the lines
$$(x^2 + y^2)(\cos^2\theta\sin^2\alpha + \sin^2\theta) = (x\tan\alpha - y\sin\theta)^2.$$

38. S is a focus and B is an end of the minor axis of an ellipse with semi-axes a, b. BS is produced to meet the curve at D. Prove that $BD = 2a^3/(2a^2 - b^2)$.

39. The normal at P to the rectangular hyperbola $xy = k^2$ meets the curve again at Q and makes angles θ, ϕ with OX and the normal at Q. Prove that $\tan^2\phi = 4\cot^2 2\theta$. Also show that the pole of PQ wo the conjugate hyperbola lies on $(x^2 - y^2)^2 + 4k^2xy = 0$.

40. A circle through the ends of a diameter of $ax^2 + by^2 = 1$ also touches the curve. Find the locus of its centre.

41. Interpret the equations:

(i) $(lx + my + 1)^2 = (la + mb + 1)^2$,

(ii) $(l^2 + m^2)\{(x-a)^2 + (y-b)^2\} = (la + mb + 1)^2$,

(iii) $(l^2 + m^2)\{(x-a)^2 + (y-b)^2\} = (lx + my + 1)^2$.

42. Show that the pedal of $y^2 = 4ax$ wo the point $(b, 0)$ can be represented by the equations $x : y : 1 = (b-a)t^2 : bt + at^3 : 1 + t^2$ and sketch the curve for $b = -a$.

43. Isosceles triangles QAB are drawn on a given base AB. Find the locus of the point of intersection of the perpendicular from A to QB and the circle on QB as diameter.

44. The tangent at P to the circle $r = 2a\cos\theta$ meets the bisector of the angle XOP at Q. Prove that Q lies on $r\cos 3\theta = 2a\cos^2 2\theta$.

45. Prove that the common diameters of concentric ellipses congruent with $ax^2 + by^2 = 1$ bisect the angles between the major axes and are of lengths $2(a\cos^2\alpha + b\sin^2\alpha)^{-\frac{1}{2}}$ and $2(a\sin^2\alpha + b\cos^2\alpha)^{-\frac{1}{2}}$, where 2α is the angle between the major axes.

46. A variable circle passes through two fixed points A, B. Prove that the loci of the points of trisection of the arcs AB are branches of two hyperbolas.

47. If the tangents from P to a hyperbola separate harmonically the lines through P parallel to the asymptotes, prove that P lies on a similar hyperbola.

48. If the diameter of $ax^2 + 2hxy + by^2 + 2gx + 2fy + c = 0$ parallel to the tangent at P passes through the origin, prove that P lies on
$$g(Cx - G) + f(Cy - F) = 0,$$
where $C = ab - h^2$, $F = gh - af$, $G = hf - bg$.

49. Find the envelope equation of the curve
$$x : y : a = t : 1 : t(1 + t^2)$$
and of the contact of the tangent $7x + y = 4a$.

50. PP' and QQ' are the parts inside a parallelogram of lines parallel to its adjacent sides. Find the locus of the point of intersection of PQ and $P'Q'$.

51. The lengths of the tangents from P to $x^2 + y^2 = r^2$ and $x^2 + y^2 + 2ax + 2by = c^2$, where $0 < r < c$, are in the ratio $\lambda : 1$. Prove that the locus of P is a circle which for all values of λ passes through two real points if $c^2 < r^2 + 2r\sqrt{(a^2 + b^2)}$, and find the value of λ for which the locus cuts $x^2 + y^2 = r^2$ orthogonally.

52. A circle centre (p, q) cuts $y^2 = 4ax$ in four points of which three form an equilateral triangle. Show that the fourth is $(p - 8a, -3q)$. Also find the locus of (p, q).

53. Find the value of k for which the parabola $kx^2 = b - y$ has the closest contact with $x^2/a^2 + y^2/b^2 = 1$.

54. Prove that a chord of $ax^2 + by^2 = 1$ which subtends a right angle at $(0, k)$, where $k \neq 0$, envelops a hyperbola if

$$(a + b)(bk^2 - 1)(abk^2 - a - b) < 0$$

and an ellipse if $(a + b)(bk^2 - 1) < 0$ and $(abk^2 - a - b) < 0$;

also discuss the cases $a + b = 0$, $k^2 = 1/b$, $k^2 = 1/a + 1/b$.

55. The normal at $P(a, b)$ to $x^2/a^2 + y^2/b^2 = 2$ meets the curve again at Q, and the circle on PQ as diameter meets the curve again at R. Show that the equations of PR and QR are

$$ax + by = a^2 + b^2 \quad \text{and} \quad x/a - y/b = 2(a^4 - b^4)/(a^4 + b^4).$$

56. Find the locus of the mid-points of chords of

$$x = at^2 + bt + c, \quad y = a't^2 + b't + c', \quad (ab' \neq ba'),$$

which are parallel to $lx + my = 0$.

57. Find the mid-point of the line joining the origin to the point of intersection of $x/a + y/b = 2$ and $x/c + y/d = 2$, and verify that it is collinear with the mid-points of the other diagonals of the quadrilateral formed by those lines and the axes.

58. What is represented by

$$(ax + by + cz)\{(px + qy + rz)^2 + (lx + my + nz)^2\} = 0$$

and by the corresponding envelope equation? Give separate answers for G_5 and G_6.

59. Verify that $ax^2 - \beta xy + \alpha y^2 + (\beta^2 - 2a\alpha)x - \alpha\beta y + a\alpha^2 = 0$ represents two lines through (α, β) touching $y^2 = 4ax$ and find the angle between them. What is the locus of the point of intersection of tangents to $y^2 = 4ax$ that intersect at a constant angle γ?

60. Find the foci and directrices of the hyperbola

$$(3x + 4y - 150)^2 - 2(4x - 3y + 25)^2 = 31250.$$

61. A triangle has orthocentre (f, g) and is self-polar wo $ax^2 + by^2 = 1$. Prove that its vertices lie on $bfy - agx = (b - a)xy$ and that the sides touch $(fx)^{\frac{1}{2}} + (-gy)^{\frac{1}{2}} = (1/a - 1/b)^{\frac{1}{2}}$.

62. If $\alpha \equiv l_1 x + m_1 y + n_1 z$, $\beta \equiv l_2 x + m_2 y + n_2 z$, $\gamma \equiv l_3 x + m_3 y + n_3 z$ and p, q, r are constants, prove that $\alpha = 0$, $\beta = 0$, $\gamma = 0$ form a triangle self-polar wo the conic $p\alpha^2 + q\beta^2 + r\gamma^2 = 0$.

63. Find the centre of the conic
$$x : y : 1 = t^2 + 4t - 1 : 4t^2 - 9t - 4 : t^2 + 1.$$

64. A_1, A_2 are $(\pm a, 0)$; B_1, B_2 are $(0, \pm b)$; C_1, C_2 are (c, d), $(c/k, d/k)$. $B_1 C_1, B_2 C_2$ meet at A; $C_1 A_1, C_2 A_2$ meet at B. Find the equation of AB and verify that AB passes through the point of intersection of $A_1 B_1, A_2 B_2$.

65. If x, y, z are rectangular cartesian coordinates in three dimensions, what type of surface is represented by an equation
$$x^2 + y^2 + z^2 = f(z)?$$

66. If the tangents at P, Q to an ellipse are at right angles and the normals at P, Q meet the curve in P', Q', prove that $PQ, P'Q'$ are parallel.

67. A circle s touches two fixed lines. Prove that the common chord of s and a fixed circle envelops two parabolas with perpendicular axes.

68. A circle touches a given line and has two given points for conjugate points. Prove that the locus of its centre is a parabola.

69. Prove that the points of contact of tangents from a fixed point to conics of the system $s = ks'$ lie on a cubic curve.

70. Prove that chords of the Folium of Descartes $x^3 + y^3 = 3axy$ which pass through $(\tfrac{3}{2}a, \tfrac{3}{2}a)$ subtend a right angle at the origin.

71. A_1, A_2, A_3 are $(x_1, 0)$, $(x_2, 0)$, $(x_3, 0)$ and B_1, B_2, B_3 are $(0, y_1)$, $(0, y_2)$, $(0, y_3)$. Find the equation of the line through the points of intersection of $A_2 B_3, A_3 B_2$ and $A_3 B_1, A_1 B_3$ and $A_1 B_2, A_2 B_1$.

72. Find the locus of the mid-points of chords of $x^2 + y^2 = 4c^2$ which are at a constant distance c from $(c, 0)$.

73. Prove that the envelope of chords of $y^2 = kx$ which subtend a right angle at $(p, 0)$ is an ellipse if p and k have the same sign. If p and $k + 4p$ have opposite signs, prove that the chords touch a hyperbola. What happens if

(i) $p = 0$,   (ii) $k + 4p = 0$,   (iii) $k(k + 4p) < 0$?

74. If the tangents to $x^2/a^2 - y^2/b^2 = 1$ at its points of intersection P, Q with $xy = k^2$ meet that hyperbola again in R, S, prove that the area of the parallelogram $PQRS$ is $2abk^{-2} \sqrt{(a^2 b^2 + 4k^4)}$.

75. A is a point outside a circle of which ABC is a diameter and AUV a secant. On CU a point P is taken so that $CP = AV$. Prove that the locus of P is an ellipse centre C.

76. Find the inverse wo $x^2 + y^2 = k^2$ of $2x = ax^2 + by^2$ and of its circle of curvature at the origin.

77. If $x : y : z = 3t^2 + t - 1 : 1 - t : 2 + t - 3t^2$, express $t^2 : t : 1$ in terms of x, y, z. Find the locus and envelope equations of the conic given by these parametric equations and find the polar of $(0, 0, 1)$ and the pole of $[0, 0, 1]$.

78. A line given in direction meets two fixed lines at A and B, and the circle on AB as diameter meets the lines again at A' and B'. Find the locus of the point of intersection of AB and $A'B'$.

79. Prove that
$$\{a(x^2 - c) - b(y^2 - c)\}^2 + 4\{axy + h(y^2 - c)\}\{bxy + h(x^2 - c)\} = 0$$
represents the sides of a rhombus, if $h^2 \neq ab$.

80. Perpendiculars PX, PY are drawn from any point P to fixed conjugate diameters of an ellipse, and the perpendicular from P to the polar of P meets XY at Z. Prove that $XZ : ZY$ is constant.

81. Find the locus of the centre of a circle which touches two given circles.

82. A rectangular hyperbola s with centre O cuts a line l in P, Q and the circle OPQ in R, S. When O and l are fixed and s varies, prove that the pole of RS wo s is fixed.

83. Find the envelope of the minor axis of an ellipse which has a given focus and two given tangents.

84. What is the nature of the curve $x^4 + y^4 = axy^2$ near the origin? Obtain the parametric equations $x : y : a = t^2 : t^3 : 1 + t^4$ of this curve and find the condition of collinearity of the points t_1, t_2, t_3.

85. A variable line through (x_1, y_1) cuts $ax^2 + 2hxy + by^2 = 0$ in A, B and the parallelogram $BOAP$ is completed. Find the locus of P.

86. Prove that the locus of the inverse of a fixed point wo a circle of a coaxal system is in general one of the orthogonal circles.

87. Show that, whatever the value of k may be, the same circle circumscribes the triangle formed by the tangents at
$$t_1, t_2, t_3 \quad \text{to} \quad x : y : a = t^2 : 2t : 1,$$
where t_1, t_2, t_3 are the roots of $(t - p)(t - q)(t - r) = k(t^2 + 1)$ and p, q, r are constants.

88. Determine the conic with asymptotes $ax^2 + 2hxy - by^2 = 0$ which cuts $ax^2 + by^2 = 1$ orthogonally.

89. A parallelogram circumscribes $x^2/a^2 + y^2/b^2 = 1$ and has two opposite vertices on $x^2/b^2 + y^2/a^2 = 1$. Prove that the others lie on

$$x^2/a^4 - y^2/b^4 = 1/(a^2 - b^2).$$

90. A, B are fixed points and c, d are fixed lines. PA, PB are separated harmonically by the lines through P parallel to c, d. Prove that P lies on a hyperbola with AB for a diameter having asymptotes parallel to c, d.

91. Show that the condition of collinearity for the points t_1, t_2, t_3 of the curve

$$x : y : 1 = a_0 t^3 + a_1 t^2 + a_2 t + a_3 : b_1 t^2 + b_2 t : c_1 t^2 + c_2 t$$

is $t_1 t_2 t_3 = -a_3/a_0$. If A, B, C and P, Q, R are two sets of collinear points on the curve, prove that AP, BQ, CR meet the curve again in collinear points.

92. P and Q lie on $\gamma = 0$. $p_1\gamma + \alpha = 0$, $p_2\gamma + \alpha = 0$ are lines a_1, a_2 through P and $q_1\gamma + \beta = 0$, $q_2\gamma + \beta = 0$ are lines b_1, b_2 through Q. Prove that the line joining $a_1 b_2$ to $a_2 b_1$ is

$$(p_1 q_1 - p_2 q_2)\gamma + (q_1 - q_2)\alpha + (p_1 - p_2)\beta = 0.$$

93. Prove that the locus of the pole of a fixed line wo a circle which touches two given lines is two conics.

94. Discuss the maximum and minimum values of KP^2, where K is a fixed point $(k, 0)$ and P is a variable point on $y^2 = 4ax$.

95. If k is such that $2xy = k^2$ touches $x^2/a^2 + y^2/b^2 = 1$, show that the points of contact lie on an equiconjugate diameter of the ellipse.

96. A part AB of the given line $lx + my = 1$ subtends a constant angle γ at the origin, O. Prove that the locus of the incentre of the triangle OAB is part of a hyperbola.

97. Three parabolas have foci A, B, C and directrices BC, CA, AB. Prove that their nine common tangents meet in sets of three at the circumcentre, incentre, and ecentres of the triangle ABC.

98. Show that $r^{\frac{1}{2}} \cos \frac{1}{3}\theta = a^{\frac{1}{2}}$ represents the same curve as

$$x : y : a = 1 - 3t^2 : 3t - t^3 : 1$$

and that the parameters of the double point are $\pm \sqrt{3}$. Sketch the curve.

MISCELLANEOUS EXERCISE C

In Nos. 1–56, sketch the cubic curve. No. 1 is an example of $y = ax^3 + bx^2 + cx + d$. Nos. 2–6 are examples of $y^2 = ax^3 + bx^2 + cx + d$. No. 7 is an example of $xy = ax^3 + bx^2 + cx + d$. Nos. 8–56 are examples of $xy^2 + 2ey = ax^3 + bx^2 + cx + d$, which may be written $xy = -e \pm \sqrt{(ax^4 + bx^3 + cx^2 + dx + e^2)}$; in some of the given equations the factor x is included. Newton showed that every cubic could be reduced to one of these four forms. See 2·31.

1. $10y = x^3 + x$ (*Cubical Parabola*).

2. $y^2 = x(x^2 + 2x + 2)$.   3. $y^2 = (x - \frac{3}{2})(x - \frac{1}{2})^2$.

4. $y^2 = (x - 2)(x - 3)(x - 4)$.   5. $y^2 = (x - 5)(x - 6)^2$.

6. $y^2 = (x - 7)^3$ (*Semi-cubical Parabola*).

7. $5xy = x^3 + x^2 + 1$ (*Trident*).   8. $4(xy + 1)^2 = (x + \frac{1}{2})(x + 2)^3$.

9. $(xy + 1)^2 = (x + 1)(x + \frac{1}{4})(x + 2)^2$.

10. $(xy + 2)^2 = (x - \frac{1}{2})(x - 1)(x - 2)(x - 4)$.

11. $(xy + 1)^2 = (x - \frac{1}{2})(x - 2)(x - 1)^2$.

12. $4(xy + 1)^2 = (x - 1)(x - 4)(x^2 + 1)$.

13. $4(xy + 1)^2 = (x - 1)(x - 4)(x + 1)^2$.

14. $(xy + 1)^2 = (x^2 - 1)(x - \frac{1}{2})(x + 2)$.

15. $xy^2 = (x - 4)^3$.   16. $xy^2 = (x - 4)(x - 5)^2$.

17. $xy^2 = (x - 4)(x - 2)^2 + 1$.   18. $xy^2 = (x - 4)(x - 2)^2$.

19. $xy^2 = (x - 4)(x - 2)^2 - 1$.   20. $xy^2 = (x - 4)(x + 3)^2 - 1$.

21. $xy^2 = (x - 4)(x + 3)^2$.   22. $xy^2 = (x - 4)(x + 3)^2 + 1$.

23. $xy^2 = (x - 4)(x - \frac{1}{2})^2$.   24. $xy^2 = (x - \frac{2}{5})(x - \frac{3}{5})(x - 4)$.

25. $xy^2 = 2x^3 + 8x^2 + 8x + 1$.   26. $xy^2 = (x - 4)(x - 1)^2$.

27. $xy^2 = 2x^3 + 4x^2 + 2x + 1$.

28. $(xy + 1)^2 = (2 - x)(x + \frac{1}{2})(x + 1)^2$.

29. $(xy + 1)^2 = (\frac{1}{2} - x)(2 + x)(1 - x)^2$.

30. $(xy + 1)^2 = (x + 1)(x - \frac{1}{2})(x - 1)(2 - x)$.

31. $(xy + 1)^2 = (\frac{1}{2} + x)(2 - x)(1 - x)^2$.

32. $(xy + 1)^2 = (1 + x)(1 - x)^3$.   33. $xy^2 = (1 - x)^3$.

34. $xy^2 = (2 - x)(1 - x)^2$.

35. $xy^2 = (1 - x)(2 - x)(3 - x)$.

36. $xy^2 = (1-x)(2-x)^2, \quad xy^2 = (8-x)(1+x)^2.$

37. $xy^2 = 2+3x^2-x^3, \quad xy^2 = (20-x)(1+x)(2+x).$

38. $2(xy+1)^2 = (1-x)(1+x)(2+x).$

39. $2(xy+1)^2 = (2-x)(1-x)^2.$ **40.** $(xy+1)^2 = (1-x)(1+x^2).$

41. $4(xy+1)^2 = (1-x)(2-x)^2.$

42. $6(xy+1)^2 = (1-x)(2-x)(3-x).$

43. $2(xy+1)^2 = (2-x)(1-x)^2.$ **44.** $(xy+1)^2 = (1-x)^3.$

45. $xy^2 = x^2+x-\frac{1}{10}, \quad xy^2 = x^2-x+\frac{1}{10}.$

46. $xy^2 = (x-1)^2.$ **47.** $xy^2 = 1+x^2.$ **48.** $xy^2 = (x+1)^2.$

49. $xy^2 = x^2+x+\frac{1}{10}.$ **50.** $x(y^2-1) = y+\frac{3}{2}.$

51. $x(y^2-1) = y.$ **52.** $x(y^2-1) = 1.$ **53.** $x(y^2+1) = y-1.$

54. $x(y^2+1) = 2$ (*Witch of Agnesi*).

55. $xy^2 = y-1.$ **56.** $xy^2 = 1.$

In Nos. 57–73, sketch the curve.

57. $y(x-2)(x-3)^3 = x(x-1)^2.$ **58.** $y(x-2)(x-4)^2 = x(x-1)^3.$

59. $y(x-1)(x-3)^4 = x(x-2)^2.$

60. $y(x-2)(x-4)^2 = x(x-1)^2(x-3)^3.$

61. $y(x-1)(x-3)^2(x-6)^3(x-8) = x(x-2)(x-4)^3(x-5)(x-7)^2.$

62. $y^2 = (x-1)(x-2)^4.$ **63.** $x^2y^2 = y-1.$

64. $x^2(xy-1)^2 = x-1.$ **65.** $(x^2+y^2-3x)^2 = 4x^2(2-x).$

66. $y^2 = x^5-x^6.$ **67.** $x^4+y^2 = 2xy.$

68. $x^4+y^4 = 2xy.$ **69.** $x^5+y^5 = 5xy^2.$

70. $x^6 = y^4-y^6.$ **71.** $x^6 = y^2-y^6.$ **72.** $y = e^{-x^2}.$

73. $y = e^{1/x}.$

In Nos. 74–78, show how the curve changes as a increases from $-\infty$ to $+\infty$.

74. $y(x^2-1) = x+a.$ **75.** $x^5+y^5 = xy(x+ay).$

76. $y^2x = (x-4)(x-a)^2.$

77. $(x^2+16y^2-16)(x^2+y^2-4) = a.$

78. $y(2x^2+ax+a) = ax^2+ax+2.$

In Nos. 79–93, sketch the curve.

79. $r = a\theta$ (*Spiral of Archimedes*).

80. $r\theta = a$ (*Reciprocal Spiral*).   **81.** $r^2\theta = a^2$.

82. $r(1 + \theta) = a\theta$.   **83.** $r = 1 - 1/\theta$.

84. $2\theta = r + 1/r$.   **85.** $r = e^{a\theta}$.

86. $r = \theta e^{a\theta}$.   **87.** $r = a \sin 5\theta$.

88. $r = a \operatorname{cosec} \theta$.   **89.** $r = a \sin 4\theta$.

90. $r^2 = \tan \theta + 4$.

91. $r = a \pm 2a \sin \frac{1}{2}\theta$ (*Freeth's Nephroid*).

92. $r = 4 - a \cos \theta$, for $a = 1, 2, 3, 4, 8$.

93. $r^n = a^n \cos n\theta$, for $n = \pm \frac{1}{2}, \pm 1, \pm 2, \frac{5}{7}$.

MISCELLANEOUS EXERCISE D

1. Q is any point on $x^2 + y^2 = 2ax$ and OQ meets $x = a$ at R. Lines through Q, R parallel to $x = 0, y = 0$ meet at P. Find the equation of the locus of P and sketch it.

2. Obtain the parametric equations $x : y : a = 2 : t + t^3 : 1 + t^2$ of the curve (*Witch of Agnesi*) in No. 1, and use them to find the coordinates of the points of inflexion.

3. Find the locus of the point of intersection of tangents to

$$x(y^2 + a^2) = 2a^3$$

at the ends of a chord through $(2a, 0)$.

4. Obtain parametric equations of $x(x^2 + y^2) = ax^2 + by^2$. Prove that the circle through the origin and two points of this curve collinear with $(a, 0)$ touches $y = 0$. Sketch the curve for $a = 3b$ and for $a = -3b$.

5. Find the inverse of $xy = k^2$ wo $x^2 + y^2 = 2k(x + y)$.

6. Obtain the equation of the locus of a point the product of whose distances from $(a, 0)$ and $(-a, 0)$ is c^2.

7. Sketch the loci in No. 6 (*Ovals of Cassini*) for

(i) $c < a$, (ii) $c = a$, (iii) $a < c < a\sqrt{2}$, (iv) $c > a\sqrt{2}$.

8. Find the polar equation of the *lemniscate* in No. 7 (ii).

9. Find the points of intersection of the lemniscate

$$(x^2 + y^2)^2 = k^2xy$$

and the circle $x^2 + y^2 = ktx$, and deduce parametric equations of the lemniscate.

10. Find the conditions for the points t_1, t_2, t_3, t_4 of the curve

$$x : y : a = t : t^3 : 1 + t^4$$

to be (i) collinear, (ii) concyclic. Also find the condition for t_1, t_2, t_3 to be collinear.

11. A circle meets a lemniscate in A, B, C, D; AB meets the curve again at P and Q; CD meets the curve again at R and S. Prove that P, Q, R, S are concyclic.

12. Find the number of tangents of the curve

$$x = zt/(1 + t^4), \quad y = zt^3/(1 + t^4)$$

which pass through a given point.

13. Tangents to a rectangular hyperbola intersect at a constant angle. Prove that the locus of the point of intersection is an Oval of Cassini.

In Nos. 14–25, prove the properties of the *Cissoid of Diocles*. A is the centre and OB, CD are fixed perpendicular diameters of a circle of radius a. G is a variable point on the circle, E is the point such that the arcs GC, CE are equal, and OE meets the lines through G and B parallel to CD at P and F. The locus of P is the cissoid.

14. With O as pole, show that the equation of the cissoid is

$$r = 2a(\sec \theta - \cos \theta).$$

15. Obtain the cartesian equation and the parametric equations

$$x : y : 2a = t^2 : t^3 : 1 + t^2.$$

16. Show that in the neighbourhood of O the curve is approximately the same as $x^3 = 2ay^2$. Also show that $x = 2a$ is an asymptote. Sketch the curve.

17. Prove that the pedal of a parabola wo the vertex is a cissoid. Find the asymptote and show in a diagram the position of the cissoid in relation to the parabola.

18. Prove that the inverse of $y^2 = -4bx$ wo $x^2 + y^2 = 4b^2$ is the same as the pedal of $y^2 = 4bx$ wo the origin.

19. Find the equation of the tangent to $x : y : 2a = t^2 : t^3 : 1 + t^2$ and the parameter of the point where it meets the curve again.

20. Show that the cissoid $x(x^2 + y^2) = 2azy^2$ passes through I and J and find the points where the tangents at I and J meet one another and the curve.

21. A rod LMN bent at right angles at M moves so that LM passes through a fixed point K, and N describes a fixed line whose perpendicular distance from K is equal to MN. Prove that the mid-point of MN describes a cissoid.

22. If a parabola rolls without sliding on an equal fixed parabola, starting with contact at the vertices, show that the locus of the vertex is a cissoid.

23. Show that the locus of the poles of tangents to

$$x : y : 2a = t^2 : t^3 : 1 + t^2$$

wo $x^2 + y^2 = k^2$ and the envelope of polars of points of the cissoid wo the same circle are both the same semi-cubical parabola.

24. Find the condition for the four points t_1, t_2, t_3, t_4 of the cissoid to be concyclic and determine the other point of intersection with the cissoid of the circle which has three-point contact at the point t.

25. Find the equation of the common chord of the cissoid and the circle in No. 24, and find the envelope of the chord when t varies.

26. A circle of radius b rolls without sliding on the outside of a fixed circle of radius a. Verify that the *epicycloid* which is the locus of a point fixed on the circumference of the rolling circle can be represented by $x = (a+b)\cos t - b\cos(a+b)t/b, \quad y = (a+b)\sin t - b\sin(a+b)t/b.$

Sketch the epicycloids given by $a = 2b$, $a = 3b$, $3a = 2b$. If $pa = qb$, where p and q are positive integers with no common factor, find the number of cusps of the epicycloid.

27. A circle of radius b rolls without sliding on the inside of a fixed circle of radius a greater than b. Find parametric equations of the *hypocycloid* which is the locus of a point fixed on the circumference of the rolling circle. Sketch the hypocycloids given by $a = 2b$, $a = 3b$, $a = 4b$.

28. Prove that the three-cusped hypocycloid given by $a = 3b$ in No. 27 is a curve of order 4 which has rational parametric equations.

29. Prove that the four-cusped hypocycloid given by $a = 4b$ in No. 27 is a curve of order 6 and class 4.

30. Find the envelope of a line of constant length whose ends are on the rectangular axes.

31. A circle of radius b rolls without sliding on the circumference of a fixed circle of radius a less than b, the contact being internal. Show that the curve obtained as the locus of a point fixed on the circumference of the rolling circle is an epicycloid which can be obtained by rolling a circle of radius $b-a$ on the outside of the circle of radius a.

32. P is a variable point on a fixed circle, centre O, and the perpendicular bisector of OP meets a fixed radius of the circle at Q. Prove that the envelope of PQ is a two-cusped epicycloid.

33. P is a variable point on a fixed circle centre O, and Q is the point on a fixed radius of the circle such that $\angle OQP = \frac{1}{2}\angle QOP$. Prove that the envelope of PQ is a three-cusped hypocycloid.

34. Show that any tangent to the three-cusped hypocycloid

$$x = a(2\cos t - \cos 2t), \quad y = a(2\sin t + \sin 2t)$$

meets the curve again in points at a distance $4a$ apart, and that the tangents at these points are at right angles and meet on the circle $x^2 + y^2 = a^2$.

In Nos. 35–47, prove the properties of the *cardioid* $r = 2a(1-\cos\theta)$.

35. The cardioid is the inverse of a parabola wo the focus, and the pedal of a circle wo a point on the circumference. Sketch the cardioid.

36. When a circle of radius a rolls without sliding on a fixed circle of radius a, the locus of a point fixed on the circumference of the rolling circle is a cardioid.

37. The locus of the ends of a rod of length $4a$ which passes through the origin and has its mid-point on the fixed circle $x^2 + y^2 - 2ax = 0$ is a cardioid.

38. Chords of a cardioid through the cusp are of constant length, and the tangents at their ends meet at right angles on a fixed circle.

39. The cardioid is a quartic curve having double intersections with the line at infinity at I and J; and I, J are cusps the tangents at which meet at $(-a, 0)$.

40. Find the points of intersection of the cardioid
$$(x^2 + y^2 + 2ax)^2 = 4a^2(x^2 + y^2)$$
and the circle $x^2 + y^2 = \lambda y$, and obtain the parametric equations
$$x : y : 4a = t^2 - 1 : 2t : (t^2 + 1)^2$$
of the cardioid. Account for the eight points of intersection of the corresponding cardioid and circle in G_6.

41. The class of the cardioid is 3.

42. Find an equation for the parameters of the contacts of tangents to the cardioid $x : y : 4a = t^2 - 1 : 2t : (t^2 + 1)^2$ that are parallel to $y = kx$.

43. Obtain the parametric envelope equations
$$4aX = 1 - 3t^2, \quad 4aY = t(t^2 - 3)$$
of the cardioid and determine the double tangent.

44. The points t_1, t_2, t_3 of $x : y : 4a = t^2 - 1 : 2t : (t^2 + 1)^2$ are collinear if $t_1 t_2 t_3 \Sigma t + \Sigma t_1^2 + \Sigma t_1 t_2 + 3 = 0$.

45. The locus of a point P such that the three tangents from P to a cardioid have collinear contacts is a circle.

46. The intercepts made on the double tangent by three parallel tangents subtend $\tfrac{1}{3}\pi$ at $(-a, 0)$.

47. The locus of the point of intersection of perpendicular tangents to a cardioid is a degenerate curve consisting of a circle and a *Pascal's limaçon* whose equation referred to a certain point as pole is
$$2r = 3a(\sqrt{3} - 2\cos\theta).$$

48. A is the fixed point $(a, 0)$ and P is the variable point $(a\cos t, a\sin t)$. PQ is the line
$$x\sin(t + s) - y\cos(t + s) = a\sin s$$
making an angle s with OP. Show that if $s = t$, the envelope of PQ is a *nephroid* (two-cusped epicycloid).

Also find the envelope when $s = 2t, \tfrac{3}{2}t, \pi - 2t, \pi - \tfrac{3}{2}t, \tfrac{1}{2}\pi - 2t$.

49. AB is a fixed diameter of a circle and N is the projection on AB of a variable point P on the circumference. Lengths PQ, PQ' equal to PN are cut off along the tangent at P. Show that the loci of Q and Q' are cardioids. If the tangents at Q, Q' to these cardioids meet at T and the normals meet at S, show that the locus of T is a nephroid and that the envelope of TS is another nephroid. Also show that the envelopes of $QN, Q'N$ are three-cusped hypocycloids, and if they touch their envelopes in R, R', show that the envelope of RR' is an *astroid* (four-cusped hypocycloid).

50. CPQ is a secant of the limaçon $r = a - b \cos \theta$ drawn through the point $C((a^2 - b^2)/2b, 0)$. Prove that $CP \cdot CQ$ is constant and show that the locus of the mid-point of PQ is another limaçon.

ANSWERS

EXERCISE 1A

5. $(x_1 + \kappa x_2)/(1 + \kappa)$.　　　　**7.** (3), (-4).

8. $(\frac{3}{2})$.　　　　　　　**9.** None.

EXERCISE 1B

1. 6.　　　　**2.** 1.　　　　**3.** 1.

4. 4.　　　　**5.** 20.　　　　**6.** 7.

7. $(\kappa_1 x_1 - \kappa_2 x_2)/(\kappa_1 - \kappa_2)$.　　　**8.** $(\frac{3}{2}b - \frac{1}{2}a)$, $(2a - b)$.

EXERCISE 1C

1. $1, -2, -\frac{1}{2}, 0$.　　　　　**2.** $B, -1$.

4. $(1, 0)$, $(0, 1)$, $(1, 1)$, $(1, -2)$, $(2, -1)$.

5. Yes.　　　　　　**6.** $(1, -1)$.

7. $(a - b)(c - d)/(a - d)(c - b)$, $(\lambda - \mu)(\nu - \rho)/(\lambda - \rho)(\nu - \mu)$.

EXERCISE 1D

1. 0, 3VG.　　　　　**4.** 0.

5. $P = 0$ or $Q = 0$ or $\theta = \frac{1}{2}\pi$.　　**6.** $2 + 3\cos\omega$.

7. 0.　　　　　　**8.** 25.

EXERCISE 1E

1. $\sqrt{(x^2 + y^2)}$.　　**2.** $\sqrt{(x^2 + xy + y^2)}$.　　**3.** $\sqrt{(x^2 - xy + y^2)}$.

4. $\sqrt{(x^2 + xy\sqrt{2} + y^2)}$.　**5.** 5.　　　**6.** $\sqrt{5}$.

7. $\sqrt{\{(a - c)^2 + (b + d)^2\}}$.　　　　**8.** $\sqrt{37}$.

9. $5\sqrt{2}$.　　**10.** $\sqrt{\{(a + c)^2 + (b - d)^2\}}$.　　**11.** 109.

12. 25.　　　**13.** $7, -17$.　　　**15.** $5, 10\sqrt{2}, 7\sqrt{5}$.

16. 272.　　　**18.** $(3\frac{1}{4}, -1\frac{1}{4})$.　　**19.** $2\sqrt{(5 - 2\sqrt{3})}$.

20. $2\sqrt{(5 - 2\sqrt{2})}$.　　**21.** $2\sqrt{7}$.　　　**22.** $\frac{2}{3}\pi$.

EXERCISE 1F

1. $(3\frac{1}{2}, 6)$.　　**2.** $(1, -1)$.　　　**3.** $(0, \frac{3}{2}b)$.

4. (a, b).　　**6.** $(7, 8)$.　　　**7.** $(\frac{7}{5}, \frac{22}{5})$.

8. $(-2, -9)$.　　**9.** $(2, -3)$.　　**10.** $(18, 3)$.

11. $((a^2 + b^2)/2a, (a^2 + 2ab - b^2)/2a)$.

12. $(-15, 1)$, $(12, -11)$.　　　　**13.** $(0, 5)$, $(-2, 2)$.

14. $(\frac{1}{2}, 63\frac{1}{2})$, $(-1, 46)$, $(-2\frac{1}{2}, 28\frac{1}{2})$. **15.** $(4, 11)$.

16. $(\frac{5}{3}, 6)$, $(-1\frac{4}{5}, 3\frac{2}{5})$. **17.** $4:3$. **18.** $-5:2$.

19. $-11:2$. **21.** $(5, 0)$. **22.** $(1, 3)$.

23. $(\frac{23}{14}, -\frac{13}{7})$. **24.** $(8, -4\frac{1}{4})$.

25. $((t_1 t_2 t_3 \Sigma t)/\Sigma t_1 t_2, 3t_1 t_2 t_3/\Sigma t_1 t_2)$. **26.** $(\frac{1}{3}, -4\frac{1}{3})$.

27. $(110, -149)$, $(-85, -14)$, $(70, 71)$.

EXERCISE 1 G

2. $(2, \frac{1}{3}\pi)$. **3.** $(2, -\frac{1}{3}\pi)$. **4.** $(2, -\frac{2}{3}\pi)$.

5. $(2, \frac{2}{3}\pi)$. **6.** $(\sqrt{(a^4+b^4)}, -\tan^{-1} b^2/a^2)$.

7. $(\sqrt{(a^4+b^4)}, \pi-\tan^{-1} b^2/a^2)$. **8.** $(\sqrt{(a^4+b^4)}, -\pi+\tan^{-1} b^2/a^2)$.

9. $(\sqrt{(x^2+y^2)}; \tan^{-1} y/x$ if $x>0, \pi+\tan^{-1} y/x$ if $x<0<y, -\pi+\tan^{-1} y/x$ if x and $y<0)$.

10. $(\sqrt{3}+1, \sqrt{3}-1)$. **11.** $(-\sqrt{3}-1, \sqrt{3}-1)$. **12.** $(\sqrt{3}+1, -\sqrt{3}+1)$.

13. $(4, 3)$. **14.** $(-4, -3)$. **15.** $(-12, -5)$.

16. $(-99, 20)$. **17.** $(\rho^2 \cos\phi, \rho^2 \sin\phi)$.

18. $(-\rho^2 \cos\phi, -\rho^2 \sin\phi)$. **19.** $\sqrt{(13-6\sqrt{3})}$.

20. 7. **21.** $|\rho|$. **22.** 8.

24. $(3\cos 20°, 90°)$.

25. $r = \frac{1}{2}\sqrt{\{r_1{}^2 + r_2{}^2 + 2r_1 r_2 \cos(\theta_1 - \theta_2)\}}$,
$\cos\theta : \sin\theta : 1 = r_1 \cos\theta_1 + r_2 \cos\theta_2 : r_1 \sin\theta_1 + r_2 \sin\theta_2 : 2r$.

EXERCISE 1 H

1. $7\frac{1}{2}$. **2.** $22\frac{1}{2}$. **3.** 3.

4. $12\frac{1}{2}$. **5.** 53. **6.** $\sqrt{3}$.

7. 142. **8.** $|ac|$.

9. $\frac{1}{2}|(b-c)(c-a)(a-b)|$. **10.** $\frac{1}{2}|(m-n)(n-p)(p-m)|$.

11. $10\frac{1}{2}$. **12.** 31. **13.** $3\frac{3}{4}$.

14. 1176. **15.** 96. **22.** 2.

23. $10, -3$. **26.** $\frac{1}{2}|ab|\sin\omega$. **27.** $\frac{1}{2}|b(a-c)|\sin\omega$.

29. $\frac{1}{2}$. **30.** $16+\frac{5}{2}\sqrt{3}$.

EXERCISE 1 I

3. $(2\triangle/3a, 2\triangle/3b, 2\triangle/3c)$. **4.** $(\frac{1}{2}a\cot A, \frac{1}{2}b\cot B, \frac{1}{2}c\cot C)$.

5. $(a\cos B\cos C \operatorname{cosec} A, b\cos C\cos A \operatorname{cosec} B, c\cos A\cos B \operatorname{cosec} C)$.

6. On BC; $(2\triangle/a, 0, 0)$; $(0, \frac{1}{2}a\sin C, \frac{1}{2}a\sin B)$.

7. $(\frac{36}{41}, \frac{42}{41}, \frac{48}{41})$. **9.** $(0, 2\triangle b/(b^2-c^2), 2\triangle c/(c^2-b^2))$.

10. $P\alpha + Q\beta + R\gamma$. **12.** $(\gamma \operatorname{cosec} B, \alpha \operatorname{cosec} B)$.

EXERCISE 1J

2. $(a, b, c) \div (a+b+c)$.

3. $(-a, b, c) \div (-a+b+c)$, etc.

4. $(\sin 2A, \sin 2B, \sin 2C) \div 4 \sin A \sin B \sin C$.

5. $(\tan A, \tan B, \tan C) \div \tan A \tan B \tan C$.

6. On BC; $(1, 0, 0)$; $(0, \frac{1}{2}, \frac{1}{2})$.

8. $2\triangle(Px_1/a + Qx_2/b + Rx_3/c)$.

EXERCISE 1K

4. $x^2 + y^2 + z^2 + 2yz \cos \lambda + 2zx \cos \mu + 2xy \cos \nu$.

$\Sigma(x_1 - x_2)^2 + 2\Sigma(y_1 - y_2)(z_1 - z_2) \cos \lambda$.

6. On plane YOZ; on line OX.

7. Cube.

8. Octahedron.

9. $\rho = \sqrt{(x^2 + y^2)}$, $\cos \phi : \sin \phi : 1 = x : y : \rho$.

10. On cylinder, axis OZ; on plane through OZ; on line parallel to OZ.

11. $r = \sqrt{(x^2 + y^2 + z^2)}$, ϕ as in No. 9, $\cos \theta : \sin \theta : 1 = z : \sqrt{(x^2 + y^2)} : r$.

12. On sphere, cone, plane, generator of cone, great circle of sphere, small circle.

EXERCISE 2A

23. $r = a \cos \theta$.

24. $(r \cos \theta - a)(r \sin \theta - b)(\sin \theta - c \cos \theta) = 0$.

25. $4r^2 = c^2 \sin 4\theta$.

26. $r = a(1 + \cos \theta)$.

27. $x^2 + y^2 = a^2$.

28. $x \sin \alpha = y \cos \alpha$.

29. $y = a$.

30. $x^2 + y^2 = ax$.

31. $(x^2 + y^2)^2 = 2a^2xy$.

32. $x \sin \{\pi/\sqrt{(x^2 + y^2)}\} = y \cos \{\pi/\sqrt{(x^2 + y^2)}\}$.

33. $x^2/a^2 + y^2/b^2 = 1$.

34. $(x^2 + y^2)(x - b)^2 = a^2y^2$.

EXERCISE 2B

1. $3, 2, 4, 6, 4$.

4. $x - y = 2$.

5. $2x + 3y = 2$.

6. $x + 7y = 11$.

7. $xy^2 = 2$.

8. $27x^4 = 16ay^3$.

9. $x^2 - y^2 = 1$.

10. $(x+3)^2/16 + (y-4)^2/9 = 1$.

11. $c^2x = ay^2 + cby$.

12. $(x^2 + y^2 - a^2)^3 + 27x^2y^2a^2 = 0$.

13. $x^2 - y^2 = 1$.

14. $x = t, y = t^3$.

15. $x = at^3, y = at^2$.

16. $x = 3 \cos t, y = \frac{3}{2} \sin t$.

17. $x = 3t - 1, y = 2 - 2t$.

18. $x = 6t - 1, y = 5t - 2$.

19. $x = bt, y = -at - c/b$.

20. $x = t, y = 1/t$.

21. $x = at, y = a/t^2$.

22. $x = 1 + t, y = t + t^2$.

EXERCISE 2c

8. Origin.   **9.** OX.   **10.** $y = x$.

11. Axes.   **12.** $x + y = 0$.   **13.** Origin.

14. Origin.   **15.** OX.   **16.** $y = n\pi$.

21. Line parallel to OY.   **22.** Two lines parallel to OY.

23. Lines parallel to OY.   **24.** Lines parallel to OX.

25. Lines through O.   **26.** Axes.

EXERCISE 2d

1. Line $8x + 14y = 33$.   **2.** Line $16x + 2y + 15 = 0$.

3. Circle $x^2 + y^2 + 10x + 9 = 0$.

4. Circle $3(x^2 + y^2) + 22x - 38y + 126 = 0$.

5. Circle $x^2 + y^2 + 14x + 13 = 0$.

6. Circle $x^2 + y^2 + 15x + 11y - 26 = 0$.

7. Parabola $y^2 = 10x - 25$.   **8.** Parabola $x^2 + 4x = 10y - 29$.

9. Ellipse $8x^2 + 9y^2 - 18x - 18y + 18 = 0$.

10. Parabola $x^2 = 6x + 12y + 15$.   **11.** Parabola $y^2 = 4ax$.

12. Circle $x^2 + y^2 + x - y - 5 = 0$.   **13.** Line $2x - 18y + 39 = 0$.

14. Two lines $16x - 12y = 107$, $16x - 12y = 93$.

15. Circle $x^2 + y^2 = c^2 - a^2$.   **16.** Line $x = -c^2/2a$.

17. Circle $x^2 + y^2 + a^2 + 2ax(c^2 + 1)/(c^2 - 1) = 0$.

18. Line $y = 0$ between $x = \pm a$, and the branch of $y^2 = (3x - a)(x + a)$ for which $x \geqslant \frac{1}{3}a$.

19. $x = \frac{1}{3}a$ and $y = 0$.

20. Parts of circles given by $x^2 + y^2 - a^2 = 2 \mid ay \mid$.

21. Circle $x^2 + y^2 = \frac{1}{4}c^2$.   **22.** Line $x + y = \frac{1}{2}c$.

23. Ellipse $(l^2x^2 + m^2y^2)(l + m)^2 = c^2l^2m^2$.

24. Hyperbola $4xy = c^2$.

25. Ellipse $9x^2 - 12xy \cos \omega + 4y^2 = 36c^2$.

26. $r(\sec \theta + \operatorname{cosec} \theta) = c$.   **27.** $x + y = \frac{1}{2}c \sec^2 \frac{1}{2}\omega$.

28. $x^2 + y^2 \pm xy\sqrt{3} = a^2$, with the fixed lines as axes.

29. Line $x + y + c = 0$.   **30.** Line $2x + 3y = 8$.

31. Circle $x^2 + y^2 = x$.   **32.** Line $x + y = 1$.

EXERCISE 2ᴇ

1. $(-1, 1), (7, -7), (a-4, b-5), (-4, -5)$.
2. $(5, 7), (0, 6), (-7, -1), (a-3, b+3)$.
3. $2x+3y+4 = 0, y = x^2+2x$.   4. $x+y = 0, x^2+(y+6)^2 = 1$.
5. $(-6, 1), (-6, 8), (a-c, b-d)$.   6. $(-3, 2), (1, -1), (-\frac{3}{2}, 1)$.
7. $(\sqrt{3}, 1), (\frac{1}{2}\sqrt{3}-\frac{1}{2}, -\frac{1}{2}\sqrt{3}-\frac{1}{2}), (\frac{1}{2}a\sqrt{3}+\frac{1}{2}b, -\frac{1}{2}a+\frac{1}{2}b\sqrt{3})$.
8. $(1, \sqrt{3}), (\frac{1}{2}\sqrt{3}+\frac{1}{2}, \frac{1}{2}-\frac{1}{2}\sqrt{3}), (\frac{1}{2}c\sqrt{3}-\frac{1}{2}d, \frac{1}{2}c+\frac{1}{2}d\sqrt{3})$.
9. $x'+y' = 0, 2x'^2-2x'y'+y'^2 = 1$.
10. $2x'-2y'\sqrt{3}+3 = 0, (x'+y'\sqrt{3})^2 = 8(x'\sqrt{3}-y')$.
11. $2y'\sqrt{2} = x'\sqrt{2}+1, -2x'y' = a^2, x'^2-y'^2 = 2c^2, 4x'^2+3y'^2 = 12$.
12. $\frac{1}{2}\cot^{-1}20$.
13. $(b, -a-b\sqrt{2}), x' = (y'+x'\sqrt{2})^2$.
14. $(1, 3), 9x^2+4y^2 = 36$.   15. $3x^2+y^2-10 = 0$.
16. Rotation $\alpha+\beta$.   22. Rotation $-\frac{1}{2}\pi$ about origin.
23. Image in $x = y$.   24. Image in OX.
25. Rotation $+\frac{1}{2}\pi$ about origin.   26. Treble scale of x.
27. Double scale of y and take image in OX.
28. Multiply scales by c.
29. Translate to (b, c) after multiplying scales by a.

EXERCISE 2ꜰ

1. Plane bisecting angle between ZOX, YOZ.
2. Planes YOZ, XOY.   3. Three coordinate planes.
4. Plane parallel to YOZ.   5. Plane parallel to XOY.
6. Planes parallel to XOZ.   7. Unit sphere, centre O.
8. Cylinder axis OZ.   9. OZ.
10. Bisector of XOY.
11. Parallel to line in No. 10, through $(0, 0, 1)$.
12. Line equally inclined to the axes.
13. Circle in plane YOZ, centre O.
14. Circle: section of No. 7 by No. 1.
15. Line OZ and plane XOY.

16. Cylinder with axis parallel to OX.

17. Planes parallel to YOZ.

18. Set of lines through O, i.e. cone, because if (x_1, y_1, z_1) satisfies the equation, so does (kx_1, ky_1, kz_1).

19. Spiral on cylinder in No. 8.   **21.** $xy = z^2$.

22. $x^2 = z + 2y$.   **23.** $x^2/a^2 + y^2/b^2 + z^2/c^2 = 1$.

24. $z = x^2$, $xz = y^2$, etc.   **25.** $4x + 3y = 17$, $y + 4z = 7$, etc.

26. $x^2 - y^2 = 1$, $xz = 1 + y$, etc.

27. $5x - 15y - z + 34 = 0$, $x = (y - 2)(y + 1)$, etc.

28 (i) Line parallel to OZ; (ii) plane $x - y = 2$.

EXERCISE 3A

1. 2.   **2.** $\frac{4}{5}$.   **3.** $-\frac{7}{8}$.

4. 0.   **5.** None.   **6.** $-a/b$.

7. $y = 2x + 1$; no.   **8.** $3y + 2x + 1 = 0$; no.

9. $3x + 5y = 2$; yes.   **10.** $2x - y = 5$; yes.

11. $x = 5$; yes.   **12.** $x + 2y = 3$; no.

13. $7x + 4y = 26$; no.   **14.** $3x = 7y$; yes.

15. $bx = ay$; no.   **16.** $x + ty = h + tk$; no.

17. $ax + by = ap + bq$; yes.   **18.** $x = 4$; yes.

19. $y = x\sqrt{3}$; no.   **20.** $x + y + 1 = 0$; no.

21. $y + x\sqrt{3} = 3 + 2\sqrt{3}$; no.   **22.** $2x - 5y = 26$; no.

23. $qx = py$; yes.   **24.** $2x + 5y = 31$; yes.

25. $x = 5$; yes.   **26.** $30x - 25y = 6$; no.

27. $6x + 8y = 5$; yes.   **28.** $4x - 11y = 8$; no.

29. $dx - cy = da - cb$; yes.   **30.** $3x + 7y = 29$; no.

31. $7y - 9x = 36$; no.   **33.** $3x - y = 2$, $x + 3y = 4$.

EXERCISE 3B

1. $(3, 5)$.   **2.** $(-1, 3)$.

3. $(\frac{10}{17}, -\frac{5}{17})$.   **4.** $(\frac{15}{17}, -\frac{14}{17})$.

5. No solution if $a = 7$; $x = t$, $y = -2 - 3t$ if $a = 2$; otherwise $x = (10 - 2a)/(a - 7)$, $y = (2a - 12)/(a - 7)$.

6. $(0, a)$.   **7.** $(a/(m_1 m_2), a/m_1 + a/m_2)$.

8. $(2at_1 t_2/(t_1 + t_2), 2a/(t_1 + t_2))$.   **11.** $-4\frac{1}{3}$.

12. $2, -4\frac{1}{7}$. **14.** $a+b+c = 0$. **15.** $9\frac{9}{11}$.

16. 37. **17.** 50. **18.** 21.

19. $\frac{1}{4}\pi$. **20.** $\frac{1}{2}\pi$. **21.** $\tan^{-1}\frac{5}{3}$.

22. $\cot^{-1}9$. **23.** $2\tan^{-1}(b/a)$.

24. $\tan^{-1}\{4m^2n^2/(m^4-n^4)\}$. **25.** $\tan^{-1}\frac{2}{3}, \tan^{-1}\frac{2}{3}, \tan^{-1}(-\frac{12}{5})$.

26. $\tan^{-1}\frac{3}{14}$. **27.** $\tan^{-1}\frac{8}{11}$.

28. $\tan^{-1}\frac{5}{3}$. **29.** $\frac{1}{2}\omega$.

30. $\tan^{-1}(\frac{5}{4}\tan\omega)$. **31.** $\tan^{-1}\{(46\sin\omega)/(3+10\cos\omega)\}$.

32. ω. **33.** $1+(t_1+t_2)\cos\omega+t_1t_2 = 0$.

34. $x+y\cos\omega = 0, \ y+x\cos\omega = 0$.

35. $(y-2)(3-\cos\omega) = (x-1)(1-3\cos\omega)$.

36. $(x-x_1)(b-a\cos\omega) = (y-y_1)(a-b\cos\omega)$.

37. $(\frac{190}{71}, -\frac{57}{71})$. **38.** $(4\frac{1}{3}, 4\frac{5}{6})$.

39. $(-a, at_1+at_2+at_3+at_1t_2t_3)$.

40. $\dfrac{\pm(a_1b_2-a_2b_1)\sin\omega}{a_1a_2+b_1b_2-(a_1b_2+a_2b_1)\cos\omega}$.

EXERCISE 3c

1. $\frac{2}{5}$. **2.** $4\frac{1}{5}$.

3. $\frac{4}{5}\sqrt{10}$. **4.** $18/\sqrt{13}$.

5. $4\frac{1}{13}$. **6.** $\frac{1}{3}a\sqrt{10}$.

7. $a\sqrt{(1+t^2)}$. **8.** $\pm(x_1\cos\alpha+y_1\sin\alpha-p)$.

9. $3\frac{2}{5}$. **12.** Same.

13. $(3, -9), (-3, 1), (0, -1)$. **14.** $c < 6\frac{4}{5}$.

15. $f+g < 5, \ 3f-2g > 6, \ f+2g > -4$.

16. $3x-11y = 20, \ 11x+3y = 30$.

17. $2x-y = 0, \ x+2y = 0$. **18.** $x = 17, \ y = 13$.

19. $220x+710y+3 = 0, \ 781x-242y-133 = 0$.

20. $11x-143y-186 = 0$. **21.** $21x-77y+188 = 0$.

22. $x+y-6 = \pm\frac{1}{3}\sqrt{6}(y-x\sqrt{2}+1)$, upper sign acute angle.

23. $47x+29y = \frac{183}{41}$ (acute), $29x-47y = -\frac{183}{37}$ (obtuse).

24. $7x-y = 15$. **25.** $63x-133y = 18$.

26. $3x + 2y = 32$, $-x + 2y = 4$, $7x - 4y = 27$; $(7, 5\frac{1}{2})$.

27. $(-\frac{5}{9}, -1\frac{4}{9})$.

28. $(10, -4\frac{1}{2})$, $(-359, 221)$, $(48\frac{1}{7}, 78\frac{1}{7})$.

30. $70 \sin \omega / \sqrt{(185 - 104 \cos \omega)}$.

31. $4x - 5y = 20$, $5x + 2y = 6$, $88x + 121y = -92$.

EXERCISE 3D

1. $x/4 + y/(-3) = 1$.   **2.** $x/(-3) + y/(-4) = 1$.

3. $x/(-3) + y/(-2) = 1$.   **4.** $x/(\frac{7}{3}) + y/(-7) = 1$.

5. $x \cos \frac{1}{3}\pi + y \sin \frac{1}{3}\pi = 2$.   **6.** $x \cos(-\frac{1}{4}\pi) + y \sin(-\frac{1}{4}\pi) = \sqrt{2}$.

7. $x \cos(-\frac{3}{4}\pi) + y \sin(-\frac{3}{4}\pi) = \frac{1}{2}\sqrt{2}$.

8. $x \cos 0 + y \sin 0 = 3$.

9. $x \cos \alpha + y \sin \alpha = \frac{12}{25}$, $\alpha = \tan^{-1} \frac{24}{7}$.

10. $x \cos(-\frac{1}{2}\pi) + y \sin(-\frac{1}{2}\pi) = 5$.

11. $x \cos(\pi - \beta) + y \sin(\pi - \beta) = 4/\sqrt{13}$, $\beta = \tan^{-1} \frac{2}{3}$.

12. $x \cos(-\beta) + y \sin(-\beta) = 3$, $\beta = \tan^{-1} \frac{99}{20}$.

13. $x \cos \beta + y \sin \beta = ab/\sqrt{(a^2 + b^2)}$, $\beta = \tan^{-1}(a/b)$.

14. $x \cos(\pi - \beta) + y \sin(\pi - \beta) = b/\sqrt{(1 + a^2)}$, $\beta = \cot^{-1} a$.

15. $x \cos(\beta - \pi) + y \sin(\beta - \pi) = c/\sqrt{(a^2 + b^2)}$, $\beta = \tan^{-1}(b/a)$.

16. $\frac{1}{5}x + \frac{1}{3}y = 1$.   **17.** $-\frac{1}{2}x + \frac{2}{3}y = 1$.

18. $\frac{1}{8}x - \frac{1}{4}y = 1$.   **19.** $x + y = 18$.

20. $x - y = 5$.

21. $x + y = x_1 + y_1 \ (x_1 y_1 > 0)$, $x - y = x_1 - y_1 \ (x_1 y_1 < 0)$.

22. $\frac{3}{2}x + \frac{4}{5}y = 7$, $2x + \frac{3}{5}y = 7$.   **23.** $y + x\sqrt{3} = 6$.

24. $12x + 5y = 13$.

25. $x \cos \frac{1}{2}(t_1 + t_2) + y \sin \frac{1}{2}(t_1 + t_2) = p \cos \frac{1}{2}(t_1 - t_2)$;
     $r = p \sec \frac{1}{2}(t_1 - t_2)$, $\theta = \frac{1}{2}(t_1 + t_2)$.

26. $r \cos \theta = c$.   **27.** $r \sin \theta = k$.

28. $\theta = \theta_1$.   **35.** $r \cos(\theta - \frac{1}{4}\pi) = \pm 1$.

36. $k/r = m \cos \theta - \sin \theta$.   **37.** $ab/r = b \cos \theta + a \sin \theta$.

38. $r = \operatorname{cosec}(\alpha - \beta) \sqrt{\{p^2 + q^2 - 2pq \cos(\alpha - \beta)\}}$,
     $\cos \theta : \sin \theta : 1 = q \sin \alpha - p \sin \beta : p \cos \beta - q \cos \alpha : r \sin(\alpha - \beta)$.

39. $\frac{1}{2}l \operatorname{cosec} \frac{1}{2}\alpha$.

40. $(\frac{1}{2}l \operatorname{cosec} \frac{1}{2}\alpha \operatorname{cosec} \frac{1}{2}\beta, \frac{1}{2}(\alpha + \beta))$.

EXERCISE 3ᴇ

1. $7x + 5y = 0.$　　　　　　**2.** $y - x = 1.$

3. $5x - 11y + 92 = 0.$　　　　**4.** $qx = py.$

5. $-x/q + y/p = 1.$　　　　**6.** $9x + 10y - 137 = 0.$

7. $15x - 88y + 56 = 0.$　　　**8.** $x - (m + n) y + amn = 0.$

9. $18x - 34y + 69 = 0.$

10. $13x + 3y = 99,\ 11x - 9y = 53,\ x + 6y = 23;\ (7, 2\tfrac{2}{3}).$

11. $r \sin \theta = a \sin \alpha.$　　　　**12.** $r \cos \theta = b \cos \beta.$

13. $cd/r = c \sin (\theta - \gamma) + d \cos (\theta - \gamma).$

14. $6\sqrt{3}/r = \sin (\theta - 10°) + 3\sqrt{3} \cos (\theta - 10°).$

15. $180/r = 141 \sin \theta - 88 \cos \theta.$

EXERCISE 3ꜰ

1. $2, -2\tfrac{1}{2}.$　　　**2.** $-3, -2.$　　　**3.** $-2\tfrac{1}{3}, \tfrac{2}{3}.$

4. $-\tfrac{1}{5}, 0.$　　　**5.** $0, \tfrac{2}{3}.$　　　**6.** $a/c, b/c.$

7. $-1/a, -1/b.$　　　**8.** $m/c, -1/c.$

9. $-\cos \alpha/p, -\sin \alpha/p.$　　　　**10.** $-100, 100.$

11. $3x + 4y + 1 = 0.$　　**12.** $-x + 5y + 1 = 0.$　　**13.** $x - y + 1 = 0.$

14. $4x + 1 = 0.$　　　**15.** $3x + 4y + 6 = 0.$　　**16.** $5y - 4 = 0.$

17. Lines through the origin.　　　　**18.** $0, 0.$

19. $4X + 5Y + 1 = 0.$　　**20.** $-2X + 3Y + 4 = 0.$

21. $(3, -7).$　　　**22.** $(-\tfrac{2}{3}, -\tfrac{5}{6}).$　　　**25.** $(-\tfrac{4}{23}, -\tfrac{5}{23}).$

26. $X_1 Y_2 = X_2 Y_1.$　　**27.** $X_1 X_2 + Y_1 Y_2 = 0.$

28. Distant a from origin.

29. $\pm (X_1 Y_2 - X_2 Y_1)/(X_1 X_2 + Y_1 Y_2).$

30. $\tfrac{7}{4}, 14;\ \tfrac{32}{9}, -\tfrac{4}{9}.$

32. $(y_1 - y_2)/(x_1 y_2 - x_2 y_1),\ (x_2 - x_1)/(x_1 y_2 - x_2 y_1).$

33. $(Y_1 - Y_2)/(X_1 Y_2 - X_2 Y_1),\ (X_2 - X_1)/(X_1 Y_2 - X_2 Y_1).$

34. $\pm (\alpha + \beta + \gamma)^2/2\alpha\beta\gamma,\ \alpha = X_2 Y_3 - X_3 Y_2,$ etc.

EXERCISE 3G

1. $4, -7.$ 2. $-\frac{2}{5}, -\frac{3}{5}.$ 3. $\frac{1}{7}, 0.$

4. $0, -\frac{2}{3}.$ 5. $3X + 4Y + 1 = 0.$ 6. $7X + 6Y - 1 = 0.$

7. $2X + 1 = 0.$ 8. $Y - 2 = 0.$ 9. $4X - 9Y - 6 = 0.$

10. The origin.   14. $\sqrt{\{[(a_1c_2 - a_2c_1)^2 + (b_1c_2 - b_2c_1)^2]/c_1^2c_2^2\}}.$

15. $(a_1c_2 + a_2c_1) X + (b_1c_2 + b_2c_1) Y + 2c_1c_2 = 0.$

16. $\frac{4}{5}, -\frac{1}{5}.$ 17. $-cX_1/(aX_1 + bY_1), -cY_1/(aX_1 + bY_1).$

18. $-\frac{4}{7}, \frac{2}{7}.$ 19. $cq/(bp - aq), -cp/(bp - aq).$

20. $a_1a_2 + b_1b_2 = 0.$ 22. $qX = pY.$ 23. $bx = ay.$

24. $x^2y = -1.$

EXERCISE 4B

1. Axes.   2. Origin.

3. Two lines parallel to OX, OY.

4. Two points on OX, OY.   5. One line $[A, B]$.

6. Points $(\pm 1, 0)$ and all lines parallel to OY.

7. Four points $(\pm 1, \pm 2)$.   8. Two lines parallel to OX, OY.

9. Two lines through the origin.   10. Nothing.

11. Two lines $x = \pm y$.   12. Points $(\frac{1}{2}, \frac{1}{2}), (-\frac{1}{2}, -\frac{1}{2})$.

13. Nothing.   14. Lines $x = \pm y, x = \pm 2y$.

15. All lines parallel to $x = y$.

16. Perpendicular bisector of $(-1, 2) (3, -4)$.

17. $(0, -a)$.   18. $[0, -1]$.   19. Two parallel lines.

20. 2, 1, 0 lines according as $a^2 + b^2 >, =, < 1$.

21. $(x + 2)(y - 3) = 0.$ 22. $(3x + y + 1)(2x - 8y - 1) = 0.$

23. $xy(x - y) = 0.$ 24. $(x - 3)^2 + (y + 5)^2 = 0.$

25. $(x^2 + y^2)(x^2 + y^2 - 2x - 2y + 2) = 0.$

26. $(a_1x + b_1y + c_1)^2 + (a_2x + b_2y + c_2)^2 = 0.$

27. $(X - 4)(Y + 3) = 0.$

28. $(4X + 7Y + 1)(3X + 2Y - 1) = 0.$

29. $X^2 + Y^2 + 2X + 2Y + 2 = 0.$

30. $\{(X - 1)^2 + Y^2\}\{X^2 + (Y - 1)^2\} = 0.$

EXERCISE 4c

1. $k < 1$; $k = 0, -1$. 2. $k < -9$ or $k > -1$; $k = 4$.

3. $\tan^{-1}\frac{3}{11}$. 4. $\tan^{-1}\frac{11}{10}$. 5. $\frac{1}{2}\pi$.

6. $\tan^{-1}\frac{2}{7}\sqrt{34}$. 7. α. 8. $\frac{1}{2}\pi - \alpha$.

9. $0, \pm 1$. 10. $\frac{1}{3}\pi$. 11. $\tan^{-1}\frac{1}{2}\sqrt{22}$.

12. $\frac{1}{2}\pi$. 13. $x^2 + xy - y^2 = 0$.

14. $11x^2 + 36xy - 11y^2 = 0$. 15. $xy = 0$.

16. $kx^2 - 4xy - ky^2 = 0$. 17. $x^2 = y^2$.

18. $x^2 = y^2$. 19. $x^2 - xy - y^2 = 0$. 20. $XY = 0$.

21. $7X^2 + 22XY - 7Y^2 = 0$. 22. $3x^2 = y^2$.

23. $5x^2 + 4xy - 7y^2 = 0$. 24. $(0, 0)$.

25. $x:y:1 = -ln(a+b): -mn(a+b): am^2 - 2hlm + bl^2$.

EXERCISE 4d

1. 15; $2x - y + 5 = 0$, $x + y + 3 = 0$.

2. $\leqslant 6\frac{1}{4}$; $\{5 \pm \sqrt{(25 - 4\lambda)}\}x + 2y + 2 = 0$.

3. 6; $x + 2y - 4 = 0$, $x + 3y - 9 = 0$.

4. $\lambda = 1$, $2x + 3y + 1 = 0$, $x - y - 1 = 0$;
   $\lambda = -5$, $2x + y + 1 = 0$, $x - 3y - 1 = 0$.

5. 14; $7x + 3 = 0$, $2y - 5 = 0$.

6. -22; $4x - 7y - 2 = 0$, $3x + 2y + 11 = 0$.

7. -2; $(1, 1), (-\frac{3}{2}, -2)$.

8. $\lambda = 0, (1, 1), (-1, \frac{1}{2})$; $\lambda = 3, (-2, 1), (\frac{1}{2}, \frac{1}{2})$.

9. $12, (-\frac{3}{5}, 0), (0, 2)$. 10. $x^2 - xy - 2y^2 = 0$.

11. $x^2 + xy - 2y^2 - 3x + 3y = 0$.

12. $x^2 + xy - 6y^2 - x - 38y - 56 = 0$.

13. $2x^2 - 3xy + y^2 = 0$.

14. $4x^2 + 3xy - 2y^2 - 25x + 6y + 16 = 0$.

15. $x^2 - xy - 3y^2 + 5x - 9y + 3 = 0$.

16. $-\frac{10}{11}$. 17. $0, 2$. 18. $-3 \pm 2\sqrt{15}$.

19. $63x^2 + 32xy - 63y^2 - 158x + 94y + 32 = 0$.

20. $x^2 - 6xy - y^2 + 14x + 18y - 41 = 0$.

21. $x^2 - y^2 + 2x + 4y - 3 = 0$.

22. $|af^2 + 2fgh + bg^2|/\sqrt{\{(a-b)^2 + 4h^2\}}$.

EXERCISE 4E

1. $(2, 4)$, $(-\frac{3}{2}, \frac{9}{4})$.

2. $(\frac{1}{4}, \frac{1}{2})$, $(\frac{1}{9}, -\frac{1}{3})$.

3. $(1, 1)$, $(\frac{1}{4}, -\frac{1}{8})$.

4. $(1, 4)$, $(3, 2)$.

5. $(1, \frac{3}{2})$, $(1, 6)$, $(2, 2)$, $(3, 2)$.

6. $(0, 0)$, $(0, \pm\sqrt{2})$, $(\pm\sqrt{2}, 0)$.

7. $(1, 1)$, $(t, -t)$.

8. $(\{2a - at^2 \pm 2a\sqrt{(1-t^2)}\}/t^2$, $\{2a \pm 2a\sqrt{(1-t^2)}\}/t)$ if $t^2 \leqslant 1$ and $t \neq 0$; $(0, 0)$ if $t = 0$; no points if $t^2 > 1$.

9. $7x - 8y + 1 = 0$.

10. $135x - 52y + 94 = 0$.

11. $18x + 31y - 227 = 0$.

12. $8x + 11y = 0$.

13. $x + 3y = 6$.

14. $x - 17y + 33 = 0$.

15. $2x + y + 4 = 0$.

16. $205(5x - 7y) + 2204 = 0$.

17. $4x + 10y + 363 = 0$.

18. $x = y$.

19. $(a_1 x + b_1 y + c_1)(a_2 f + b_2 g + c_2) = (a_2 x + b_2 y + c_2)(a_1 f + b_1 g + c_1)$.

20. $(a - c)y = ad - bc$.

21. $(ad - bc)(x + my) = e(b - ma)$.

22. $5x - 15y - 61 = 0$.

23. $244x - 23y + 287 = 0$.

25. $(1, 1)$.

26. $(-\frac{1}{2}, \frac{1}{2})$.

28. $(\pm 1, 0)$, $(\pm 1, 4)$, $(0, 0)$, $(0, 4)$.

29. $(\frac{3}{2}, \frac{1}{4})$, $(-\frac{3}{2}, -\frac{1}{4})$, $(\frac{1}{2}, \frac{3}{4})$, $(-\frac{1}{2}, -\frac{3}{4})$.

30. $(x - y)(a - b) + 2a = 0$, $(x + y)(a + b) - 7a = 0$.

31. $3x^2 + xy - 2y^2 = 0$.

32. $30x^2 - 19xy + 2y^2 = 0$.

33. $4x^2 + 12xy + 3y^2 = 0$.

34. $4x^2 + 9xy + 3y^2 = 0$.

35. $4x^2 + 11xy + 5y^2 = 0$.

EXERCISE 4F

1. $[1, 4]$, $[3\frac{4}{5}, -1\frac{3}{5}]$.

2. $[\frac{1}{2}, \pm\frac{1}{3}]$, $[-\frac{1}{2}, \pm\frac{1}{3}]$.

3. $[1, -2]$, $[-1, \frac{1}{2}]$.

4. $[1/a, \pm 1/a]$.

5. $[0, \pm\sqrt{3}/(3c)]$.

6. $[\pm 1, 1]$, $[0, t]$.

7. $3X + 7Y = 4$.

8. $7X - Y = 31$.

9. $(2, -5)$.

10. $(p_1 X + q_1 Y + r_1)(p_2 A + q_2 B + r_2)$
$$= (p_2 X + q_2 Y + r_2)(p_1 A + q_1 B + r_1).$$

12. $[1, 0]$.

13. $3x - 4y = 8$.

14. $\pm x \pm y = 1$.

EXERCISE 4G

1. Point and line.

2. $-x\cos\frac{1}{12}\pi + y\sin\frac{1}{12}\pi = 1$, $x\cos\frac{5}{12}\pi - y\sin\frac{5}{12}\pi = 1$.

3. $17x^2 - 14xy - 17y^2 - 90x - 122y - 199 = 0$.

5. $\frac{1}{2}\sqrt{2}$. **6.** $x^2 - y^2 - 3x - y + 2 = 0$.

7. $1\frac{1}{5}$. **8.** $ab = h^2$, $2fgh = af^2 + bg^2$, $g^2 > ac$.

9. $\tan^{-1}(\frac{3}{13}\sqrt{3})$. **10.** $3x^2 - 4xy - 5y^2 = 0$.

11. $h = a\cos\omega$. **12.** $a + b = 2h\cos\omega$.

13. $kx^2 - 2xy = \lambda(ky^2 + 2xy)$. **14.** $\sqrt{(h^2 - ab)/(bX^2 - 2hXY + aY^2)}$.

16. $(\{(b-a)l - 2hm\}/2d, \{(a-b)m - 2hl\}/2d)$, $d \equiv bl^2 - 2hlm + am^2$,
$(\frac{2}{3}(bl - hm)/d, \frac{2}{3}(am - hl)/d)$.

17. $x : y : 1 = m(fm - gl) - ln : l(gl - fm) - mn : l^2 + m^2$.

20. $(a_1 b_2 - a_2 b_1)^2 - 4(a_1 h_2 - a_2 h_1)(h_1 b_2 - h_2 b_1) = 0$.

21. $(a_1 a_2 - b_1 b_2)^2 + 4(a_1 h_2 + b_2 h_1)(h_1 a_2 + h_2 b_1) = 0$.

22. $G^2 + 4H^3 = 0$. **23.** $a = c$ and $b = d$.

24. $(b+d)(be+ad) + (a+c+e)(a-e)^2 = 0$.

25. $b(x^3 + y^3) = 3axy(x-y)$; $\pm a\sqrt{(2\sqrt{3} - 3)}$.

EXERCISE 5A

1. $x^2 + y^2 + 4x - 6y - 3 = 0$. **2.** $x^2 + y^2 + 2x - 3 = 0$.

3. $x^2 + y^2 - 8x + 10y - 8 = 0$. **4.** $x^2 + y^2 + 3x + 4y = 0$.

5. $x^2 + y^2 - 2ax + 2by - 2ab = 0$. **6.** $(-1, 3)$, 5.

7. $(-1, 0)$, 3. **8.** $(1\frac{1}{2}, -2)$, $\frac{1}{2}\sqrt{5}$.

9. $(-2, -\frac{1}{6})$, $2\frac{1}{6}$. **10.** $(0, -\frac{3}{5})$, $\frac{1}{5}\sqrt{14}$.

11. $(a, 0)$, $a\sqrt{3}$. **12.** $(b/2a, c/2a)$, $\sqrt{(b^2 + c^2)/(2a)}$.

17. $x^2 + y^2 + 8x - 10y + 7 = 0$. **18.** $x^2 + y^2 - 38x + 74y = 0$.

19. $x^2 + y^2 - 2cy + c^2 - 169 = 0$. **20.** $19(x^2 + y^2) - 142x - 44y = 0$.

21. $9(x^2 + y^2) - 29x - 37y - 340 = 0$.

22. $3(x^2 + y^2) - 85x + 61y - 52 = 0$.

23. $b(x^2 + y^2) = (a^2 + b^2)(x - a + b)$.

24. $(x-a)^2+(y-b)^2 = 85^2$. **25.** $x^2+y^2-10x+5 = 0$.

26. $5(x^2+y^2)-22x-2y-30 = 0$.

27. $x^2+y^2-6x+4y+9 = 0$, $x^2+y^2+10x+20y+25 = 0$.

28. $x^2+y^2+2c(x-y)+c^2 = 0$, $c = 5 \pm 2\sqrt{3}$.

29. $x^2+y^2-3x+4y = 0$. **30.** $c(x^2+y^2-a^2) = y(b^2+c^2-a^2)$.

31. $x^2+y^2-b^2 = \pm 2y\sqrt{(a^2-b^2)}$. **34.** $c(c-a)+d(d-b) = 0$.

35. $x^2+y^2-7x+2y-23 = 0$. **36.** $x^2+y^2-ax-by = 0$.

37. $x^2+y^2-7x-21y+120 = 0$. **38.** $x^2+y^2-10x-12y+51 = 0$.

39. $(l^2+m^2)(x^2+y^2+c)$
$$= 2x(l+flm-gm^2)+2y(m+glm-fl^2)+2gl+2fm+2.$$

EXERCISE 5B

1. $3x+4y = 25$. **2.** $5x-12y = 32$.

3. $x(x_1-f)+y(y_1-g) = f(x_1-f)+g(y_1-g)+h^2$.

4. $5x-2y = 29$. **5.** $xx_1+yy_1 = a^2$. **6.** $x+10y = 39$.

8. $3x-2y = 13$. **9.** $x+2y = \frac{5}{3}\sqrt{3}$. **10.** $5x+13y = 13$.

11. $7x-11y = 148$. **12.** $k(xx_1+yy_1)+u(x+x_1)+v(y+y_1)+w = 0$.

13. $x^2+y^2+10x+8y+25 = 0$. **14.** $x^2+y^2-4x-6y+12 = 0$.

15. $(x^2+y^2)(a^2+b^2) = c^2$. **16.** $x^2+y^2 = 4$ or 144.

17. $x^2+y^2 = 9$ or 529. **18.** $1/a^2+1/b^2 = 1/c^2$.

19. $6, -12$. **20.** $(-\frac{3}{10}a\sqrt{10}, \frac{1}{10}a\sqrt{10})$.

21. $(lf+mg-1)^2 = h^2(l^2+m^2)$. **22.** $y = x\sqrt{3}\pm 2$.

23. $4x-3y = \pm 10$. **24.** $x = 0, -4$.

25. $mx-ly = \pm a\sqrt{(l^2+m^2)}$. **26.** $2x-3y+13 = 0$.

27. $3x+y+1 = 0$. **28.** $2\sqrt{5}$.

29. $2\sqrt{\{c^2-(a^2b^2)/(a^2+b^2)\}}$.

EXERCISE 5C

1. 16. **2.** 0. **3.** 81.

4. $-c^2$. **5.** $2\sqrt{15}$. **6.** $\frac{1}{2}\sqrt{285}$.

7. 6. **8.** $2\sqrt{14}$. **9.** $\sqrt{(a^2+b^2-c^2)}$.

10. $\sqrt{(f^2+g^2+2uf/a+2vg/a+w/a)}$. **11.** 36.

12. 50. **13.** -15. **14.** -2.

15. $\frac{1}{3}\pi$. **16.** $\cos^{-1}\frac{2}{9}$.

19. $x^2+y^2+2\kappa x-b = 0$.

EXERCISE 5D

1. $x+y+1 = 0.$

2. $6x - 5y + 4 = 0.$

3. $26x - 23y - 292 = 0.$

4. $2gx + 2fy + c + kr^2 = 0.$

5. $2x = 5.$

6. $(-2, -5).$

7. $3:13, 13:23, 3:7, -9:11.$

8. $(3, 2), (2, 3).$

9. $22(x^2 + y^2) - 9x - 115y - 146 = 0.$

10. $\frac{1}{10}\sqrt{149}.$

11. $(0, 0), 2.$

12. $65\frac{19}{85}.$

13. $(-\frac{3}{2} \pm \frac{1}{2}\sqrt{5}, 0).$

14. $(2\frac{2}{5}, -3\frac{1}{5}).$

15. $(\frac{12}{25}(9 \pm \sqrt{6}), \frac{2}{25}(-3 \pm 8\sqrt{6})).$

16. $(0, \pm\sqrt{(b/a)}).$

17. $x^2 + y^2 = 100,\ x^2 + y^2 + 26x - 56 = 0.$

18. $(\frac{28}{25}, \frac{4}{25}).$

21. $\sqrt{(h^2 - k^2)}.$

22. $\lambda(x^2 + y^2 + 8x + 6y + 25) = \mu(x^2 + y^2 - 4x - 10y + 29).$

24. $(x^2 + y^2 + a^2)(1 - \lambda^2) = 2ax(1 + \lambda^2).$

25. $(\kappa_1^2 - \kappa_2^2)(x^2 + y^2) = 2x(\kappa_1^2 x_1 - \kappa_2^2 x_2) + 2y(\kappa_1^2 y_1 - \kappa_2^2 y_2)$
$$- \kappa_1^2(x_1^2 + y_1^2) - \kappa_2^2(x_2^2 + y_2^2).$$

EXERCISE 5E

1. $r^2 + 2r\sin\theta - 3 = 0.$

2. $r = 2a\cos\theta.$

3. $k^2 c/(c^2 - r_1^2).$

4. Circle of radius ka through $O.$

5. Circle $k^2(a^2 - b^2) = r^2 - 2bkr\cos\theta.$

6. $x = 0.$

7. $c(x^2 + y^2) = k^2 y.$

8. $ar = k^2\cos(\theta - \alpha).$

9. $(-cg/(g^2 + f^2), -cf/(g^2 + f^2)).$

10. $4.$

11. $x':y':1 = t^2:\frac{1}{2} + t^3:\frac{1}{4} + t^3 + t^4 + t^6.$

12. $r = \frac{1}{2}(1 - \cos\theta).$

13. $a = r(1 + \cos\theta),\ r = -4a\cos\theta.$

14. $y(x^2 + y^2) = k^2(ax^2 + bxy + cy^2).$

EXERCISE 5F

1. $89x^2 - 157xy + 29y^2 = 0.$

2. $(1 - 2gX + cX^2)x^2 - 2(fX + gY - cXY)xy + (1 - 2fY + cY^2)y^2 = 0;$
$(f^2 - c)X^2 - 2fgXY + (g^2 - c)Y^2 + 2gX + 2fY - 1 = 0.$

3. $t_1 t_2 t_3 t_4 = 1.$

4. $a^2 x^2 - (b^2 - a^2)y^2 - 2a^2 bx + a^2 b^2 = 0.$

5. $(g_1 f_2 - g_2 f_1)^2 - c_1(f_2^2 + g_2^2) - c_2(f_1^2 + g_1^2) + (c_1 + c_2)(g_1 g_2 + f_1 f_2)$
$$= \tfrac{1}{4}(c_1 - c_2)^2.$$

6. $3x \pm 4y = 75,\ 9x \pm y\sqrt{19} = 150.$

7. $120,\ 100.$　　　　　　**8.** $\cos^{-1}\tfrac{3}{5}.$

9. $\sqrt{\{2(h_1^2 + h_2^2) - d^2 - (h_1^2 - h_2^2)^2/d^2\}},\ d^2 = (f_1 - f_2)^2 + (g_1 - g_2)^2.$

10. $x^2 + xy + y^2 - 7x - 8y + 10 = 0.$

12. $c\sin^2\omega < f^2 + g^2 - 2fg\cos\omega;$
$((f\cos\omega - g)\operatorname{cosec}^2\omega,\ (g\cos\omega - f)\operatorname{cosec}^2\omega);$
$\sqrt{\{\operatorname{cosec}^2\omega(f^2 + g^2 - 2fg\cos\omega) - c\}}.$

13. $x^2 + 2xy\cos\omega + y^2 = \pm 2kx\sin\omega.$

14. $x^2 + 2xy\cos\omega + y^2 = ax + by.$

15. $x(x_1 + y_1\cos\omega) + y(y_1 + x_1\cos\omega) = k^2.$

16. $(y - 2x)\sin\omega = \pm\sqrt{(5 + 4\cos\omega)}.$

17. $(a\sin\alpha - b\sin\beta)\sin\theta = (b\cos\beta - a\cos\alpha)\cos\theta;$
$(ab/2r)\sin(\alpha - \beta) = a\sin(\alpha - \theta) - b\sin(\beta - \theta).$

18. Circle: $2r = a\cos(\theta - \alpha) + b\cos(\theta - \beta).$

22. $r = a - b\cos\theta$ with pole at $(b, 0).$

23. $(x^2 + y^2 + gx + fy)^2 = (g^2 + f^2 - c)(x^2 + y^2).$

24. $(x - a)^2(ax - c^2)^2 = y^2\{a^2 c^2 - (ax - a^2 - c^2)^2\}.$

25. Semicircle on $AB.$

26. $(a_1 - a_2)^2 - (r_1 - r_2)^2;$ equation of the common tangents.

28. $\{x(r_1 \mp r_2) - (a_1 r_1 \mp a_2 r_2)\}^2 + \{y(r_1 \mp r_2) - (b_1 r_1 \mp b_2 r_2)\}^2$
$$= \{x(b_1 - b_2) - y(a_1 - a_2) + (a_1 b_2 - a_2 b_1)\}^2.$$

30. $(x - a)^2 + (y - b)^2 + (z - c)^2 = r^2.$

31. $(-u, -v, -w),\ \sqrt{(u^2 + v^2 + w^2 - d)}.$

MISCELLANEOUS EXERCISE A

1. 6. **2.** $\frac{1}{4}x^2 + y^2 = c^2$, $x^2 + \frac{1}{4}y^2 = c^2$.

3. $\tan^{-1}\frac{17}{31}$, $\tan^{-1}\frac{4}{3}$, $\pi - \tan^{-1}7$. **4.** $2r\cos\theta + k = 0$.

5. $-\frac{1}{3}, -\frac{1}{3}$. **6.** $\frac{1}{2}\pi$.

7. $a(x - a - b) = \pm y\sqrt{(2ab + b^2)}$. **8.** $\sqrt{93}$.

10. $x^2 - xy - y^2 = 0$. **11.** 36.

12. $61x - 79y - 908 = 0$. **13.** $y = 5 \pm 2(x+2)\sqrt{2}$.

14. $(c^2 - b^2)(x^2 + y^2 + a^2) = 2ax(c^2 + b^2)$.

15. $(2\frac{1}{3}, 1\frac{2}{3})$, $(10\frac{36}{37}, 7\frac{31}{37})$. **16.** 150.

17. $y = x\tan\{\sqrt{(x^2 + y^2)} - 1\}$. **18.** $38x - 19y = 2$.

19. $71x^2 + 20xy - 11y^2 = 0$.

21. $bx + ay + 2ab = 0$; $x^2 + y^2 + x/a - y/b - c = \mu(bx + ay + 2ab)$.

22. $((\Sigma x)/n, (\Sigma y)/n)$.

23. $(19, -2), (1, 1), (-3, 9), (-7, -15)$.

24. $-2x/\sqrt{7} + 5y/2\sqrt{7} = 3/\sqrt{7}$.

25. $x^2 - 7xy + y^2 + 43x - 38y + 181 = 0$.

26. $11X + 7Y = 3$.

27. $(G/C, F/C), \Delta/C$; notation of 0·6.

29. $(\frac{1}{4}, 9\frac{1}{4})$.

31. $y - k = (x - h)\tan\frac{1}{2}(\alpha + \beta)$; $y - k = (x - h)\tan\frac{1}{2}(\alpha + \beta + \pi)$.

32. $\tan^{-1}(\frac{1}{3}\sqrt{17})$. **33.** $c^2 - 2cp\cos\alpha + 2p^2 = a^2$.

34. $(10, 7), 10$.

35. $x(7 + 4\cos\omega) - y(4 - 6\cos\omega) = 13 + 24\cos\omega$.

36. $\sec^{-1}3$. **37.** $(48\frac{1}{7}, 78\frac{1}{7})$.

38. $x - y = 2a$, $7x - y = 14a$. **39.** $\tan^{-1}2$.

41. $72/\sqrt{73}$. **42.** $(-ac/(a^2 + b^2), -bc/(a^2 + b^2))$.

44. $x^2 + 2xy\cos\omega + y^2 = c^2\mathrm{cosec}^2\omega$.

45. $(uv, -uv)$. **46.** $x^2 + y^2 = 1$.

48. $x'^2 + y'^2 + x'\sqrt{2} - 3y'\sqrt{2} + 1 = 0$.

49. $3dx^2 - 2cxy + dy^2 = 0$; $\pm\frac{1}{3}c\sqrt{3}$.

50. $(2, 3), \tan^{-1}\frac{1}{2}(1 \pm \sqrt{5})$. **51.** $x + 2 = 0$, $x + 2 = (y - 3)\sqrt{3}$.

52. $(7, 5\frac{1}{2})$. **53.** $208x + 13y = 15$.

54. $x^2 - 91xy - 36y^2 = 0$. **55.** $35 \pm 5\sqrt{65}$.

56. Circle, centre mid-point of AB.

58. $\frac{1}{2}\pi$.   **59.** $232x - 116y + 1 = 0$.

60. $2x^2 + 17xy - y^2 - 38x - 13y + 32 = 0$.

61. $X : Y : 1 = (X_1 Y_2 - X_2 Y_1)\, b + X_1 - X_2 : (X_2 Y_1 - X_1 Y_2)\, a + Y_1 - Y_2$
$\qquad\qquad\qquad\qquad\qquad\qquad\qquad : (X_2 - X_1)\, a + (Y_2 - Y_1)\, b$.

62. $(3 + \sqrt{17})\, x^2 + (3 - \sqrt{17})\, y^2 + 28 = 0$.

63. Cuts orthogonally the circle orthogonal to $s_1,\, s_2,\, s_3$.

64. $-2 : 5$.   **65.** Surface of revolution.

66. 7.   **67.** $\tan^{-1}\frac{6}{5}$.

69. $x^2 + y^2 - 22x - 8y - 81 = 0$.

71. $Px_1/a_1 + Qx_2/a_2 + (P + Q)\, x_3/a_3 = 0$; $(a_1,\, a_2,\, -a_3)$.

72. $138x - 53y - 63 = 0$.   **73.** $|c - d|/\sqrt{(1 + m^2)}$.

74. $13X + 9Y + 4 = 0,\ 2X + Y + 1 = 0$.

75. $7x = 9y,\ 9x - 28y + 70 = 0$.

76. $2\sqrt{\{(af - bg)^2 - c(a^2 + b^2)\}}/\sqrt{(a^2 + b^2)}$.

78. $(a^2 + b^2)\,(x^2 + y^2) \pm 4abxy = (a^2 - b^2)^2$.

79. $(m + n)\, k^2 + 2(l^2 - mn)\, k - (m + n)\, l^2 = 0$.

80. $15x - 8y = 100$.   **81.** $x^2 - 2xy - y^2 + 2x - 2y + 1 = 0$.

82. $b(x - x_1)^2 - 2h(x - x_1)\,(y - y_1) + a(y - y_1)^2 = 0$.

83. $(x - a)\cos t + (y - b)\sin t = c$.   **84.** $\cos^{-1}\frac{5}{13}$.

87. $[-\frac{3}{4}, -\frac{1}{2}],\ [-\frac{3}{8}, +\frac{1}{4}]$.   **88.** 2.

89. $(l_1 m_2 - l_2 m_1)\,(ax - by) + a(m_1 - m_2) + b(l_1 - l_2) = 0$.

90. $2(x^2 + y^2) - 6x - 2y - 21 = 0$.

91. $x_1^2 + 2x_1 y_1 \cos\omega + y_1^2 + 2gx_1 + 2fy_1 + c$.

92. $(\kappa_1 + \kappa_2)\, r = \kappa_1^2 r_1^2 + \kappa_2^2 r_2^2 - 2\kappa_1 \kappa_2 r_1 r_2 \cos(\theta_1 - \theta_2)$;
$\qquad \cos\theta : \sin\theta : 1 = \kappa_1 r_1 \cos\theta_1 + \kappa_2 r_2 \cos\theta_2$
$\qquad\qquad\qquad\qquad\qquad : \kappa_1 r_1 \sin\theta_1 + \kappa_2 r_2 \sin\theta_2 : (\kappa_1 + \kappa_2)\, r$.

93. $a\cos\frac{1}{2}(\alpha - \beta)$.

94. $x + y = 9,\ 5x + 2y = 34,\ 2x + 5y = 29$.

95. $8x^2 - 12xy + 3y^2 + 4x + 6y - 13 = 0$.

96. 2α.   **97.** $x'y' = -k^2,\ x'^2 - 3y'^2 = 3k^2$.

98. $(-k^2 g/c,\ -k^2 f/c),\ (k^2/c)\sqrt{(g^2 + f^2 - c)}$.

99. $5\frac{1}{4}$.   **100.** $\tan^{-1}\{(a^2 - b^2)/2ab\}$.

102. $x^2 - xy - y^2 - x + 18y - 61 = 0$.

103. $(1, 2)$.   **104.** $\frac{9}{13}\sqrt{23}$.

105. $(-1, +1),\ (\frac{1}{5}, \frac{8}{5})$; $x^2 + y^2 + 102x + 48y - 23 = 0$,
$\qquad x^2 + y^2 - 2x - 4y + 3 = 0$.

EXERCISE 6A

1. $y^2 = ax$ except the origin.

2. $y^2 = ax$ in first quadrant, every point twice.

3. $y = x^2$ in first quadrant. **4.** $(4, 2)$, $(9, 3)$.

5. $(-\frac{1}{2}, -\frac{1}{2})$, $(1\frac{3}{4}, \frac{1}{7})$. **6.** $(1, 2)$, $(\frac{1}{4}, \frac{5}{8})$, $(\frac{1}{9}, \frac{10}{27})$.

7. $(\frac{1}{2}k, \frac{1}{2}k)$, $(\frac{4}{9}k, \frac{2}{9}k)$, $(\frac{2}{7}k, -\frac{4}{7}k)$. **8.** $(a, 0)$, $(0, -b)$.

9. $x - \frac{1}{2}y(t_1 + t_2) + at_1 t_2 = 0$, $x - yt + at^2 = 0$, $tx + y = at^3 + 2at$.

10. $x + yt_1 t_2 = k(t_1 + t_2)$, $x + yt^2 = 2kt$, $t^2 x - y = kt^3 - k/t$.

11. $x(t_1 t_2 - t_1 - t_2) + 2y = a(t_1 t_2 + t_1 + t_2)$, $x(t^2 - 2t) + 2y = a(t^2 + 2t)$,
$2x - (t^2 - 2t)y = a(t^4 - 2t^3 - 2t - 2)/(1 - t)$.

12. $x(1 - t_1 t_2) + y(1 + t_1 t_2) = 2a(t_1 + t_2)$, $x(1 - t^2) + y(1 + t^2) = 4at$,
$x(1 + t^2) - y(1 - t^2) = 2a(t^4 - 1)/t$.

13. $x(t_1^2 + t_1 t_2 + t_2^2) - y(t_1 + t_2) = at_1^2 t_2^2$, $3tx - 2y = at^3$,
$3tx - 2y = at^3$, $2x + 3ty = at^2(2 + 3t^2)$.

14. $x(t_1 + t_2) + yt_1^2 t_2^2 = k(t_1^2 + t_1 t_2 + t_2^2)$, $2x + yt^3 = 3kt$,
$t^3 x - 2y = k(t^4 - 2/t^2)$.

15. $a_1 c_2 = a_2 c_1$, $a_1 a_2 + c_1 c_2 = 0$.

18. $a(1 + t^2)$. **22.** $-a$, $at_1 + at_2$.

24. $\frac{1}{2}a(t_1^2 + t_2^2)$, $a(t_1 + t_2)$; $t_1 + t_2$ constant; $y = a(t_1 + t_2)$.

EXERCISE 6B

2. $(h, k \sec \theta)$.

3. $y'^2 = kx' \cos^2 \theta$, $y'^2 = kx' \sec \theta$.

4. OX vanishing line, $\theta = \frac{1}{4}\pi$; OY vanishing line, $\theta = \frac{1}{3}\pi$.

5. $x'^2/a^2 + y'^2/b^2 = 1$. **6.** $x'^2 + y'^2 = b^2$.

EXERCISE 6C

1. $x \cos \frac{1}{2}(\phi_1 + \phi_2) + y \sin \frac{1}{2}(\phi_1 + \phi_2) = a \cos \frac{1}{2}(\phi_1 - \phi_2)$,
$x \cos \phi_1 + y \sin \phi_1 = a$.

2. $(x/a) \cos \frac{1}{2}(\phi_1 + \phi_2) + (y/b) \sin \frac{1}{2}(\phi_1 + \phi_2) = \cos \frac{1}{2}(\phi_1 - \phi_2)$,
$(x/a) \cos \phi_1 + (y/b) \sin \phi_1 = 1$.

3. $(a \sec \beta \cos \alpha, a \sec \beta \sin \alpha)$, $(a \sec \beta \cos \alpha, b \sec \beta \sin \alpha)$.

4. Chord $\phi_1 \phi_2$; see No. 2.

5. $ax \sec \phi_1 - by \operatorname{cosec} \phi_1 = a^2 - b^2$.

6. Parameters are roots of $t^2(n - al) + 2tbm + (n + al) = 0$.

7. $(x/a)(1 - t_1 t_2) + (y/b)(t_1 + t_2) = (1 + t_1 t_2)$.

8. $(x/a)(1 - t^2) + 2ty/b = 1 + t^2$; 2, 2 coincident, or 0.

11. 4.

EXERCISE 6D

1. $(x/a)(1+t_1t_2)-(y/b)(t_1+t_2) = 1-t_1t_2$;
   $(x/a)(1+t^2)-2(y/b)t = 1-t^2$;
   $2atx+by(1+t^2) = 2t(a^2+b^2)(1+t^2)/(1-t^2)$.

2. $(x/a)\operatorname{ch}\tfrac{1}{2}(\phi_1+\phi_2)-(y/b)\operatorname{sh}\tfrac{1}{2}(\phi_1+\phi_2) = \operatorname{ch}\tfrac{1}{2}(\phi_1-\phi_2)$;
   $(x/a)\operatorname{ch}\phi-(y/b)\operatorname{sh}\phi = 1$; $ax/(\operatorname{ch}\phi)+by/(\operatorname{sh}\phi) = a^2+b^2$.

3. $(x/a)\cos\tfrac{1}{2}(\theta_1-\theta_2)-(y/b)\sin\tfrac{1}{2}(\theta_1+\theta_2) = \cos\tfrac{1}{2}(\theta_1+\theta_2)$;
   $(x/a)-(y/b)\sin\theta = \cos\theta$; $ax\sin\theta+by = (a^2+b^2)\tan\theta$.

4. $(x/a)(1+t_1t_2)-(y/b)(1-t_1t_2) = t_1+t_2$;
   $(x/a)(1+t^2)-(y/b)(1-t^2) = 2t$;
   $ax(1-t^2)+by(1+t^2) = (a^2+b^2)(1-t^4)/(2t)$.

5. $m = (1-t)/(1+t)$.   6. $a^2b^2/(a^2+b^2)$.

10. $c, 0; a^2/c, b^2/c$; where $c^2 = a^2+b^2$.

EXERCISE 6E

2. (i) $t_1+t_2 = 0$; (ii) t_1t_2 constant.

5. $t_1t_2t_3t_4 = -1$.   7. $t_1^2x-y = k(t_1^4-1)/t_1$.

8. $x(t_1^2+t_1t_2+t_2^2)-y = at_1t_2(t_1+t_2)$; $3t_1^2x-y = 2at_1^3$;
   $4a^2X^3+27Y = 0$.

EXERCISE 6F

1. $cx-2aty = cb-2adt-act^2$; $X:Y:1 = c:-2at:act^2+2adt-cb$.

2. $x(t-1)^2-y(t^2+1)+1 = 0$; $X = (t-1)^2$, $Y = -t^2-1$.

3. $x(3t+t^3)-2y = at^3$; $X:Y:1/a = -3t-t^3:2:t^3$.

4. $xt(t^2-3)-y(3t^2-1)+2at^3 = 0$; $X:Y:1/a = t^3-3t:1-3t^2:2t^3$.

5. $(2+t^2+t^4)x+2ty = (2+t^2)^2$; $X:Y:-1 = 2+t^2+t^4:2t:(2+t^2)^2$.

6. $x\cos\tfrac{1}{2}t-y\sin\tfrac{1}{2}t = a(t\cos\tfrac{1}{2}t-2\sin\tfrac{1}{2}t)$;
   $X:Y:1/a = 1:-\tan s:2(\tan s-s)$.

7. $x+yt_1t_2(t_1+t_2) = a(t_1^2+t_1t_2+t_2^2)$.

8. $3x-4(t_1+t_2)y = 3-8(t_1+t_2)-12t_1t_2$.

9. $2\Sigma t_1t_2 = 1$; $(\tfrac{8}{7}a, \pm\tfrac{4}{21}a\sqrt6)$; $(0, 0)$ isolated.

10. $t_1t_2t_3 = -1$.   11. $(11\tfrac{1}{4}, 1\tfrac{7}{20})$.   12. $(4\tfrac{1}{2}, 1\tfrac{1}{4})$.

13. $t_1t_2 = -1$.   15. $(-a/2t, 4at^2)$.   16. ±2.

17. $x:y:1 = 3+2t:3t+2t^2:t^3$.

18. $y = t$, $x = 1/(1+t^2)$; $(4x_1^2y_1^2, (1-y_1^2)/2y_1)$.

19. $x:y:1 = 1-t^2:t-t^3:(1+t^2)(9-7t)$; $(\tfrac{1}{15}, \tfrac{1}{5})$.

20. $\sin\theta_1+\sin\theta_2+\sin\theta_3 = 0$.

23. $x:y:1 = t^2:2+t^3:t^4+(2+t^3)^2$, or $x^3 = (x^2+y^2)(2x^2+2y^2-y)^2$.

EXERCISE 7A

1. O and, if $ab > 0$, $(\pm \sqrt{(a^3/b^3)}, \pm \sqrt{(a/b)})$.

2. $(k^2 - 2k, k - 2)$; line is part of curve.

3. $O, O, (\pm \sqrt{(1 - k^2)/(1 + k^2)}, \pm k\sqrt{(1 - k^2)}/(1 + k^2))$; O, O, O, O; O, O.

4. $a = -b$, O, O; $a = \pm 2b$, O.

5. O and, if $b/a < 1$, $t = \pm \sqrt{\{(a - b)/a^3\}}$; $x = y$.

6. O, O and, if $b \ne 0$, $t = a(a - 2b)/b^3$; $x = 0$, $x = 2y$.

7. O, O and, if $b^2 \geqslant 4a^2$ and $b \ne 0$, $t = \pm \sqrt{\{(b^2 - 4a^2)/b^4\}}$; $y = \pm 2x$.

8. O, O and, if $a \ne 0$, $t = (a^2 + b^2)/a^3$; isolated.

9. O, O and, if $a + b \ne 0$, $t = (a - b)^2/(a + b)^3$; $x = y$.

10. O, O, O and, if $a \ne 0$, $t = b(b^2 - a^2)/a^4$; $y = 0$, $y = \pm x$.

11. O, O, O, O and, if $b \ne 0$, $t = (b^2 - a^2)/(b^2 - 4a^2)/a^5$;
    $y = \pm x$, $y = \pm 2x$.

12. O, O, O and, if $ab \ne 0$, $t = (a^3 + b^3)/a^2b^2$; $x = 0$, $y = 0$.

13. O, O, O and, if $a \ne 0$, $t = -\{ab \pm b\sqrt{(a^2 + ab)}\}/a^3$; $y = 0$.

14. O, O, O and, if $a \ne 0$ and $b^8 + 4a^7b \geqslant 0$,
    $t = \{-b^4 \pm \sqrt{(b^8 + 4a^7b)}\}/(2a^5)$; $x = 0$, $y = 0$.

15. $(0, 0)$, $(0, \pm 1)$, $(\pm 1, 0)$, $(1, \pm 1)$, $(-1, \pm 1)$; $a = 0$, $b = 0$; $a = 4b$.

16. $\frac{3}{2}$, $-\frac{31}{2}$, -15, 30; cusped like $x^3 = 30y^2$.

17. $y = (1 \pm \sqrt{2})x^2$.   18. $y = x^2$, $y = 2x^2$, $y = -3x^2$.

19. $x = y^3$, $y = x^2 \pm \sqrt{x^9}$.   20. $y^2 = x$, $y^3 = -x^2$, $y = \pm x^2$.

EXERCISE 7B

1. $16x^2 + 23xy + 6y^2 - 64x - 36y + 48 = 0$.

2. $(3x - y - 13)(3x - 2y + 1) = 0$.

3. $3X^2 + XY + 2Y^2 - 4X - 11Y + 9 = 0$.

5. $x^2 - 2xy + 3y^2 = 6y$.   6. $(x + y)(3x - y - 5) = 0$.

7. $t = \pm 1, \pm 2, \pm 4$.   8. Conic is part of cubic.

10. $(2/(a + b), 0)$.   11. $(k + kt^2, -kt)$.

13. $x^3 + 10x^2y - 4y^3 = x + 6y$.

14. l and conic; conic is a line-pair.

17. $(0, 0)$, $(0, 1)$, $(0, -1)$, $(1, 1)$, $(1, 2)$, $(1, 0)$, $(2, 2)$, $(2, 3)$, $(2, 1)$;
    $y^3 - 3y^2x + 3yx^2 - 2x^3 + 3x^2 - y - x = 0$.

18. (i) l and a quartic; (ii) s and a cubic; (iii) c and any one of ∞^1
    conics; (iv) l, s and any one of ∞^2 conics.

19. $14, 4, 8$.   20. $\frac{1}{6}n(n^2 + 6n + 11)$.   21. $\frac{1}{2}n(n + 3)$.

EXERCISE 8 A

4. 2, 2 coincident, 0, as $|k| <, =, > |a|$.

5. 2, 2 coincident, 0, as $|k| >, =, < |a|$.

6. 2, 2 coincident, 0, as $|k| <, =, > 5$.

7. 3, 3 (2 coincident), 1, as $k^2 <, =, > 4\sqrt{2}$.

8. 0, 2 pairs coincident, 4, 2 and 2 coincident, 2, according as $k < 0$, $k = 0$, $0 < k < 1$, $k = 1$, $k > 1$.

9. 2, 2 and 2 coincident, 4, 2 pairs coincident, 0, according as $k < 0$, $k = 0$, $0 < k < \frac{2}{9}\sqrt{3}$, $k = \frac{2}{9}\sqrt{3}$, $k > \frac{2}{9}\sqrt{3}$.

EXERCISE 8 B

1. $x : y : 1 = -ac \pm b\sqrt{(a^2 + b^2 - c^2)} : -bc \mp a\sqrt{(a^2 + b^2 - c^2)} : a^2 + b^2$; $a^2 + b^2 = c^2$.

2. $((c^2 + r^2)/(2c), (c^2 - r^2)/(2ci))$.

3. None; $(2, \pm i)$. 4. $(4, 3)$, $(3, 4)$ in both.

5. $(0, 0)$, $(2, 4)$; $(0, 0)$, $(2, 4)$, $(2\omega, 4\omega^2)$, $(2\omega^2, 4\omega)$.

6. $(1, 1)$; $(1, 1)$, $(\frac{1}{2}(-1 \pm i\sqrt{7}), \frac{1}{2}(5 \mp i\sqrt{7}))$.

7. $(2, -1)$; $(2, -1)$, $(\frac{1}{8}(3 \pm i\sqrt{423}), \frac{1}{8}(-21 \pm i\sqrt{423}))$.

8. G_4 only: $(0, i)$, $(0, i)$, $(\pm\sqrt{3}, i)$.

9. $(p \pm \sqrt{(p^2 - q)}, 2p^2 - q \pm 2p\sqrt{(p^2 - q)})$, unless, in G_3, $p^2 < q$.

10. $(4, i)$. 11. $x + yi = 8 + 6i$, $x - yi = -2 - 2i$. 12. Zero.

13. 1 if $t = 0$; 2 coincident if $t = \{b \pm \sqrt{(b^2 - a)}\}/2a$ or if $a = 0$ and $t = 1/4b$; otherwise 2 distinct.

EXERCISE 8 c

1. $(b, \pm 2\sqrt{ab}, 1)$ if $ab > 0$; $(c^2, 4ac, 4a)$; $(1, 0, 0)$, $(1, 0, 0)$.

2. $(k, k, 1)$, $(-k, -k, 1)$; $(1, k^2, 1)$, $(0, 1, 0)$; $(1, 0, 0)$, $(0, 1, 0)$.

3. $(1, t, 0)$, $(1, 0, 0)$, $(0, 1, 0)$, $(b, -a, 0)$, $(a, -b, 0)$.

4. Two coincident intersections at infinity.

5. $(1, 1, 0)$, $(1, 1, 0)$, $(2, 2, 5)$; $(2, 1, 0)$, $(2, 1, 0)$, $(2, 1, 3)$; $(3, 1, 0)$, $(3, 1, 0)$, $(6, 2, 7)$; $(4, 1, 1)$, $(8, 2, 3 \pm \sqrt{21})$.

6. $4, -5, 0$; $0, 1, 0$; $a, 0, c$.

7. $Z = 0$, $X = 0$. 8. $(h \pm k)X - aY = 0$.

9. $X \pm Y = 0$. 10. $4X + (11 \pm \sqrt{65})Y = 0$.

12. $(1, 1, 0)$, $(1, 1, 0)$, $(1, 0, 1)$, $(0, 0, 1)$.

EXERCISE 8 D

1. $(a, \pm bi, 0)$.   2. $(0, 1, 0), I, J$.

3. $\pm ib/a; (b, \pm ai, 0)$.   4. $(k, \pm \sqrt{(3-k^2)})$.

5. $(-1, -1); (-1, -1), (\tfrac{1}{2}(1 \pm i\sqrt{7}), \tfrac{1}{2}(-5 \mp i\sqrt{7}))$.

6. $(0, 1, 0), (0, 1, 0), (1, 1, 1)$.

7. $(1, 1); (1, 1), (1, \omega), (1, \omega^2); (1, 1, 1); (1, 1, 1), (1, \omega, 1), (1, -\omega^2, 1)$.

8. $(0, 0), (9, 3); (0, 0), (9, 3), (9\omega^2, 3\omega), (9\omega, 3\omega^2); (0, 0, 1), (9, 3, 1);$
   $(0, 0, 1), (9, 3, 1), (9\omega^2, 3\omega, 1), (9\omega, 3\omega^2, 1)$.

9. $(k, \pm \tfrac{2}{3}\sqrt{(9-k^2)})$ but not in G_3 if $k^2 > 9$.

10. $((a^2+k^2)/2k, (a^2-k^2)/2k)$ if $k \neq 0$; $(1, 1, 0), (a^2+k^2, a^2-k^2, 2k)$.

11. $(0, 0, 1), (a^2+ab, ab+b^2, a^2-4b^2)$; yes.

12. $(0, 0, 1), (a\sqrt{(a-b)}, b\sqrt{(a-b)}, \pm a\sqrt{a})$; same unless $a(a-b) < 0$.

13. $(0, 0, 1)$ thrice, $(ab(b^2-a^2), b^2(b^2-a^2), a^4)$; yes.

14. $(0, 0)$ four times; $(0, 0, 1)$ four times and $(1, 0, 0)$.

15. $X = 0, Y = 0$.   16. $aX \pm ibY = 0$.

17. $aX^2 + bXY + aY^2 = 0$.

18. $x_1{}^2 X^2 + 2(x_1 y_1 - 2k^2) XY + y_1{}^2 Y^2 = 0$.

20. I, I, J, J.   21. $(x^2+y^2)(ax+by) = xy$.

22. $(x \pm iy)^2 + 2gxz + 2fyz + cz^2 = 0; I, I$ or J, J.

23. $32; 2\sqrt{2}, 2\sqrt{3}, 4$.

EXERCISE 9 A

6. $(ac/(b-c), 0)$ if $b \neq c; (ca, 0, b-c)$.

7. $(ac/(b-c), bd/(b-c))$ if $b \neq c; (ac, bd, b-c)$.

8. $ay' = \lambda(bx' - cx' - ac); ay' = \lambda(bx' - cx' - acz')$.

9. $ay' - dx' - da = -\lambda ab; ay' - dx' - daz' = -\lambda abz'$.

10. $\{c^2 x' - a(b^2-c^2)\}^2 + c^2 a^2 y'^2 = a^2 b^2 (b^2-c^2);$
    $\{c^2 x' + a(b^2+c^2)\}^2 - c^2 a^2 y'^2 = a^2 b^2 (b^2+c^2)$.

11. $(x' + \tfrac{3}{2}a)^2/a^2 - y'^2/b^2 = \tfrac{5}{4}; ay'^2 = b^2 x'; (x'-3a)^2/8a^2 + y'^2/b^2 = 1$.

12. $(b(b-c-a)/(b-c), 0)$ if C is $(c, 0)$.

13. $ay'(x'+a)^2 = (bx')^3; (ay')^2(x'+a) = (bx')^3$.

14. $ay'(x'+az')^2 = (bx')^3; (ay')^2(x'+az') = (bx')^3$.

15. $(a_1 b^2 + 2g_1 b + c_1) x'^2 + 2(h_1 b + f_1) x'y' + b_1 a^2 y'^2$
    $\qquad + 2(g_1 b + c_1) x' + 2f_1 a^2 y' + c_1 a^2 = 0;$
    $(a_2 b^2 + 2g_2 b + c_2) x'^2 + 2(h_2 b + f_2) x'y' + b_1 a^2 y'^2$
    $\qquad + 2(g_2 b + c_2) x'z' + 2f_1 a^2 y'z' + c_2 a^2 z'^2 = 0$.

16. Points on $x' = -a$; points on $z = 0$.

17. Parallel lines $by + cbz - cx = kabz$.

EXERCISE 10 A

1. $-\frac{8}{27}$.   **2.** $AP = PB$.   **3.** $CP = BA$.

5. Both or neither between A and B.

6. x_3 if $x_1 \neq x_2$; x_2 if $x_1 \neq x_3$; if $(ABCD) = 0$, $D \equiv C$ if $A \not\equiv B$; if $(ABCD) = 1$, $D \equiv B$ if $A \not\equiv C$.

7. $-q(p+q+r)/rp$.   **11.** $-\frac{1}{3}$, -3, $\frac{4}{3}$, 4, $\frac{3}{4}$, $\frac{1}{4}$.

12. (i) to (vi) 5, $\frac{1}{5}$, -4, $\frac{4}{5}$, $-\frac{1}{4}$, $\frac{5}{4}$.

     (vii) 100, $\frac{1}{100}$, -99, $\frac{99}{100}$, $-\frac{1}{99}$, $\frac{100}{99}$.

     (viii) $-\frac{9}{10}$, $-\frac{10}{9}$, $\frac{19}{10}$, $\frac{19}{9}$, $\frac{10}{19}$, $\frac{9}{19}$.

16. $1/\lambda$, $1-\lambda$, $1/(1-\lambda)$, $\lambda/(\lambda-1)$, $1-1/\lambda$.

EXERCISE 10 B

2. $t = 0$ gives no point; $(1, 0, 0)$ is given by no value of t.

3. No value of t gives $(-a, 0)$.

4. $(t_1 - t_2)(t_3 - t_4)/(t_1 - t_4)(t_3 - t_2)$.   **5.** a/b.

6. $(p - pq)/(q - pq)$; $-p/(1-p)$; $(ap - ab)/(bp - ab)$; a/b.

7. $k_2 k_1'/k_1 k_2'$.   **8.** t/t'; 2; $2k/(k-1)$; 2.

9. $x : y : c = -1 : -t_r : a + bt_r$; $(t_1 - t_2)(t_3 - t_4)/(t_1 - t_4)(t_3 - t_2)$.

11. $(Y_p - Y)/(YX_r - XY_r)$; $(X_1 - X_2)(X_3 - X_4)/(X_1 - X_4)(X_3 - X_2)$.

12. $(c/(ad - bc), -a/(ad - bc))$; -3.

EXERCISE 11 A

1. 12.   **2.** $\frac{1}{2}$, 2, $\frac{1}{2}$.   **4.** $k + k' = 0$.

8. $(a_1 h_2 - a_2 h_1) x^2 + (a_1 b_2 - a_2 b_1) xy + (h_1 b_2 - h_2 b_1) y^2 = 0$.

EXERCISE 11 B

1. $bx = ay$.   **2.** 2, -10, 3; 2λ, -10μ, 3ν, $-\rho$.

5. $XB_2 C_2$.   **10.** $11 \cdot 62$.

11. Four determinants like $\begin{vmatrix} a_2 & a_3 & a_4 \\ b_2 & b_3 & b_4 \\ c_2 & c_3 & c_4 \end{vmatrix}$ must not all be zero.

EXERCISE 11c

2. $ax + hy = 0.$ **3.** $(Y_1, X_1, 0).$

4. $AXX_1 + H(XY_1 + YX_1) + BYY_1 = 0.$

5. $(x_1X_1 + y_1Y_1 + 1)(x_2X + y_2Y + 1)$
$$+ (x_1X + y_1Y + 1)(x_2X_1 + y_2Y_1 + 1) = 0.$$

8. $(l_1x_1 + m_1y_1 + n_1z_1)(l_2x + m_2y + n_2z)$
$$+ (l_1x + m_1y + n_1z)(l_2x_1 + m_2y_1 + n_2z_1) = 0;$$
$$\Sigma(axx_1) + \Sigma f(yz_1 + zy_1) = 0.$$

EXERCISE 12a

1. $-2:1.$ **2.** $-(ax_1 + by_1 + c):(ax_2 + by_2 + c).$

3. $2:1, 1:2.$ **4.** $7 \pm \sqrt{145}:24.$

11. $s_{11} + 2s_{12} + s_{22} = 0.$ **12.** $1:4, 3:2, 114:211.$

13. Opposite. **14.** $b^2 < 4ac.$

15. $ax_1^2x_2 + b(x_1^2y_2 + 2x_1x_2y_1) + c(y_1^2 + 2y_1y_2) + d(2x_1 + x_2) + e;$
$$ax_1x_2^2 + b(2x_1x_2y_2 + x_2^2y_1) + c(2y_1y_2 + y_2^2) + d(x_1 + 2x_2) + e;$$
line; conic; P_1 lies on curve.

EXERCISE 12b

1. $ax_1^2 + by_1^2 = c; \ ax_1x + by_1y = c.$

2. $x_1y_1 = 1; \ xy_1 + yx_1 = 2.$

3. $3x_1^2 + 4x_1y_1 = 5; \ 3xx_1 + 2(xy_1 + yx_1) = 5.$

4. $x - 3y = 4.$ **5.** $3x - 4y = 6.$

6. $4(y_1^2 - ax_1 - by_1)(y^2 - ax - by) = \{2yy_1 - a(x + x_1) - b(y + y_1)\}^2.$

7. $x^2 - 10xy + y^2 + 98x - 10y + 1 = 0.$

8. $(b + c)x^2 = a(y + b)^2.$

9. $24x^2 - 60xy - 25y^2 - 72x - 160y - 196 = 0.$

10. $xx_1 + yy_1 = a^2.$ **11.** $xx_1/a^2 - yy_1/b^2 = 1.$

12. $y + y_1 = 2xx_1.$

13. $(4x_1 + y_1 - 1)x + (x_1 - 6y_1 + 2)y - (x_1 - 2y_1 - 2) = 0.$

14. $14y = 5(x + 4).$ **15.** $2x - 7y = 2.$ **16.** $13x + 24y = 3.$

17. $(\frac{1}{2}, \frac{3}{4}).$ **18.** $(-5, 3).$ **19.** $(\frac{1}{5}, \frac{2}{5}).$

20. $(1, 1).$ **21.** $(-a/2b, c/b).$ **22.** $(\frac{1}{18}, \frac{16}{9}).$

23. $(0, -2a/b)$ if $b^2 \neq 4ac.$ **24.** $xy_1 + yx_1 = 2k^2.$

25. (i) $a(x-b)^2 + by^2 = 0$;

(ii) $ax^2 - bxy - ay^2 + (2a^2 + b^2)x + aby + a^3 = 0$.

28. (i) $(0, 0)$; (ii) points on $x = 0$. **29.** $x + yt^2 = 2kt$.

31. $gx + fy + c = 0$, $x^2 + y^2 + gx + fy = 0$. **34.** $(3\frac{3}{5}, \frac{1}{5})$.

35. $a(by_1{}^2 - 1)x^2 - 2abx_1y_1xy + b(ax_1{}^2 - 1)y^2 = 0$.

37. $x^2 + y^2 = a^2 - b^2$, $x + a = 0$, $z(x + az) = 0$.

40. $\begin{vmatrix} x & y & 1 \\ x_1 & y_1 & 1 \\ G & F & C \end{vmatrix}^2 = 0$.

EXERCISE 12c

3. $xx_1 - yy_1 = a^2zz_1$. **4.** $xy_1 + yx_1 = 2k^2zz_1$.

5. $k(xx_1 + yy_1) + g(xz_1 + zx_1) + f(yz_1 + zy_1) + czz_1 = 0$.

6. $x + y + az = 0$. **7.** $12x + 23y - 17z = 0$.

8. $x_1x_2/a^2 - y_1y_2/b^2 = 1$; $x_1x_2/a^2 - y_1y_2/b^2 = z_1z_2$.

9. $x_1y_2 + x_2y_1 = 2k^2$; $x_1y_2 + x_2y_1 = 2k^2z_1z_2$.

11. $ax_1x_2 + h(x_1y_2 + x_2y_1) + by_1y_2 - k$.

12. $k(x_1x_2 + y_1y_2) + g(x_1z_2 + x_2z_1) + f(y_1z_2 + y_2z_1) + cz_1z_2$.

13. $(b^2, 4ab, 4a)$, $(1, 0, 0)$; $(b, 2a, 0)$.

14. $(c, 0, ak)$. **15.** $(30, 8, -21)$.

16. $((hm - bl)/nC, (hl - am)/nC)$ if $nC \neq 0$, and no point if $n = 0$; no point unless $l : m = a : h$, and then any point on
$$a(ax + hy) = -l/n \text{ if } n \neq 0.$$

17. $(hm - bl, hl - am, nC)$ if $C \neq 0$; no point unless $l : m = a : h$, and then any point on $an(ax + hy) + lz = 0$.

18. $(lx + my + n_1 + n_2)(lx_1 + my_1 + n_1 + n_2) = (n_1 - n_2)^2$, unless P_1 is midway between the lines. In G_6, P_1 may be any point except $(-m, l, 0)$.

19. $y^2 - 2ax = gy - 2af$. **20.** $2xy = gx + fy$.

21. $amx = bly$. **22.** $(0, 0, 1)$, $(0, 0)$. **23.** $(0, 0, 1)$, $(0, 0)$.

24. $(1, 0, 0)$, none. **25.** $(1, -2)$. **26.** $(7\frac{23}{31}, -4\frac{6}{31})$.

27. $(12, -6, 7)$. **28.** Any point on $3x - 4y = 2$.

29. Any point on $z = 0$.

30. $k(ax + hy + g) + mk(hx + by + f) + (gx + fy + c) = 0$;

$(ax + hy + g) + m(hx + by + f) = 0$.

31. $(1, 0, 0)$, $(0, 1, 0)$, $(0, 0, 1)$; conic is degenerate.

EXERCISE 12D

1. $X^2/a + Y^2/b = 1/c.$ **2.** $x^2/a + y^2/b = z^2/c.$

3. $4XY = 1.$ **4.** $x^2 + 2yz = 0.$

5. $c^2X^2 + 2(2ad - bc)XY + b^2Y^2 - 2acX - 2abY + a^2 = 0.$

6. $a^2X^2 - b^2Y^2 = 1.$ **7.** $4k^2XY = 1.$

8. $kc(X^2 + Y^2) - (fX - gY)^2 - 2kZ(gX + fY) + k^2Z^2 = 0.$

9. $8X^2 - 4XY + 5Y^2 - 24X + 6Y + 9 = 0.$

10. $19X^2 + 27Y^2 - 2Z^2 - 2YZ - 2ZX + 14XY = 0.$ **11.** $x^2 + y^2 = k^2.$

12. $(3x - y - 2)^2 = 0$; conic is degenerate. **13.** $(X - Y)^2 = 0.$

14. $\{(m - m')X - (l - l')Y + (lm' - l'm)\}^2 = 0.$ **15.** $0 = 0.$

16. $y + 2 = 0$ joins the points $(3X + 2Y - 1)(X - 2Y + 1) = 0.$

17. The pole of $[g/c, f/c]$ has no equation and lines through $(G/C, F/C)$ have no poles.

EXERCISE 12E

1. $[3, 0], [\frac{3}{53}, \frac{234}{53}].$ **2.** $[11, -8, 2], [0, 3, 2].$

3. $PXX_1 + QYY_1 = 1.$ **4.** $X/X_1 + Y/Y_1 = 2Z/Z_1.$

5. $(-1, 0).$ **6.** $(15, 3, -7).$ **7.** $\frac{3}{2}, -\frac{7}{2}.$ **8.** $1, 2.$

9. $(X + 1)(11X + 12Y - 13) = 0.$

10. $(aX + 2aY + 1)(aX - 2aY + 1) = 0.$

11. $(0, c, b), (c, 0, a), (b, a, 0).$

12. $2aYY_1 = X + X_1;\ 2aYY_1 = XZ_1 + ZX_1.$

13. $X = 1.$ **14.** $X - 7Y + 8Z = 0.$

15. $GX + FY + CZ = 0.$ **16.** $[2, -1].$

17. $(F(FL - GM - HN), G(GM - HN - FL), H(HN - FL - GM)).$

18. $2(X_1X_2 + Y_1Y_2) = Y_1Z_2 + Y_2Z_1.$

19. $2p_1p_2 + q_1q_2 = 2r_1r_2.$ **20.** $(0, 0).$

21. $(1, 1, -1).$ **22.** $(1, -\frac{3}{2}).$ **23.** $(b, a, 0).$

24. $(-c/a, -b/a).$ **25.** $x^2/a + y^2/b = 1/c.$ **26.** $ky^2 = 2zx.$

27. $(x - az)^2 + (y - bz)^2 = r^2z^2.$ **28.** $x^2 + 12xy + 2y^2 + 10x + 8y + 3 = 0.$

29. The meet $p_1p_2.$ **30.** $c(X^2 + Y^2) - 2gX - 2fY + a + b = 0.$

31. $SS_{11} = S_1^2;\ a + b = 0.$ **32.** $(bX^2 + aY^2)Z_1^2 = (XY_1 - YX_1)^2.$

34. $x^2 : y^2 : 1 = bc(b - d) : ad(c - a) : ab(bc - ad);$

on $ac(b + d)x^2 + bd(c + a)y^2 = bc + ad.$

EXERCISE 12f

[The answers to Nos. 1–4 are not unique]

1. $x:y:1 = 1:t:t+t^2$.

2. $x:y:1 = 1-t:t-t^2:(1+t)^2$.

3. $x:y:1 = 4-5t:4t-5t^2:1+2t+3t^2$.

4. $x:y:1 = 5t^2-17t-12:-16t^2-10t+2:5t^2+4t+2$.

5. $(x-y)^2 = 25(y-4)$.　　　　**6.** $(2x-y)(2y-x) = 9$.

7. $202x^2-135xy+19y^2-40x+5y+7 = 0$;
$3X^2+20XY+24Y^2+10X+40Y-17 = 0$.

8. $102x^2-70xy+12y^2+90zx-31yz+20z^2 = 0$;
$X^2-20XY-60Y^2-20XZ-48YZ+4Z^2 = 0$.

9. $[2, 15, -10]$.

11. $x(t_1t_2-t_1-t_2-3)+y(3t_1t_2-2t_1-2t_2-6) = t_1t_2-t_1-t_2-2$.

12. $[4u-7v+6w, -7u+12v-11w, 6u-11v+10w]$.

13. $(1, 1, -1)$.　　　　　　**14.** $(1, 1, -1)$.

15. $[1, -1, 1], [-1, 2, 2]$.　　**16.** $\sqrt{2}$.

17. $(\frac{19}{49}, -\frac{6}{49})$; $(x-5y-1)(9x+4y-3) = 0$;
parallel to $41x^2+58xy-41y^2 = 0$.

18. $(-\frac{2}{3}, -\frac{1}{3})$; parallel to $x^2-12xy-y^2 = 0$.

19. $t = -4$; $12x+4y = 7$.

20. The point (a_1, a_2, a_3) or (b_1, b_2, b_3) or (c_1, c_2, c_3).

22. $(dy+2ef)^2 = 4(ax+cy+e^2)(bx+f^2)$.

23. $y^2 = 4x^3$, $X^3 = -27Y^2$.　　**24.** Four points $(\pm 1, \pm 1)$.

25. $y^4 = 4ax^2$, $aY^4 = 4X^2$.　　**26.** $(x/2a)^{\frac{2}{3}}+(y/2b)^{\frac{2}{3}} = 1$.

27. $(x-y+1)^2 = 4x$, $XY+X+Y = 0$.

28. $x^2+y^2 = 1$.　　　　　　**29.** $27x^4+y^4 = 27x^2$.

30. Touches four lines $[\pm 1, \pm 1]$.

EXERCISE 13a

1. $ax-by+b^2 = 0$.　　　　**2.** $(x-y+a)(x-2y+4a) = 0$.

3. $g^2 = a^2+6af+f^2$.　　　　**4.** $2y^2 = kx$.

5. t_1+t_2 constant.

7. (at_1t_2, at_1+at_2); $Xt_1^2+2Yt_1+1/a = 0$.

9. $\pm\frac{1}{4}\pi$.

EXERCISE 13B

1. $x \pm y + a = 0.$　　　　**3.** $4a \operatorname{cosec}^2 \alpha.$

4. $(7x - y)^2 = 276x + 32y - 506.$

5. $(x + 2y)^2 = 56x + 12y - 184.$

6. $(4x - 3y)^2 + 148x + 164y - 964 = 0.$

13. $r = a(1 + \cos \theta)$ with pole S and initial line SA.

14. $r \cos \theta = a \cos 2\theta$ with pole at $(-a, 0)$.

EXERCISE 13C

1. $|y^2 - 4ax|/4a.$　　　　**4.** $b, \ -2b \cos \omega.$

6. $(4ax - y^2)(4a^2 + y^2) = a^2 l^2.$　　　**8.** $y^2 = 4bx - k^2$ (axes of 13·75).

EXERCISE 13D

1. $3x - y = 33.$

2. $(b - 2y_1)x - ay = (b - 2y_1)x_1 - ay_1.$

3. $x + y = 2a.$　　　　**5.** $km(2l^2 + m^2) + l^2 n = 0.$

6. (i) $y - 3x + 33a = 0;$

(ii) $y + 3x = 33a, \ 2y + (-3 \pm \sqrt{5})x = 2a(-12 \pm 5\sqrt{5}).$

7. $(a(2 + t_1{}^2 + t_1 t_2 + t_2{}^2), \ -at_1 t_2(t_1 + t_2)).$

9. $2a \operatorname{cosec} \theta, \ 4a \sec^2 \theta \operatorname{cosec} \theta.$　　**10.** $y^2(x + 2a) + 4a^3 = 0.$

11. $y^4 - 2ay^2 x + 4a^2 y^2 + 8a^4 = 0.$

12. $y^2 = 16a(x - 2a \pm 4a).$　　　**19.** $2x + ty + 4a = 0.$

EXERCISE 13E

1. $x = 0, y = 0, 3.$　　　　**2.** $y = 0, x = -2, 5.$

3. $y = 2, x = 1, 4.$　　　　**4.** $x = -a, 2by = a^2 - c, 2|b|.$

5. $(-1, 0), x = -2.$　　　　**6.** $(0, -1), y = 3.$

7. $(-b/a + a/4, 0), x = -b/a - a/4.$

8. $(a - b^2/c + 1/4c, -b/c), x = a - b^2/c - 1/4c.$

9. $x + y = 0, (0, 0).$　　　　**10.** $x - 2y + 1 = 0, (-\frac{1}{5}, \frac{2}{5}).$

11. $x + 2y - 8 = 0, (4, 2).$　　**12.** $5x + 12y + \frac{1}{2} = 0, (-2\frac{1}{2}, 1).$

13. $7x - y - 19 = 0, (3, 2).$　　**14.** $7x - 24y + 1 = 0, (-\frac{7}{625}, \frac{24}{625}).$

15. $x = (u^2/g)\sin\alpha\cos\alpha$; $((u^2/g)\sin\alpha\cos\alpha, (u^2/2g)\sin^2\alpha)$; $(2u^2/g)\cos^2\alpha$.

16. $gx^2 = -2yu^2\cos^2\alpha$.

19. $x(1-t_1)/a + yt_1/b = t_1 - t_1^2$.   **20.** $13a$.

21. $gx - 2bt_1y = ga - 2bt_1f - bgt_1^2$; $4bx = 4ab - g^2$.

22. $x:y:1 = bc^2 - 2acd - ba^2 : da^2 - 2abc - dc^2 : 4b^2 + 4d^2$.

23. $r\sin^2\theta = 4a\cos\theta$; $r\cos\theta = a\sin^2\theta$.

24. Parabola with focus the pole.

25. See No. 24: $2ar = k^2(1+\cos\theta)$.

EXERCISE 13F

1. $a\operatorname{cosec}^2\theta$, $a\sec^2\theta$, $2a\operatorname{cosec}\theta$, $4a\sec^2\theta\operatorname{cosec}\theta$.

3. Parabola with focus at centre of given circle.

11. $2x + 15 = \pm y\sqrt{5}$.   **12.** $2x + 3595 = \pm(y+171)\sqrt{221}$.

16. $bx - 4ay + 16a^2 = 0$.   **19.** $1 + (t_0+t_1)(t_0+t_2) = 0$.

20. $(t_2-t_3)(1+t_1^2) : (t_3-t_1)(1+t_2^2) : (t_1-t_2)(1+t_3^2)$.

EXERCISE 14A

1. $12x - 7y = 85$.   **2.** $x = 0, y = -1$.

3. $x + 3y = \pm\sqrt{4\cdot1}$.   **4.** $3x - 56y = 3$.

5. $b^2cx - a^2dy = b^2c^2 + a^2d^2 - a^2b^2$.

6. $x = -ka^2$.   **7.** $3x + 4y + 5 = 0$.

8. $b^2lx + a^2my = 0$.

9. $249x^2 + 300xy + 80y^2 - 6x + 100y - 259 = 0$.

10. $16x^2 + 12xy - 59y^2 - 76x + 94y + 29 = 0$.

11. $(-\frac{4}{35}, \frac{3}{10})$.   **12.** $(p - r^2/p, 0)$.   **13.** $(-15, -12)$.

14. $(a^2u, b^2v, 0)$.   **15.** $(X-1)(11X - 12Y + 13) = 0$.

16. $a^2b^2(mX - lY)^2 = (a^2l^2 + b^2m^2)Z^2$.

18. $(-a^2ln/(b^2m^2 + a^2l^2), -b^2mn/(b^2m^2 + a^2l^2))$.

19. $n^2(x^2/a^2 + y^2/b^2) = (lx + my)^2$.

20. $x^2 + y^2 - 6x - 4y + 11 = 0$.   **21.** $\sqrt{(a^2\sin^2\psi + b^2\cos^2\psi)}$.

23. $9a^2 + 36b^2 = 4$.   **24.** $(-a^2l/n, -b^2m/n)$.

26. Perpendicular tangents from (x, y); no tangents from (x, y).

27. $t_1t_2 = -1$.   **28.** $(t_1 + t_2)/(1 - t_1t_2)$ constant.

29. $b^2(1-t_1t_2)(1-t_3t_4)+a^2(t_1+t_2)(t_3+t_4) = 0$.

30. $(a(1-t_1^2t_2^2)/\{(1+t_1^2)(1+t_2^2)\},\ b(t_1+t_2)(1+t_1t_2)/\{(1+t_1^2)(1+t_2^2)\})$.

31. $\phi_1-\phi_2 = (2n+1)\pi,\ b^2\cos\phi_1\cos\phi_2+a^2\sin\phi_1\sin\phi_2 = 0$.

32. $\phi_1+\phi_2$ is constant. **33.** $4x^2y^2 = b^2x^2+a^2y^2$.

35. $(x/a)\cos\frac{1}{2}\phi+(y/b)\sin\frac{1}{2}\phi = \cos\frac{1}{2}\phi$,

$\quad -(x/a)\sin\frac{1}{2}\phi+(y/b)\cos\frac{1}{2}\phi = \sin\frac{1}{2}\phi$;

$\quad (a,\ b\tan\frac{1}{2}\phi),\ (-a,\ b\cot\frac{1}{2}\phi)$.

40. $x^2/\{a^4(b^2+d^2)\}+y^2/\{b^4(a^2+c^2)\} = 1/(b^2c^2+a^2d^2)$.

EXERCISE 14B

1. $2\sqrt{3}$ in. **2.** 4 in. **3.** $\frac{5}{8}$.

4. $(\pm4\sqrt{2},\ 0);\ x = \pm\frac{9}{2}\sqrt{2};\ \frac{2}{3}\sqrt{2}$.

5. $(0,\ \pm2\sqrt{10});\ y = \pm\frac{49}{20}\sqrt{10};\ \frac{2}{7}\sqrt{10}$.

6. $(\frac{1}{2},\ -\frac{1}{2}),\ (-\frac{1}{2},\ -\frac{3}{2});\ x+y = 3,\ x+y = -5;\ \frac{1}{2}$.

7. $(0,\ 0),\ (-\frac{2}{3}p\cos\alpha,\ -\frac{2}{3}p\sin\alpha)$;

$\quad x\cos\alpha+y\sin\alpha = p,\ x\cos\alpha+y\sin\alpha = -\frac{5}{3}p;\ \frac{1}{2}$.

8. $\frac{1}{2}\sqrt{2},\ \sqrt{3}$. **9.** $\frac{1}{5}\sqrt{21},\ \frac{8}{5}$. **10.** $\frac{1}{2}\sqrt{2},\ 2\sqrt{2}$.

11. $\frac{1}{9}\sqrt{77},\ \frac{8}{9}$. **12.** $\pm x\sqrt{7}\pm4y = 16$.

13. $1+t_1t_2 = \pm e(1-t_1t_2)$. **15.** $l/(1-e^2),\ l/\sqrt{(1-e^2)}$.

27. $x^2+y^2 = 2ax+b^2;\ y^2 = 2x,\ 2x+1 = 0$.

EXERCISE 14C

1. $b^2x+a^2ty = 0$; parallels to $b^2x+a^2sy = 0$.

2. $bl_1l_2+am_1m_2 = 0$. **6.** $bl^2+am^2 = 2abn^2$.

8. $n(mx-ly)+(a^2-b^2)\,lm = 0$. **13.** $(a^2x^2+b^2y^2)^3 = (a^4x^2+b^4y^2)^2$.

16. $x^2/a^2+y^2/b^2 = \pm ex/a$.

17. $(a^2+b^2)(x^2/a^2+y^2/b^2)^2 = b^2x^2/a^2+a^2y^2/b^2$.

21. Find C by mid-point loci of 14·71, and use 14·81.

22. $-2\phi_1-\phi_2$. **23.** -3θ. **24.** -3ϕ.

EXERCISE 14D

11. $a^2x^2/(2a^2-b^2)^2+y^2/b^2 = \frac{1}{4}$.

14. $(x^2/a^2+y^2/b^2)^2(a^6/x^2+b^6/y^2) = (a^2-b^2)^2$.

20. $2b;\ 3a^2b^2\sqrt{3}/\sqrt{(a^2+b^2)^3}$. **23.** $x:y:a^2+b^2 = f:-g:a^2-b^2$.

EXERCISE 14E

1. $4x^2 + 9y^2 - 64x + 54y + 301 = 0.$

2. $16x^2 + 4xy + 19y^2 - 34x - 23y + 4 = 0.$ **3.** $\frac{1}{3}\sqrt{3}, \frac{8}{87}\sqrt{58}.$

4. $2x + y = 7,\ 2\sqrt{5},\ x^2 + y^2 - 6x - 2y + 5 = 0.$

5 and 6. $(x+11)\sqrt{3} = \pm 8y,\ 3x - 11 = \pm 4y.$

7. $(x^2 + y^2)^2 = k^4(x^2/a^2 + y^2/b^2).$

8. $2ab^2x(x^2 + y^2) = k^2(b^2x^2 + a^2y^2).$ **10.** Circle.

20. $(x\cos\phi + y\sin\phi)^2/a_1^2 + (x\sin\phi - y\cos\phi)^2/b_1^2 = 1.$

23. $a^2x^2 - xy(a^2\tan\gamma + b^2\cot\gamma) + b^2y^2$
$$= (a^2 - b^2)^2(a^2\sin^2\gamma - b^2\cos^2\gamma)^2/(a^2\sin^2\gamma + b^2\cos^2\gamma)^2.$$

25. $x^2/(pc)^2 + y^2/(qd)^2 = r^2,$ where
$$p:q:r = b^2c^2 - a^2d^2 - a^2b^2 : a^2d^2 - b^2c^2 - a^2b^2 : b^2c^2 + a^2d^2 - a^2b^2.$$

EXERCISE 15A

1. $2x + 12y = 1.$ **2.** $x/a = \pm y/b.$ **3.** $z = 0.$

4. $a^2X = kb^2Y.$ **5.** $z = 0.$ **6.** $5x - 2y - 3 = 0.$

7. $(2x - y)(5x - 2y - 3z) = 0.$ **8.** $a^2X^2 = b^2Y^2.$

9. $(x - a)\{(b^2 + k^2)x - 2aky - (b^2 - k^2)a\} = 0.$

10. $1/a^2 - 1/b^2$; no such diameters exist if $b^2 < a^2.$

11. $(1 - t_1^2)(1 - t_2^2) > 0.$ **12.** $t_1 t_2 = 1.$

13. $(t_1 + t_2)/(1 + t_1 t_2)$ constant.

14. $a^2(t_1 + t_2)(t_3 + t_4) + b^2(1 + t_1 t_2)(1 + t_3 t_4) = 0.$

15. $x:y:1 = a(1 - t_1^2 t_2^2):b(t_1 + t_2)(1 - t_1 t_2):(1 - t_1^2)(1 - t_2^2).$

16. $x^2 + y^2 = a^2 - b^2.$

17. $(1 + t_1 t_2)x/a + (1 - t_1 t_2)y/b = t_1 + t_2;\ (1 + t_1^2)x/a + (1 - t_1^2)y/b = 2t_1;$
$$a(1 - t_1^2)x - b(1 + t_1^2)y = (a^2 + b^2)(1 - t_1^4)/(2t_1).$$

18. $(la + mb)t^2 + 2nt + (la - mb) = 0.$

19. $(u + w)x/a + (u - w)y/b + 2v = 0.$

20. $x:y:1 = a(t_1 t_2 + 1):b(t_1 t_2 - 1):t_1 + t_2.$

21. $x:y:1 = -a(u + w):b(u - w):2v.$

22. $(x/a)\operatorname{ch}\alpha - (y/b)\operatorname{sh}\alpha = 1.$

23. $(x/a)\cos\frac{1}{2}(\phi_1 - \phi_2) - (y/b)\sin\frac{1}{2}(\phi_1 + \phi_2) = \cos\frac{1}{2}(\phi_1 + \phi_2);$
$$(x/a) = (y/b)\sin\phi + \cos\phi.$$

24. $r^2 = a^2\cos 2\theta.$ **25.** $x^2y^2 = a^2y^2 - b^2x^2.$

EXERCISE 15B

7. Except when $APB = BP + PC$. **8.** Half a hyperbola.

12. $(\pm\sqrt{5}, 0)$, $x = \pm\frac{4}{5}\sqrt{5}$, $\frac{1}{2}\sqrt{5}$. **13.** $(\pm\sqrt{13}, 0)$, $x = \pm\frac{4}{13}\sqrt{13}$, $\frac{1}{2}\sqrt{13}$.

14. $(\pm 13, 0)$, $x = \pm 11\frac{1}{13}$, $1\frac{1}{12}$. **15.** $(1\pm\sqrt{2}, -2)$, $x = 1\pm\frac{1}{2}\sqrt{2}$, $\sqrt{2}$.

16. 3, $\frac{1}{2}\sqrt{7}$. **17.** $2\sqrt{2}$, $\sqrt{2}$. **18.** $2\frac{2}{3}$, $\frac{1}{3}\sqrt{13}$.

19. $\frac{1}{2}$, $\frac{1}{2}\sqrt{5}$. **20.** $a^2b^2/(a^2+b^2)$.

21. $(x^2-y^2)^3 + 4a^2x^2y^2 = 0$. **22.** $a^6/x^2 - b^6/y^2 = (a^2+b^2)^2$.

23. $x^2/(a^2+b^2) - y^2/b^2 = 1$. **24.** $((a+2b^2/a)\,\mathrm{ch}\phi, -b\,\mathrm{sh}\phi)$.

EXERCISE 15C

1 and 2. $x/a \mp y/b = 0$. **3.** $(x-a)(bx-ay) = 0$.

4. $(x+3y)(x-3y) = 0$. **5.** $(2x+y)(2x-y) = 0$.

6. $(x-y+11)(x+y+1) = 0$. **7.** $(x+y-3)(x-y-1) = 0$.

8. $(7x-y-6)(x+7y-8) = 625$.

9. $x^2 - 4y^2 = \pm 11$; $\frac{1}{2}\sqrt{5}$, $\sqrt{5}$. **10.** $1/e_1^2 + 1/e_2^2 = 1$.

EXERCISE 15D

2. $(a\,\mathrm{sh}\phi, b\,\mathrm{ch}\phi)$, $(-a\,\mathrm{sh}\phi, -b\,\mathrm{ch}\phi)$. **9.** $Aa^2 = Bb^2$.

EXERCISE 15E

3. $t_1^2 + t_2^2 = 0$. **4.** $h = 0$.

5. $x\{e^2x_1 - (2-e^2)y_1\} - y\{e^2y_1 - (2-e^2)x_1\} = e^2(x_1^2 - y_1^2)$.

6. $(x-2)(y-3) = 0$. **7.** $(4x+1)(3y-2) = 0$.

8. $(2x-3)(2y-3) = 0$. **9.** $(cx+d)(cy-a) = 0$.

10. $x+y \pm 2k/e$. **14.** A similar hyperbola.

16. $2axy + (b-af)x + (c-ae)y - (be+cf) = 0$.

21. $abc < 0$.

36. $m^2(mx+y)(x-my) = k^2(1-m^2)(1+m^2)^2$.

40. $(p_1/(1+p_4), p_3/(1+p_4))$.

EXERCISE 16A

1. $441xy = 11(4x+5y-20)(3x+2y+6)$.

2. $3x^2+12xy+8y^2+6x-8y-24 = 0$.

3. $(2x+3y)^2 = 72x+93y-230$, $(2x-3y)^2 = 24x+69y-182$.

4. $y^2 = 4x+2y+11$, $(x-2y)(x-2y+8) = 0$.

5. $(x-1)(y+1) = 0$.   **6.** $5(x^2-y^2) = 3$.

7. $11x^2-25xy+11y^2 = 0$, $(x+y)^2 = 47$, $(x-y)^2 = 3$.

8. $\{(a+2h+b)r^2-1\}(x^2+y^2-r^2) = r^2(ax^2+2hxy+by^2-1)$.

9. $4x^2-3y^2 = 4ax$.

10. $\{x+(1\pm\sqrt{2})y\}^2 = 7\pm 3\sqrt{2}$.

11. $4x^2-4y^2-ax+2a^2 = 0$.   **12.** $(x+2)(y+2) = 0$.

14. $12x^2+13xy+4y^2 = 54$.

15. $x^2/a^2+y^2/b^2-1 = kxy$; degenerate parabolas if $k = \pm 2/ab$.

17. $Y = 1+X+X^2$.   **18.** $b^2X^2+(c^2-a^2)Y^2 = b^2c^2$.

19. $X^2-3YZ+2ZX = 0$.

20. $(a^2+b^2)X^2 = 1$, $(a^2+b^2)Y^2 = 1$.

21. $(X\pm Y)^2 = 1$.   **22.** $2\frac{2}{5}$.

23. $X^2+5XY+Y^2+2X-2Y = 0$.

24. $(X_1{}^2+Y_1{}^2)/(a^2X_1{}^2+b^2Y_1{}^2-Z_1{}^2)$; only $X^2+Y^2 = 0$.

EXERCISE 16B

1. $(x/a+y/b-1)(x/a+y/c+1) = kxy$; $a^2 = bc$;
$\quad x^2+y^2+(c-b)y-bc = 0$.

2. $X^2-1 = k(Y^2-1)$; $x^2/\lambda+y^2/(1-\lambda) = 1$.

3. $(am^2+bl^2)y^2+2alx-l^2-a = 0$, $(am^2+bl^2)x^2-2blx-m^2+b = 0$.

4. $8kXY+X+Y = 0$.

7. (i) $(y^2-4ax) = (x-ty+at^2)\{p(x-at^2)+q(y-2at)\}$,
$\quad\quad (aY^2-X) = (aX-2atY+t^2)\{p(aX-t^2)+q(aY-t)\}$;
$\quad$ (ii) $(y^2-4ax) = p(x-ty+at^2)^2$, $(aY^2-X) = q(aX-2atY+t^2)^2$.

9. $t^3(x^2+y^2)-k(3t^4+1)x-kt^2(t^4+3)y+3k^2t(t^4+1) = 0$.

10. $x^2-2txy-y^2+2a(3t^2+2)x-2at^3y+a^2t^4 = 0$.

11. $21x^2+19xy-21y^2+19x+73y-71 = 0$.

14. Touches at B, meets at A and A'; $\kappa b^2/(\kappa b^2-2)$.

15. Double contact along $x = a$; hyperbola, centre O.

16. Focus O, directrix $x\cos\alpha + y\sin\alpha = a$, eccentricity κ.

17. Double contact with $y^2 + 2kx = 0$ along P_1P_2.

18. Four-point contact with $xy = k^2$ at $(kt, k/t)$.

19. A diameter and an asymptote in common with $xy = k^2$.

20. Not meeting $xy = k^2$; asymptote $x = 0$.

21. Four-line contact with $aY^2 = X$ on $[1/a, 1/a]$.

22. Circle of curvature at vertex of $aY^2 = X$.

23. Double contact with ellipse along minor axis.

24. Double contact with ellipse along $x + y = 0$.

EXERCISE 16 c

8. $(a - b\cos\omega)/a^2b$.

9. $(x/a + y/b - 1)(x/a' + y/b' - 1) = kxy$; if $aa' = bb'$.

EXERCISE 16 D

1. $ax^2 + 2hkx + bky = 0$.   **2.** $5x^4 + 19x^2y^2 - 40y^4 = 0$.

3. $x^2 + y^2 + 3x + 3y = 0$.   **4.** $x^2 + y^2 - x - 3y = 0$.

6. $20x^2 + 20y^2 + 13x + 34y - 11 = 0$.

8. $(a + \tfrac{1}{2}f, \tfrac{1}{4}g)$, $(\tfrac{2}{3}f - \tfrac{4}{3}a, 0)$, $(f - 6a, -\tfrac{1}{3}g)$.

20. Plane collinear with $\alpha = 0$, $\beta = 0$.

21. Sphere through common circle of $\sigma = 0$, $\sigma' = 0$.

22. Sphere through common circle of $\sigma = 0$, $\alpha = 0$.

23. Sphere concentric with $\sigma = 0$.

24. Conicoid through common curve of $s = 0$, $s' = 0$.

25. Conicoid meeting $s = 0$ in two conics (double contact).

26. Conicoid having ring contact with $s = 0$.

27. Two planes collinear with $\alpha = 0$, $\beta = 0$.

28. Two conicoids through common curve of $s = 0$, $s' = 0$.

EXERCISE 16E

2. $(ax+by-ab)(bx+ay) = kxy$; $2(x^2-y^2) = bx-ay$.

5. $(3, 1), (-21, 25)$.

6. $62x^2+62y^2-3x-17y+1 = 0$.

8. $(bx+ay-ab)^2\cos\omega + (a^2+b^2-2ab\cos\omega)xy = 0$.

9. $(p/r, 0), (0, p/q)$.

10. Conic touching joins of $(a, 0, 1)$ $(0, b, 1)$ to $(0, 0, 1)$ and $(c, d, 1)$; centre of conic; join of mid-points of diagonals.

11. Circle touching $y^2 = 4ax$ at $(at^2, 2at)$; $\lambda = -3at^2$.

12. Product of perpendiculars from P_1, P_2 to $[X, Y, Z]$ is constant; conic with foci at the given points.

MISCELLANEOUS EXERCISE B

1. $(2x-a)(c-d) = (2y-c)(a-b)$.

7. $(\frac{3}{2}a, \pm\frac{1}{3}a\sqrt{3})$; $xy^2 = a^3$.

8. Parabola touching the lines.

11. Parabola touching the axes.

12. When $\lambda\lambda' = 1$, QR is parallel to OY.

18. Hyperbola with focus at the foot of the perpendicular from A to BC.

20. $xy(ak+bh-hk)^2 = ab(kx+hy-hk)^2$.

21. $bt_1t_2 = a$.   **22.** $n = 2m$.

23. $2|f_1g_2-f_2g_1|/\sqrt{\{(f_1-f_2)^2 + (g_1-g_2)^2\}}$.

24. $x^2/a^2+y^2/4b^2 = 1$.   **26.** $(2a, \pm 2a\sqrt{2}), (0, 0)$.

28. $8x^3 - 2a^2y + 3a^2y_1 = 0$.

29. $\{b(c-a)+dc(b-a)\}x = \{(c-a)+d(b-a)\}y$.

30. $4; 6$.   **32.** $\sqrt{2}; \sqrt{10}$.   **37.** 2α.

40. $4ab(bx^2+ay^2) = (a-b)^2$.

41. (i) Parallel lines, one through (a, b), $lx+my+1 = 0$ midway between them; (ii) Circle, centre (a, b) touching $lx+my+1 = 0$; (iii) Parabola, focus (a, b), directrix $lx+my+1 = 0$.

43. Circle, centre B.

49. $4(aX+1)^3 + 27a^2Y^2 = 0$; $4aX-8aY+5 = 0$.

50. The diagonals.   **51.** $(r\sqrt{2})/\sqrt{(r^2+c^2)}$.

52. $9y^2 = 4a(x - 8a)$.　　　**53.** $b/(2a^2)$.

54. $a + b = 0$, parabola unless $k^2 = -1/a$; Frégier's point if $k^2 = 1/b$; there is only one chord if $k^2 = 1/a + 1/b$.

56. $(al + a'm)\{a'(x - c) - a(y - c')\} = \frac{1}{2}(bl + b'm)(ab' - a'b)$.

57. $(ac(b - d)/(bc - ad), \; bd(c - a)/(bc - ad))$.

58. G_5: Line and point, point and line. G_6: Three lines, three points.

59. $\tan^{-1}[\{\sqrt{(\beta^2 - 4a\alpha)}\}/(a + \alpha)]$; $(y^2 - 4ax)\cos^2\gamma = (x + a)^2\sin^2\gamma$.

60. $(14 \pm 15\sqrt{3}, \; 27 \pm 20\sqrt{3})$; $3x + 4y = 150 \pm \frac{250}{3}\sqrt{3}$.

63. The origin.

64. $(bx + ay)(k + 1) = 2(bc + ad) + ab(k - 1)$.

65. Surface of revolution.

71. $\Sigma[x_1 y_1\{x(y_2 - y_3) - y(x_2 - x_3) + (x_2 y_3 - x_3 y_2)\}] = 0$.

72. $r = c + c\cos\theta$.

73. (i) Chords pass through $(k, 0)$; (ii) there is only one chord; (iii) there are no chords.

76. $2x(x^2 + y^2) = k^2(ax^2 + by^2)$; $2x = k^2 b$.

77. $x + y : x - y + z : x + 2y + z$; $(x - y + z)^2 = (x + y)(x + 2y + z)$.
$(X - Y + Z)^2 + 12(X - Z)(X - Y - 2Z) = 0$;
$[1, -3, 2]$; $(-17, 5, 25)$.

78. A line through the meet of the given lines.

81. Two conics with foci at the centres.

83. A parabola.

84. Ordinary branch and cusp; $t_2 t_3 + t_3 t_1 + t_1 t_2 + t_1^2 t_2^2 t_3^2 = 0$.

85. $ax^2 + 2hxy + by^2 = 2x(ax_1 + hy_1) + 2y(hx_1 + by_1)$.

88. $ax^2 + 2hxy - by^2 = (b - a)/(b + a)$.

94. $k > 2a$, max k^2 and min $4a(k - a)$; $k \leqslant 2a$, min k^2.

MISCELLANEOUS EXERCISE D

1. $x(y^2 + a^2) = 2a^3$.　　**2.** $(\frac{3}{2}a, \; \pm\frac{1}{3}a\sqrt{3})$.　　**3.** $xy^2 = a^3$.

4. $x : y : 1 = a + bt^2 : at + bt^3 : 1 + t^2$.

5. $\{x^2 + y^2 - 2k(x + y) + 2k^2\}(x + y - 2k) + 2kxy = 0$.

6. $(x^2 + y^2)^2 - 2a^2(x^2 - y^2) + a^4 - c^4 = 0$.

8. $r^2 = 2a^2 \cos 2\theta$.　　　**9.** $x : y : k = t : t^3 : 1 + t^4$.

10. $\Sigma t_1 t_2 = 0 = t_1 t_2 t_3 t_4 - 1$; $t_1 t_2 t_3 t_4 = 1$; $t_1 t_2 t_3 \Sigma t_1 t_2 + \Sigma t_1 = 0$.

12. Six.   **15.** $x(x^2 + y^2) = 2ay^2$.   **17.** The directrix

19. $x(t^3 + 3t) - 2y = 2at^3$; $-\frac{1}{2}t$.   **20.** $(-a, 0, 1)$, $(-2a, \pm ai, 3)$.

24. $t_2 t_3 t_4 + t_1 t_3 t_4 + t_1 t_2 t_4 + t_1 t_2 t_3 = 0$; $-\frac{1}{3}t$.

25. $(7t + t^3) x - 6y = 2at^3$; $243y^2(2a - x) = 343x^3$.

26. q.

27. $x = (a - b)\cos t + b\cos (a - b) t/b$, $y = (a - b)\sin t - b\sin (a - b) t/b$.

30. $x = a\cos^3 t$, $y = a\sin^3 t$.

40. The origin and $(4a\lambda^2(4a^2 - \lambda^2)/(4a^2 + \lambda^2)^2, \; 16a^2\lambda^3/(4a^2 + \lambda^2)^2)$; two intersections at I and at J and three at O.

42. $kt^3 - 3t^2 - 3kt + 1 = 0$.   **43.** $x = \frac{1}{2}a$.

48. Four-cusped epicycloid; three-cusped epicycloid; four-cusped hypocycloid; three-cusped hypocycloid; four-cusped hypocycloid.

INDEX